Lecture Notes in Artificial Intelligence 8578

Subseries of Lecture Notes in Computer Science

Tim Dwyer Helen Purchase
Aidan Delaney (Eds.)

Diagrammatic Representation and Inference

8th International Conference, Diagrams 2014
Melbourne, VIC, Australia, July 28 – August 1, 2014
Proceedings

Springer

Volume Editors

Tim Dwyer
Monash University
PO Box 197, Caulfield East, VIC 3145, Australia
E-mail: tim.dwyer@monash.edu

Helen Purchase
University of Glasgow
University Avenue, Glasgow, G12 8QQ, UK
E-mail: helen.purchase@glasgow.ac.uk

Aidan Delaney
University of Brighton
Lewes Road, Brighton, BN2 4GJ, UK
E-mail: aidan@ontologyengineering.org

ISSN 0302-9743 e-ISSN 1611-3349
ISBN 978-3-662-44042-1 e-ISBN 978-3-662-44043-8
DOI 10.1007/978-3-662-44043-8
Springer Heidelberg New York Dordrecht London

Library of Congress Control Number: 2014942442

LNCS Sublibrary: SL 7 – Artificial Intelligence

Typesetting: Camera-ready by author, data conversion by Scientific Publishing Services, Chennai, India

Printed on acid-free paper

Springer is part of Springer Science+Business Media (www.springer.com)

Preface

The 8th International Conference on the Theory and Application of Diagrams (Diagrams 2014) was held in Melbourne, Australia, during July and August 2014. For the second time, Diagrams was co-located with the IEEE Symposium on Visual Languages and Human-Centric Computing. Prior to Diagrams 2014, the Diagrams 2008 co-location with the VL/HCC produced lively interaction between the delegates of both conferences. At the 2014 conference we again found that co-location stimulated inter-community debate.

Contributers to Diagrams 2014 continued to interpret the meaning of "diagram" broadly, as befits an intentionally interdisciplinary conference. This led to the wide spectrum of papers found in this proceedings volume, covering the areas of diagram notations, diagram layout, diagram tools, diagrams in education, empirical studies of diagrams, and diagrammatic logic.

Submissions to Diagrams 2014 were soclicited under the categories of full papers, short papers, and posters. The peer-review process entailed each paper being reviewed by three members of the Program Committee or a nominated sub-reviewer. For the first time in this conference series, authors were invited to comment on the reviewers' feedback in a rebuttal phase. Both the original review and the author rebuttal were considered when making acceptance decisions.

The process of selecting high-quality papers for the conference and for this proceedings volume would have been impossible without the commitment and efforts of the 39 members of the Program Committee and the eight additional reviewers — we are grateful to them. A total of 40 submissions were received, of which 15 were accepted as full papers. A further nine papers were presented at the conference as short papers, together with six posters.

In addition to the research program, Diagrams 2014 included a graduate symposium and two tutorial sessions. The two invited talks were keynote events shared between Diagrams and VL/HCC, and reflected the Oceaniac location of the conference — we are greatful to our invited speakers Peter Eades of the University of Sydney and Kim Marriot of Monash University.

We acknowledge and appreciate the considerable assistance of the administration team at Monash University for their help in organizing Diagrams 2014. We thank the Monash Immersive Analytics project who have contributed sponsorship, furthermore the graduate symposium ran with the continued support of the US NSF to whom we are also thankful. Finally, we are immensely grateful to John Grundy for all of his help in bringing Diagrams and VL/HCC together.

April 2014

Tim Dwyer
Helen C. Purchase
Aidan Delaney

Organization

Executive Commitee

General Chair

Tim Dwyer	Monash University, Australia

Program Chairs

Aidan Delaney	University of Brighton, UK
Helen C. Purchase	University of Glasgow, UK

Workshops Chair

Karsten Klein	University of Sydney, Australia

Tutorials Chair

Michael Wybrow	Monash University, Australia

Graduate Symposium Chair

Stephanie Schwartz	Millersville University, USA

Treasurer

Cagatay Goncu	Monash University, Australia

Publicity Chair

Jim Burton	University of Brighton, UK

Program Committee

Gerard Allwein	US Navy Research Lab, USA
Dave Barker-Plummer	Stanford University, USA
Lisa Best	University of New Brunswick, Canada
Alan Blackwell	Cambridge University, UK
Paolo Bottoni	University of Rome, Italy
Sandra Carberry	University of Delaware, USA
B. Chandrasekaran	Ohio State University, USA
Peter Cheng	University of Sussex, UK
Lopamudra Choudhury	Jadavpur University, India
James Corter	Columbia University, USA
Phil Cox	Dalhousie University, Canada

Richard Cox	Monash University, Australia
Frithjof Dau	SAP
Jim Davies	Carleton University, Canada
Stephanie Elzer-Schwartz	Millersville University, USA
Jacques Fleuriot	University of Edinburgh, UK
Jean Flower	Autodesk
Ashok Goel	Georgia Institute of Technology, USA
Mary Hegarty	University of California, Santa Barbara, USA
John Howse	University of Brighton, UK
Mateja Jamnik	University of Cambridge, UK
John Lee	University of Edinburgh, UK
Emmanuel Manalo	Waseda University, Japan
Kim Marriott	Monash University, Australia
Nathaniel Miller	University of Northern Colorado, USA
Mark Minas	Universität der Bundeswehr München, Germany
Ian Oliver	HERE.com
Luis Pineda-Cortes	Universidad Nacional Autónoma de México, Mexico
Beryl Plimmer	University of Auckland, New Zealand
Peter Rodgers	University of Kent, UK
Frank Ruskey	University of Victoria, Canada
Atsushi Shimojima	Doshisha University, Japan
Sun-Joo Shin	Yale University, USA
Gem Stapleton	University of Brighton, UK
Nik Swoboda	Universidad Politécnica de Madrid, Spain
Ryo Takemura	Nihon University, Japan
Barbara Tversky	Columbia and Stanford, USA
Yuri Uesaka	The University of Tokyo, Japan
Michael Wybrow	Monash University, Australia

Additional Reviewers

Andrew Blake
Jim Burton
Peter Chapman
Veronika Irvine
Sonja Maier
Petros Papapanagiotou
Sahand Saba
John Taylor

Keynote Talks (Abstracts)

What Is a Good Diagram? (Revisited)

Peter Eades

School of Information Technologies, University of Sydney
peter.eades@sydney.edu.au

Graphs have been broadly used to model binary relations since the beginning of Computer Science. Nodes represent entities, and edges represent relationships between entities. Such models become more useful when the graph model is represented as a diagram, because visualization of a graph enables humans to understand the underlying model.

A quality metric assigns a number $q(D)$ to each diagram D such that $q(D)$ is larger than $q(D')$ when D is a higher quality diagram than D'. Quality metrics for graph visualization have been discussed since the 1970s. Sugiyama et al. [6] wrote lists of quality metrics and his subsequent book [5] contains an extensive discussion. The seminal paper What is a good diagram? by Batini et al.[1] presented guidelines for database diagrams. These early works were entirely based on intuition and introspection; later Purchase et al. [4] began the scientific investigation of quality metrics with a series of experiments that validated some of the metrics.

Of course, the quality of a diagram is "a hopeless matter to define formally" [1]: quality depends on specific users, specific tasks, and rather informal and subjective notions of aesthetics. Nevertheless, formal quality metrics are helpful if not essential in the design of automatic graph visualization methods, because such methods are optimisation algorithms with quality metrics as objective functions. As an example, it is well established that edge crossings in a diagram inhibit human understanding, and edge crossings form the basis of so-called "planarity-based" quality metrics. Methods that reduce the edge crossings have received considerable attention in the literature (see, for example, [2]).

In this talk we review the history of quality metrics for graph visualization, and suggest a new approach.

The new approach is motivated by two observations: (1) the size of data sets is much larger now than ever before, and it is not clear that established quality metrics are still relevant, and (2) there is a disparity between methods used in practice and methods used in academic research.

Using a pipeline model of graph visualization, we classify quality metrics into "readability" metrics and "faithfulness" metrics. Readability metrics measure how well the human user perceives the diagram; these metrics have been extensively investigated and they are (at least partially) understood. Faithfulness metrics (see [3]) measure how well the diagram represents the data; these metrics are not well developed and they are poorly understood.

We argue that faithfulness metrics become more relevant as the data size increases, and suggest that the commercial dominance of stress-based methods over of planarity-based methods is somewhat due performance on faithfulness metrics.

We introduce some specific faithfulness metrics aimed at large graphs. In particular, we suggest that metrics based on proximity graphs (see [7]) may play a key role.

Much of this talk is based on joint work and discussions with Karsten Klein, SeokHee Hong, and Quan Nguyen, among others.

References

1. Batini, C., Furlani, L., Nardelli, E.: What is a good diagram? a pragmatic approach. In: Proceedings of the Fourth International Conference on Entity-Relationship Approach, pp. 312–319. IEEE Computer Society, Washington, DC (1985), `http://dl.acm.org/citation.cfm?id=647510.726382`
2. Jünger, M., Mutzel, P.: Maximum planar subgraphs and nice embeddings: Practical layout tools. Alogrithmica 16, 33–59 (1996)
3. Nguyen, Q.H., Eades, P., Hong, S.H.: On the faithfulness of graph visualizations. In: 2013 IEEE Pacific Visualization Symposium (PacificVis), pp. 209–216 (February 2013)
4. Purchase, H.C., Cohen, R.F., James, M.: Validating graph drawing aesthetics. In: Brandenburg, F.J. (ed.) GD 1995. LNCS, vol. 1027, pp. 435–446. Springer, Heidelberg (1996), `http://dx.doi.org/10.1007/BFb0021827`
5. Sugiyama, K.: Graph drawing and applications for software and knowledge engineers. World Scientific (2002); Japanese language version 1993
6. Sugiyama, K., Tagawa, S., Toda, M.: Methods for visual understanding of hierarchical system structures. IEEE Transactions on Systems, Man and Cybernetics 11(2), 109–125 (1981)
7. Toussaint, G.T.: A graph-theoretical primal sketch. In: Computational Morphology, pp. 229–260. North-Holland (1988)

Seeing with Sound and Touch

Kim Marriott and Cagatay Goncu

Caulfield School of IT, Monash University, Australia,
Kim.Marriott@monash.edu

Information graphics – diagrams, plans, maps, plots and charts – are widespread in written communication. The ability to comprehend, use and create these graphics is an important skill that most of us take for granted. However, for those of us who are blind or have severe vision impairment access to such graphics is severely limited, restricting effective participation in the workplace, limiting educational opportunities, especially in art and design, mathematics, and science and technology, and constraining enjoyment of popular media including the web. Lack of access to maps and plans of public spaces and buildings also contributes to one of the most disabling consequences of being blind: the difficulty of leaving the safety of home and having to find one's way in a new environment.

Tactile graphics are currently the best way of providing access to graphical material for those who are blind. However they are usually produced manually by skilled transcribers and so are time consuming and expensive to create. For this reason technologies that provide effective, low-cost, on-demand accessible graphics have been a "holy grail" of accessibility research for several decades. Fortunately, new technologies developed for mobile devices, gaming and graphics recognition now make this goal achievable.

In the GraVVITAS (Graphics Viewer using Vibration Interactive Touch and-Speech) project (http://gravvitas.infotech.monash.edu/) we are developing new computer technologies that are designed to work on touch screens and to provide people who are blind or have severe vision impairment with fast, inexpensive access to a wide variety of information graphics at home, at school and at work. The main components are:

1. An iPad app for information graphics presentation. As the user explores the screen with their fingers, a mixture of speech and non-speech audio allows them to hear the graphic as they touch its elements. We are also developing inexpensive "data rings" for providing haptic feedback during this exploration.
2. A web-based authoring tool that allows teachers, friends or family members of a blind person to easily create accessible graphics suitable for display with the presentation app.

We are also exploring alternatives to manual creation of accessible graphics. One way is direct generation of the graphic from applications like statistical software, the other is to use a combination of graphics recognition and automatic graphics transformation to create an accessible graphic from an on-line image such as a floor plan.

Table of Contents

Diagram Layout

Diagram Notations

Diagramming Tools

Diagrams in Education

Empirical Studies

Logic and Diagrams

Octilinear Force-Directed Layout with Mental Map Preservation for Schematic Diagrams

Daniel Chivers and Peter Rodgers

The University of Kent, Canterbury, Kent, CT2 7NF
{dc355,P.J.Rodgers}@kent.ac.uk

Abstract. We present an algorithm for automatically laying out metro map style schematics using a force-directed approach, where we use a localized version of the standard spring embedder forces combined with an octilinear magnetic force. The two types of forces used during layout are naturally conflicting, and the existing method of simply combining these to generate a resultant force does not give satisfactory results. Hence we vary the forces, emphasizing the standard forces in the beginning to produce a well distributed graph, with the octilinear forces becoming prevalent at the end of the layout, to ensure that the key requirement of line angles at intervals of 45° is obtained. Our method is considerably faster than the more commonly used search-based approaches, and we believe the results are superior to the previous force-directed approach. We have further developed this technique to address the issues of dynamic schematic layout. We use a Delaunay triangulation to construct a schematic "frame", which is used to retain relative node positions and permits full control of the level of mental map preservation. This technique is the first to combine mental map preservation techniques with the additional layout criteria of schematic diagrams. To conclude, we present the results of a study to investigate the relationship between the level of mental map preservation and the user response time and accuracy.

Keywords: force directed, schematic layout, Delaunay triangulation, mental map preservation.

1 Introduction

The layout of metro maps and other schematics has been the focus of much recent research effort. The goal of these systems is to take a network diagram, for instance a public transport map, and draw the network as a schematic, so allowing the information contained in the network to be more easily analyzed. One of the key features of a schematic is octilinearity (all edges at angles which are multiples of 45°).

Much research has been undertaken to improve the readability of schematics by automated layout, such as hill-climbing [1], simulated annealing [2], linear programming [3] and path simplification [4]. Although these search-based methods can already compute layouts of a reasonable quality, a force-directed method that could produce comparable results would provide the following benefits: 1. Fast optimisation relative to search-based methods 2. This layout method has been studied in great detail from a number of perspectives and so will be able to piggyback on considerable related work in, for

T. Dwyer et al. (Eds.): Diagrams 2014, LNAI 8578, pp. 1–8, 2014.

example, performance improvements and scalability 3. Node movements during layout can be viewed in real time as a transition animation.

Our algorithm uses modified versions of two force based techniques: a spring embedder [5] and a magnetic spring model [6] which encourages octilinearity (Section 2). At the beginning of layout only the spring embedder forces are applied and this ensures an even layout of stations. We then transition between the two force types; the octilinear forces are introduced with increasing strength as the spring embedder forces are reduced in strength. Eventually the spring embedder forces reduce to zero and only octilinear forces are present - this final stage ensures that octilinearity is enforced. We then developed a mental map preservation technique using a Delaunay triangulation to construct a proximity graph; the edges of which are modelled as springs in order to constrain node movement to a variable degree by altering spring strength (Section 3). Using this mental map preservation method, we then perform a study to test for significance between level of mental map preservation and user response time and accuracy (Section 4). Finally, Section 5 concludes.

The contribution of this paper is to: 1. Describe a force-directed metro map layout technique that has considerably improved the previous attempt for producing octilinear lines and well distributed stations; and which we believe produces results of a comparable quality to search-based methods in much less time 2. Present a new, fully tuneable, technique for mental map preservation 3. Perform a study on how mental map preservation affects diagram usability in order to augment the current understanding of this potentially useful concept. The technique has been implemented in java and is called FDOL (Force-Directed Octilinear Layout). The application, example schematic files and data from the study are available for download at `http://www.cs.kent.ac.uk/projects/fdol`.

2 Octilinear Force-Directed Layout

As is common in schematic layout, we model the schematic as a graph by treating stations as nodes and line segments between stations as edges. We do not differentiate between different lines, and multiple lines between stations are treated as a single edge. In addition, where some systems remove two degree nodes from the graph before layout [7], we keep them in the graph. This helps ensure a reasonable separation between nodes on the same line and avoids having to use individually weighted lines. The algorithm uses modified versions of two force based techniques: a spring embedder [5] and a magnetic spring model [6].

We use the geographic positions of stations to define our starting layout, as can be seen in Fig. 2a, with the entire schematic scaled to have a mean edge length equal to the length of an edge spring at equilibrium – this prevents the schematic quickly expanding/contracting in the first iteration by minimising large initial forces. The standard spring embedder has two types of force which act upon the nodes to produce the layout [5]. Standard spring embedder forces are intended to produce an aesthetically pleasing graph layout; however, these forces do nothing to ensure edges are aligned to octilinear angles. In order to achieve this, additional forces are required that will cause edges to rotate to desired angles. We use a technique similar to that explained in [6] in

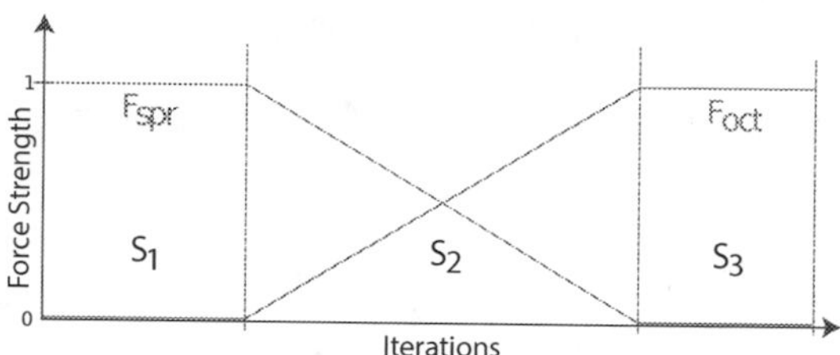

Fig. 1. The application of forces over method duration

which equal and opposite (perpendicular to the edge) forces are applied to the nodes connected to each edge in order to rotate them around the midpoint.

Our method works by applying both force methods in each iteration, but varying the strength of each so as to perform a smooth switch from spring to rotational forces. Variables F_{spr} and F_{oct} are used as respective coefficients for this purpose and are varied throughout the layout process as shown in Fig. 1. At the start only spring embedder forces are applied (S_1) - this stage runs until the energy level of the schematic falls below a set threshold; then there is a switchover in S_2 until only octilinear forces are applied (S_3). The switchover of force types in S_2 is required to allow the octilinear force to have a gradually increasing effect whilst the spring embedder forces still ensure a well distributed layout. If the two force types are applied one after the other, with no switchover period, the octilinearity stage would rotate edges without consideration for features that the spring embedder forces prevent (such as node-node occlusions and preservation of edge lengths). This final stage of layout without any spring embedder forces is necessary in order to achieve the highest possible compliance with the octilinearity criterion. We then perform a post-processing step to straighten periphery line sections. A schematic showing examples of these steps is shown in Figure 2.

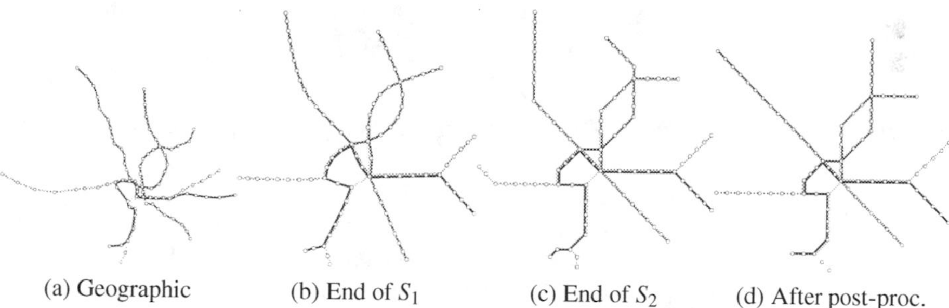

(a) Geographic (b) End of S_1 (c) End of S_2 (d) After post-proc.

Fig. 2. Washington metro map at key stages throughout the layout method

On a range of ten real-world metro maps, our method achieved a layout speedup factor in the range of 2.23 to 27.67 times ($\bar{x} = 10.12$, $\sigma = 8.3$) that of our hill-climbing technique; the speedup factor for Sydney was 20.08 times, with an optimisation time of 0.59s against 11.85s. We have included for comparison images of the Sydney schematic as laid out using our force-directed method (Fig. 3a) and the previous force-directed

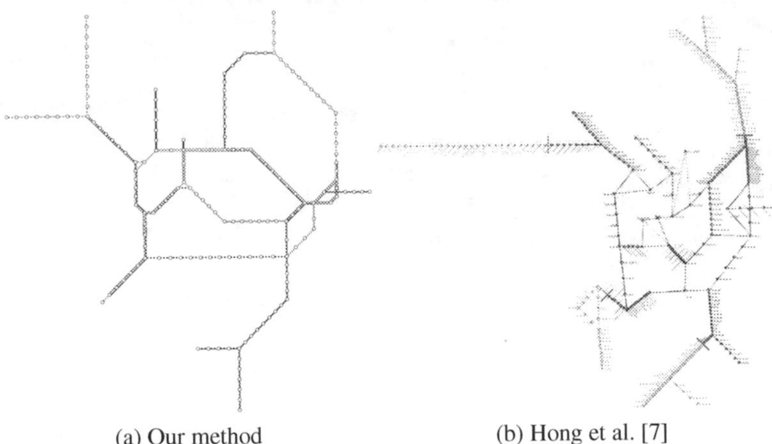

(a) Our method (b) Hong et al. [7]

Fig. 3. Comparison of our octilinear force-directed method (3a) against the previous method by Hong et al. (3b). Note: Fig 3b contains additional stations demarcated by short perpendicular red lines and slightly faded.

method by Hong et al. [7] (Fig. 3b). Sydney optimisations using alternate methods can be seen in [3][1][8]. Our method, Fig. 3a, has octilinear lines throughout and fairly even node spacing, achieving our main design goals. The previous force-directed attempt, Fig. 3b, fails on both of these counts; moreover some odd node positioning is evident, for instance the blue line is very jagged. We believe these problems are due to the combination of removing 2 degree nodes and because of the way the forces are combined. Each type of force will have different ideal positions for nodes to move to, and combining the two into a composite function creates lowest energy positions which fully satisfy neither. It should be noted that the previous method is drawing a network that is larger than the one we show. Although the schematic used is slightly different, in terms of layout the extra stations should not affect the quality.

3 Dynamic Layout and Mental Map Preservation

This section explains how a Delaunay triangulation is used to constrain node movement by proximity during optimisation and thus help preserve the users' mental map. A Delaunay triangulation is first constructed over the entire schematic. Figure 4 shows an example of this applied to the Vienna metro map; the underlying schematic can be made out under the thinner, red edges of the triangulation. During construction of the triangulation, four "anchor" nodes are created which surround the schematic. These anchor nodes are left in the triangulation but are not represented with nodes, and so cannot move. This has the effect of slightly anchoring the schematic to the underlying canvas, preventing effects such as rotation. The generated triangulation edges are then used during layout as a frame to hold nodes in place. Each edge is modelled as a Hookean spring, with a length at rest equal to its initial length, and a spring strength equal to a user defined value, k. This spring strength value can be varied in order to affect the level

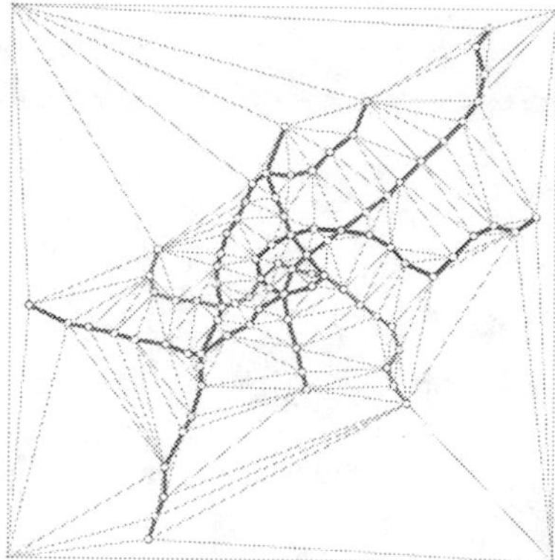

Fig. 4. Delaunay triangulation of the Vienna metro map (including triangulation frame)

of mental map preservation; using a low k value will create weak springs, and will not hold nodes in place. In our algorithm Delaunay frame forces are calculated after the calculation of spring embedder and edge rotation forces, this is to ensure that all node movements from other forces can be counteracted by the mental map preserving frame. In order to allow the Delaunay frame forces to be used in conjunction with octilinear forces, it is required that the Delaunay frame uses a force coefficient similar to that of the standard spring embedder, which decreases over time. This force constant, F_{Del}, mirrors the value of F_{Spr} used for the standard spring embedder. Our system provides a scalable slider from $0-100\%$ which provides a linear mapping to the average resultant mental map preservation level, measured using a similarity metric based upon the change in frame edge lengths.

4 Mental Map Preservation Study

This section describes the procedure for a study to answer our research question: "Can mental map preservation improve diagram understanding in dynamic schematic layout?". Our hypothesis was that different levels of mental map preservation would change response time and accuracy. Previous research into the effect of mental map preservation has not shown conclusive benefit [9][10], and we hoped to augment the research in this area. We used three sizes of map as follows: Small (S) 4 lines, 28 nodes; Medium (M) 5 lines, 35 nodes; and Large (L) 6 lines, 42 nodes. There were three modification types as follows: Line addition (A) $+1$ line, $+7$ nodes; Line removal (R) -1 line, -7 nodes; Line addition and removal (A-R) $+1$ line, $+7$ nodes, -1 line, -7 nodes. These variations create nine map cases. Each map was reoptimized with three levels of mental map preservation (MMP); these are 0%, 50% and 100% preservation. We used a between-subject methodology, and therefore each subject viewed only one MMP variant. In order to help alleviate the effect of a learning curve when answering questions, the user completed five training maps. Map data was generated using the Flickr API which we used to retrieve a number of photos, represented as nodes, and their associated tags, represented by coloured lines between nodes. Nodes that shared the same

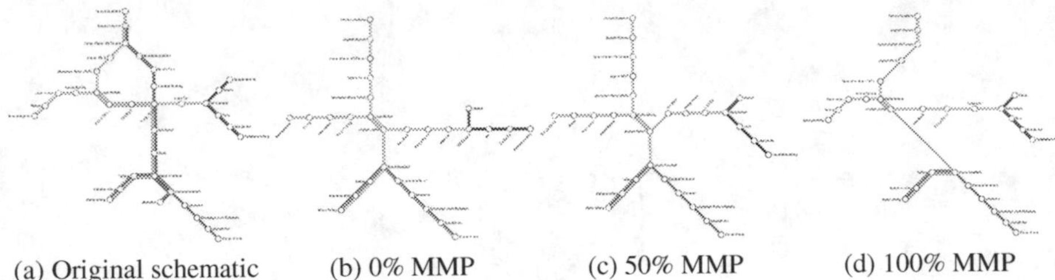

(a) Original schematic (b) 0% MMP (c) 50% MMP (d) 100% MMP

Fig. 5. Comparison of MMP variants used in the study. 5b-5d show the modified schematic, in which the magenta line has been removed.

tag appeared along the same line, much like LineSets [11]. Figure 5 shows an example of a diagram used in the study. There were two stages for each map: five stage one questions, followed by a map modification and a single stage two question. There were three types of stage two question (X, Y and Z) which were assigned to maps so that each type of question was asked once for each map size, and each modification type. Only stage two questions were used in the data analysis as stage one questions had the purpose of familiarising the participant with the schematic before the modification was made. Example stage two questions were as follows:

X. How many photos contain the "___" or "___" tags and not the "___" tag?
Y. What is the minimum number of tag changes required to travel between the highlighted nodes?
Z. Which tag contains the most photos?

We first performed a pilot study to identify and fix any testing issues. The main study used a total of 60 participants recruited from the University of Kent Job Centre. These participants are from across all disciplines, ranging from 16 to 34 years of age ($\bar{x} = 20.83$, $\sigma = 2.88$), 20 male and 40 female.

In terms of results, the statistical analysis techniques were performed following guidelines in [12]. In our analysis, data from questions which were answered incorrectly are also included as response time is intended to be an indication of cognitive effort required, independent of whether or not the effort resulted in a correct answer. Table 1 shows mean time and mean error against each condition. In the following tests, we used a p-value of < 0.05 to indicate significance. There were no significant differences in

Table 1. Mean response time and mean number of errors for the three mental map conditions, over all non-test maps and all post-modification questions

	Time		Error %	
MMP	$\bar{x}$	σ	$\bar{x}$	σ
0%	19.95	4.31	14.44	13.54
50%	20.27	4.81	10.56	12.73
100%	23.96	7.07	12.78	13.13

response time ($p = 0.207$) or error rate ($p = 0.593$) as represented by the error data according to condition under a non-parametric independent-measures Kruskal-Wallis test. These results indicate that the difference between the conditions can likely be attributed to random chance, rather than being due to the differing nature of the conditions.

This result provides evidence against our hypothesis, for which a possible explanation is due to multiple impacts on diagram comprehension. One impact occurs when the mental map is preserved, the layout is compromised, so making analysis of the diagram difficult because of features such as increased line bends and less effective station spacing. The alternative extreme is that the mental map is not preserved, so the diagram changes a great deal, impacting comprehension because the participant needs to re-examine parts of the diagram that were previously held in their memory. It may be that these two conflicting impacts on comprehension are broadly equal, and so it is not important which approach is taken for dynamic data. There have been a number of studies on the effect of mental map preservation on user readability, e.g. [9][10]; however, none have found conclusive evidence of any effect, supporting this view.

5 Conclusion

We have presented a force-directed method for automatically laying out schematic diagrams whilst enforcing octilinearity. Our method uses a three stage process to transition between types of forces. We have explained how a force based approach can be more beneficial than methods using an explicit target function, and demonstrated that our methodology allows us to produce superior results than those from the previous octilinear force based method. We believe our results are also of comparable quality to those produced by slower techniques that are known for producing high-quality layouts. We have also shown a new technique for the preservation of mental map between two graph states. This technique uses a Delaunay triangulation to preserve node positions relative to connected nodes. This varies from the more common approach of limiting nodes by absolute distance, and allows more node position flexibility with minimal mental map expense, such as moving clusters of nodes whilst maintaining their internal structure.

References

1. Stott, J., Rodgers, P.J., Martinez-Ovando, J.C., Walker, S.G.: Automatic Metro Map Layout Using Multicriteria Optimization. TVCG 6, 101–114 (2011)
2. Anand, S., Avelar, S., Ware, M.J., Jackson, M.: Automated Schematic Map Production using Simulated Annealing and Gradient Descent Approaches. In: 15th Annual GISRUK (2007)
3. Nöllenburg, M., Wolff, A.: A Mixed-Integer Program for Drawing High-Quality Metro Maps. In: Healy, P., Nikolov, N.S. (eds.) GD 2005. LNCS, vol. 3843, pp. 321–333. Springer, Heidelberg (2006)
4. Merrick, D., Gudmundsson, J.: Path Simplification for Metro Map Layout. In: Kaufmann, M., Wagner, D. (eds.) GD 2006. LNCS, vol. 4372, pp. 258–269. Springer, Heidelberg (2007)
5. Eades, P.: A Heuristic for Graph Drawing. Congressus Numerantium 42, 149–160 (1984)
6. Sugiyama, K., Misue, K.: A Simple and Unified Method for Drawing Graphs: Magnetic-Spring Algorithm. In: Tamassia, R., Tollis, I.G. (eds.) GD 1994. LNCS, vol. 894, pp. 364–375. Springer, Heidelberg (1995)

7. Hong, S.-H., Merrick, D., Nascimento, H.A.D.: Automatic Visualisation of Metro Maps. IJVLC 17, 203–224 (2006)
8. Wang, Y.-S., Chi, M.-T.: Focus+Context Metro Maps. TVCG 17, 2528–2535 (2011)
9. Archembault, D., Purchase, H.C., Pinaud, B.: Animation, Small Multiples, and the Effect of Mental Map Preservation in Dynamic Graphs. TVCG 17, 539–552 (2011)
10. Purchase, H.C., Samra, A.: Extremes are Better: Investigating Mental Map Preservation in Dynamic Graphs. In: Stapleton, G., Howse, J., Lee, J. (eds.) Diagrams 2008. LNCS (LNAI), vol. 5223, pp. 60–73. Springer, Heidelberg (2008)
11. Alper, B., Riche, N.H., Ramos, G., Czerwinski, M.: Design Study of LineSets, a Novel Set Visualization Technique. TVCG 17, 2259–2267 (2011)
12. Purchase, H.C.: Experimental Human-Computer Interaction. Cambridge University Press (2012)

Counting Crossings for Layered Hypergraphs

Miro Spönemann, Christoph Daniel Schulze,
Ulf Rüegg, and Reinhard von Hanxleden

Department of Computer Science, Christian-Albrechts-Universität zu Kiel
{msp,cds,uru,rvh}@informatik.uni-kiel.de

Abstract. Orthogonally drawn hypergraphs have important applications, e. g. in actor-oriented data flow diagrams for modeling complex software systems. Graph drawing algorithms based on the approach by Sugiyama et al. place nodes into consecutive layers and try to minimize the number of edge crossings by finding suitable orderings of the nodes in each layer. With orthogonal hyperedges, however, the exact number of crossings is not determined until the edges are actually routed in a later phase of the algorithm, which makes it hard to evaluate the quality of a given node ordering beforehand. In this paper, we present and evaluate two cross counting algorithms that predict the number of crossings between orthogonally routed hyperedges much more accurately than the traditional straight-line method.

Keywords: edge crossings, hypergraphs, graph drawing, layered graphs.

1 Introduction

Many kinds of diagrams, e. g. data flow diagrams and circuit diagrams, can be formalized as *directed hypergraphs.* A directed hypergraph is a pair $G = (V, H)$ where V is a set of nodes and $H \subseteq \mathcal{P}(V) \times \mathcal{P}(V)$ is a set of hyperedges. Each $(S, T) \in H$ has a set of *sources* S and a set of *targets* T.

The layer-based approach to graph drawing proposed by Sugiyama et al. [1] has been extended for drawing hypergraphs orthogonally [2,3]. This approach can be structured in five phases: eliminate cycles by reversing edges, assign nodes to *layers*, reorder the nodes of each layer such that the number of crossings is low, determine concrete positions for the nodes, and finally route the hyperedges by computing bend points and junction points. The reordering phase is usually done with the *layer sweep* heuristic, where layers are processed one at a time with several iterations until the number of crossings is not further improved.

In order to determine whether an iteration of the the layer sweep heuristic has reduced the number of edge crossings, they need to be counted. A fundamental problem with this approach is that the actual number of crossings in orthogonal drawings does not depend only on the order of nodes in each layer, but also on the routing of edges. This routing in turn depends on the concrete positions of the nodes, which are unknown at the time the layer sweep heuristic is executed.

Standard algorithms for counting crossings in layered graphs assume that edges are drawn as straight lines [4]. We call this standard approach STRAIGHT,

T. Dwyer et al. (Eds.): Diagrams 2014, LNAI 8578, pp. 9–15, 2014.

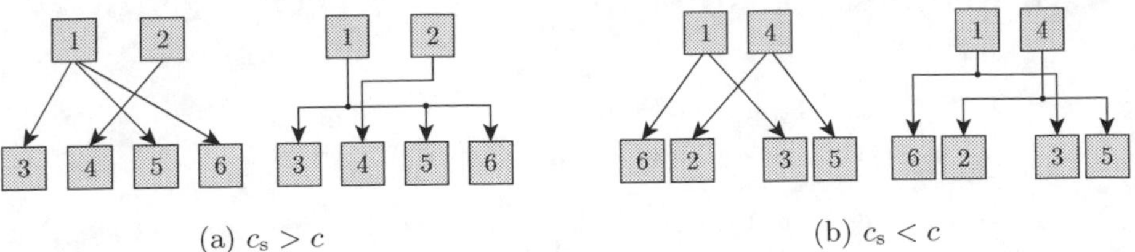

(a) $c_s > c$

(b) $c_s < c$

Fig. 1. The number of crossings c_s resulting from a straight-line drawing can be (a) greater or (b) less than the actual number of crossings c resulting from an orthogonal hyperedge routing.

and denote its result as c_s. As noted by Eschbach et al. [5], there are simple examples where c_s is always different from the actual number of crossings c obtained after routing edges orthogonally (see Fig. 1). In order to quantify this difference, we measured c and c_s for a number of data flow diagrams from the Ptolemy project (see Sect. 3). The difference $c - c_s$ averaged -34 with a standard deviation of 190. As a general observation, the STRAIGHT approach tends to overestimate the crossing number, possibly compromising the quality of the resulting drawings.

Contributions. The number of crossings between straight-line edges is a bad predictor for the number of crossings between orthogonal hyperedges in the final drawing. We propose two methods for counting crossings that predict the number of crossings much more accurately, as our evaluation shows.

Related Work. Eschbach et al. [2] and Sander [3] proposed methods for the orthogonal routing of edges in the layer-based approach. They both noted that the number of crossings determined during the node ordering phase is only an approximation, but gave no proposals on how to solve this problem. In this paper, we will present algorithms that give much more accurate approximations.

Several authors have addressed the problem of counting straight-line crossings in layered graphs [4,6,7]. These methods produce exact results for normal graphs, but not for hypergraphs, as explained above. The "STRAIGHT approach" which we refer to in this paper represents any exact straight-line method. For our experiments we implemented the method of Barth et al. [4].

2 Counting Crossings

The main concept for representing a hyperedge in the layer-based approach is to replace the hyperedge by regular edges. Let (V, H) be a hypergraph and $h = (S, T) \in H$ be a hyperedge; for each $v \in S$ and each $v' \in T$ we generate an edge $e = (v, v')$. We call e a *representing edge* of h and define E_h to be the set of all representing edges of h. For instance, the hyperedge connected to node 1 in Fig. 1(a) would be represented by three edges $(1, 3)$, $(1, 5)$, and $(1, 6)$. These edges may partly overlap each other in the final drawing.

2.1 Lower Bound Method

Since counting straight-line crossings tends to yield rather pessimistic estimates when hyperedges are involved, we assumed that a more accurate approach might be to use a lower bound of the number of crossings.

In the following, let $G = (V, H)$ be a hypergraph with a set $E = \bigcup_{h \in H} E_h$ of representing edges and two layers L_1, L_2, i. e. $V = L_1 \cup L_2$, $L_1 \cap L_2 = \emptyset$, and all $h \in H$ have their sources in L_1 and their targets in L_2.

We propose an optimistic method MinOpt and denote its result as c_{m}. This method counts the minimal number of crossings to be expected by evaluating each unordered pair $h_1, h_2 \in H$: if any edge $e_1 \in E_{h_1}$ crosses an edge $e_2 \in E_{h_2}$ if drawn as a straight line, h_1 and h_2 are regarded as crossing each other once, denoted as $h_1 \bowtie h_2$. The result is $c_{\mathrm{m}} = |\{\{h_1, h_2\} \subseteq H : h_1 \bowtie h_2\}|$.

Observation 1. $c_{\mathrm{m}} \leq c_{\mathrm{s}}$.

Observation 2. *Let c be the number of hyperedge crossings in a layer-based drawing of G. Then $c_{\mathrm{m}} \leq c$.*

Theorem 1. *Let $q = |H|$ and $H = \{h_1, \ldots, h_q\}$. The time complexity of* MinOpt *is $\mathcal{O}\left(\sum_{i=1}^{q-1} \sum_{j=i+1}^{q} |E_{h_i}| \cdot |E_{h_j}|\right)$. If $|S| = |T| = 1$ for all $(S, T) \in H$, the complexity can be simplified to $\mathcal{O}(|H|^2)$.*

Proof. The result of MinOpt is $|\{\{h_i, h_j\} \subset H : h_i \bowtie h_j\}|$, which requires to check all unordered pairs $U = \{\{h_i, h_j\} \subset H\}$. This is equivalent to $U = \{(i, j) \in \mathbb{N}^2 : 1 \leq i < q,\ i < j \leq q\}$, hence $|U| = \sum_{i=1}^{q-1} \sum_{j=i+1}^{q} 1$. Whether $h_i \bowtie h_j$ is determined by comparing all representing edges of h_i with those of h_j, which requires $|E_{h_i}| \cdot |E_{h_j}|$ steps. In total we require $\sum_{i=1}^{q-1} \sum_{j=i+1}^{q} |E_{h_i}| \cdot |E_{h_j}|$ steps. The simplification follows immediately. □

2.2 Approximating Method

Theorem 1 shows that MinOpt has a roughly quadratic time complexity. In this section we propose a second method with better time complexity, which we call ApproxOpt. The basic idea is to approximate the result of MinOpt by checking three criteria explained below, hoping that at least one of them will be satisfied for a given pair of hyperedges if they cross each other in the final drawing.

Let again $G = (V, H)$ be a hypergraph with layers L_1, L_2. Let $\pi_1 : L_1 \to \{1, \ldots, |L_1|\}$ and $\pi_2 : L_2 \to \{1, \ldots, |L_2|\}$ be the permutations of L_1 and L_2 that result from the layer sweep heuristic for crossing minimization. We denote the result of ApproxOpt as c_{a}.

The ApproxOpt method is based on the four *corners* of a hyperedge: for each $h = (V_{h1}, V_{h2}) \in H$ and $i \in \{1, 2\}$, we define the left corners $\kappa_i^{\leftarrow}(h) = \min\{\pi_i(v) : v \in V_{hi}\}$ and the right corners $\kappa_i^{\rightarrow}(h) = \max\{\pi_i(v) : v \in V_{hi}\}$. The *virtual edges* are defined by $E^* = \{(\kappa_1^{\leftarrow}(h), \kappa_2^{\leftarrow}(h)) : h \in H\}$. The method consists of three steps:

Algorithm 1. Counting crossings with the ApproxOpt method

```
Input: L_1, L_2 with permutations π_1, π_2, hyperedges H with arbitrary order ϑ
for each h ∈ H do                                                    // Step 1
 └ Add (κ_1^←(h), κ_2^←(h)) to E*
c_a ← number of crossings caused by E*, counted with a straight-line method
for i = 1...2 do                                               // Steps 2 and 3
 │ for each h ∈ H do
 │  └ Add (κ_i^←(h), κ_i^→(h), ϑ(h), −1)) and (κ_i^→(h), κ_i^←(h), ϑ(h), 1)) to C_i
 │ Sort C_i lexicographically
 │ d ← 0
 │ for each (x, x', j, t) ∈ C_i in lexicographical order do
 │  │ d ← d − t
 │  │ if t = 1 then
 │  │  └ c_a ← c_a + d
 └  └
return c_a
```

1. Compute the number of straight-line crossings caused by virtual edges between the left corners.
2. Compute the number of overlaps of ranges $[\kappa_1^{\leftarrow}(h), \kappa_1^{\rightarrow}(h)]$ in the first layer for all $h \in H$.
3. Compute the number of overlaps of ranges $[\kappa_2^{\leftarrow}(h), \kappa_2^{\rightarrow}(h)]$ in the second layer for all $h \in H$.

The result c_{a} is the sum of the three numbers computed in these steps. A more detailed description is given in Alg. 1.

Step 1 aims at "normal" crossings of hyperedges such as h_1 and h_2 in Fig. 2. The hyperedge corners used in Steps 2 and 3 serve to check for overlapping areas, as shown in Fig. 2(c). For instance, the ranges spanned by h_4 and h_5 overlap each other both in the first layer and in the second layer. This is determined using a linear pass over the hyperedge corners, which are sorted by their positions. The sort keys are constructed such that the overlapping of two ranges is counted only if it actually produces a crossing. The variable d is increased whenever a left-side corner is found and decreased whenever a right-side corner is found. This variable indicates how many ranges of other hyperedges surround the current corner position, hence its value is added to the approximate number of crossings.

While MinOpt counts at most one crossing for each pair of hyperedges, ApproxOpt may count up to three crossings, since the hyperedge pairs are considered independently in all three steps. Fig. 3(a) shows an example where MinOpt counts a crossing and ApproxOpt counts none, while Fig. 3(b) shows an example where ApproxOpt counts a crossing and MinOpt counts none. Thus neither $c_{\mathrm{m}} \leq kc_{\mathrm{a}}$ nor $c_{\mathrm{a}} \leq kc_{\mathrm{m}}$ hold in general for any $k \in \mathbb{N}$. However, as shown in Sect. 3, the difference between c_{m} and c_{a} is rather small in practice.

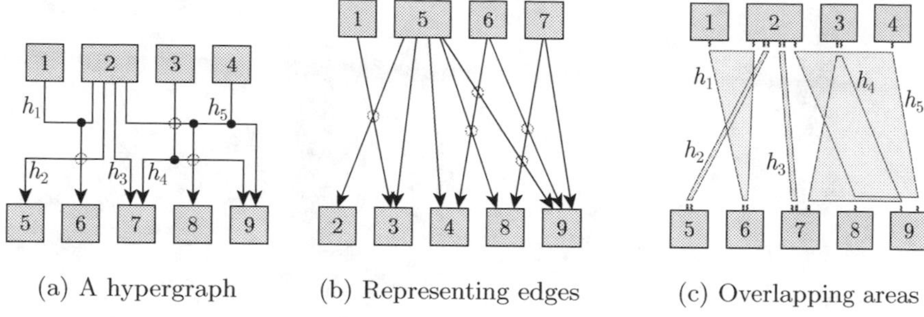

(a) A hypergraph (b) Representing edges (c) Overlapping areas

Fig. 2. The hypergraph (a) can be drawn orthogonally with $c = 3$ crossings. The straight-line crossing number (b) is $c_{\mathrm{s}} = 5$, the result of MinOpt is $c_{\mathrm{m}} = 2$, and the result of ApproxOpt is $c_{\mathrm{a}} = 4$. ApproxOpt counts three crossings between h_4 and h_5 (c) because the virtual edges $(2, 8)$ and $(3, 7)$ cross (Step 1 in Alg. 1) and the ranges spanned by the corners overlap both in the top layer and in the bottom layer (Steps 2 and 3).

(a) $c_{\mathrm{m}} = 1$, $c_{\mathrm{a}} = 0$ (b) $c_{\mathrm{m}} = 0$, $c_{\mathrm{a}} = 1$

Fig. 3. Differences between the MinOpt and ApproxOpt methods: (a) $c_{\mathrm{m}} = 1$ due to the crossing of $(1, 4)$ and $(2, 3)$, but $c_{\mathrm{a}} = 0$ since none of the three steps of ApproxOpt is able to detect that. (b) $c_{\mathrm{m}} = 0$ because $(2, 4)$ crosses neither $(1, 4)$ nor $(3, 4)$; $c_{\mathrm{a}} = 1$ because one crossing is detected in Step 2 of Alg. 1.

Theorem 2. *Let* $b = \sum_{(S,T) \in H}(|S| + |T|)$. *The time complexity of* ApproxOpt *is* $\mathcal{O}(b + |H|(\log|V| + \log|H|))$.

Proof. In order to determine the corners $\kappa_i^{\leftarrow}(h), \kappa_i^{\rightarrow}(h)$ for each $h \in H$, $i \in \{1, 2\}$, all source and target nodes are traversed searching for those with minimal and maximal index π_i. This takes $\mathcal{O}\left(\sum_{(S,T) \in H}(|S| + |T|)\right) = \mathcal{O}(b)$ time. The number of virtual edges created for Step 1 is $|E^*| = |H|$. Counting the crossings caused by E^* can be done in $\mathcal{O}(|E^*| \log|V|) = \mathcal{O}(|H| \log|V|)$ time [4]. Steps 2 and 3 require the creation of a list C_i with $2|H|$ elements, namely the lower-index and the upper-index corners of all hyperedges. Sorting this list is done with $\mathcal{O}(|C_i| \log|C_i|) = \mathcal{O}(|H| \log|H|)$ steps. Afterwards, each element in the list is visited once. The total required time is $\mathcal{O}(b + |H| \log|V| + |H| \log|H|) = \mathcal{O}(b + |H|(\log|V| + \log|H|))$. □

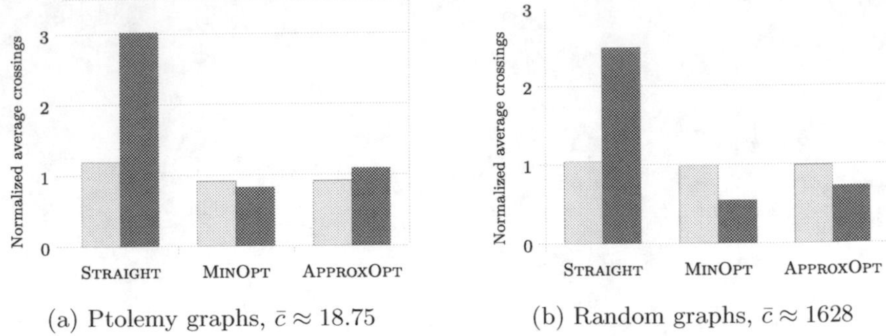

(a) Ptolemy graphs, $\bar{c} \approx 18.75$

(b) Random graphs, $\bar{c} \approx 1628$

Fig. 4. Average number of crossings when using the given cross counting algorithm (light) and average number of crossings predicted by the algorithm (dark).

3 Experimental Evaluation

For evaluating our algorithms, we used 171 diagrams taken from the set of demo models shipping with the Ptolemy open source project [8].[1] Since Ptolemy allows models to be nested using *composite actors* that represent subsystems composed of other actors, we eliminated the hierarchy by moving nested elements out of their composite actors. Diagrams unsuitable for evaluations were left out, e. g. those with very few nodes.

We executed our drawing algorithm once for each cross counting algorithm on each of the selected Ptolemy diagrams. For each execution, the actual number of crossings in the final diagram as well es the number predicted by the cross counting algorithm were measured. The results can be seen in Fig. 4(a). Our proposed methods are by far more accurate at predicting the number of crossings compared to the straight-line method. While the average difference of the actual and predicted numbers of crossings was 35.6 for Straight, the difference averaged 5.3 for MinOpt and 5.7 for ApproxOpt. A further important observation is that the average number of actual crossings is reduced by 23.6% when using MinOpt and by 23.8% when using ApproxOpt instead of Straight. These differences of mean values are significant: the p-values resulting from a t-test with paired samples are 4.5% for MinOpt and 4.0% for ApproxOpt.

We performed a second experiment with randomly generated bipartite graphs with 5 to 100 nodes and 2 to 319 hyperedges each. We performed the same measurements as for the Ptolemy diagrams, the results of which are shown in Fig. 4(b). They confirm the general observations made before. The average number of actual crossings is reduced by 5.6% when using MinOpt and by 4.6% when using ApproxOpt instead of Straight. Although the relative difference of mean values is lower compared to the Ptolemy diagrams, their significance is much higher: $p \approx 7.8 \times 10^{-43}$ for MinOpt and $p \approx 3.2 \times 10^{-32}$ for ApproxOpt.

[1] http://ptolemy.eecs.berkeley.edu

Performance evaluations conducted on a set of 100 randomly generated large bipartite graphs (500 nodes, 3 edges per node) confirmed our theoretical results: STRAIGHT (mean time 0.3ms) was significantly faster than APPROXOPT (1.7ms), which in turn was significantly faster than MINOPT (24ms).

4 Conclusion

We proposed two methods for counting crossings in orthogonal hypergraph drawings more accurately. Our experiments indicate that the algorithms lead to significantly fewer edge crossings both with real-world and with random diagrams.

We see two main areas for future research. First, the number of crossings between orthogonal hyperedges depends not only on the results of the crossing minimization, but also on the exact placement of nodes. However, current node placement algorithms only try to minimize either edge length or the number of bend points. And second, limiting the routing of each hyperedge to one horizontal segment reduces the number of bend points at the expense of edge crossings. Future research could address routing algorithms that reduce the number of edge crossings as well by creating multiple horizontal segments.

References

1. Sugiyama, K., Tagawa, S., Toda, M.: Methods for visual understanding of hierarchical system structures. IEEE Transactions on Systems, Man and Cybernetics 11(2), 109–125 (1981)
2. Eschbach, T., Guenther, W., Becker, B.: Orthogonal hypergraph drawing for improved visibility. Journal of Graph Algorithms and Applications 10(2), 141–157 (2006)
3. Sander, G.: Layout of directed hypergraphs with orthogonal hyperedges. In: Liotta, G. (ed.) GD 2003. LNCS, vol. 2912, pp. 381–386. Springer, Heidelberg (2004)
4. Barth, W., Mutzel, P., Jünger, M.: Simple and efficient bilayer cross counting. Journal of Graph Algorithms and Applications 8(2), 179–194 (2004)
5. Eschbach, T., Guenther, W., Becker, B.: Crossing reduction for orthogonal circuit visualization. In: Proceedings of the 2003 International Conference on VLSI, pp. 107–113. CSREA Press (2003)
6. Nagamochi, H., Yamada, N.: Counting edge crossings in a 2-layered drawing. Information Processing Letters 91(5), 221–225 (2004)
7. Waddle, V., Malhotra, A.: An $E \log E$ line crossing algorithm for levelled graphs. In: Kratochvíl, J. (ed.) GD 1999. LNCS, vol. 1731, pp. 59–71. Springer, Heidelberg (1999)
8. Eker, J., Janneck, J.W., Lee, E.A., Liu, J., Liu, X., Ludvig, J., Neuendorffer, S., Sachs, S., Xiong, Y.: Taming heterogeneity—the Ptolemy approach. Proceedings of the IEEE 91(1), 127–144 (2003)

Evolutionary Meta Layout of Graphs

Miro Spönemann, Björn Duderstadt, and Reinhard von Hanxleden

Department of Computer Science, Christian-Albrechts-Universität zu Kiel
{msp,bdu,rvh}@informatik.uni-kiel.de

Abstract. A graph drawing library is like a toolbox, allowing experts to select and configure a specialized algorithm in order to meet the requirements of their diagram visualization application. However, without expert knowledge of the algorithms the potential of such a toolbox cannot be fully exploited. This gives rise to the question whether the process of selecting and configuring layout algorithms can be automated such that good layouts are produced. In this paper we call this kind of automation "*meta layout.*" We propose a genetic representation that can be used in meta heuristics for meta layout and contribute new metrics for the evaluation of graph drawings. Furthermore, we examine the use of an evolutionary algorithm to search for optimal solutions and evaluate this approach both with automatic experiments and a user study.

Keywords: graph drawing, layout algorithms, evolutionary algorithms, meta layout, readability metrics, user study.

1 Introduction

There are many different approaches for drawing graphs, and all have their specific strengths and weaknesses. Therefore successful graph drawing libraries include multiple algorithms, and usually they offer numerous configuration options to allow users to tailor the generated layouts to their needs. However, the proper choice of a layout algorithm as well as its configuration often require detailed knowledge of the background of these algorithms. Acquiring such knowledge or simply testing all available configuration options is not feasible for users who require quick results.

An inspiring idea was communicated by Biedl et al. [1]: by displaying multiple layouts of the same graph, the user may select those that best match her or his expectations. In this paper we build on that idea and apply meta heuristics for generating a variation of layouts using existing layout libraries.

We introduce the notion of *abstract layout*, that is the annotation of graphs with directives for layout algorithm selection and configuration. *Concrete layout* is a synonym for the *drawing* of a graph and is represented by the annotation of graph elements with position and size data. When a layout algorithm is executed on a graph, it transforms its abstract layout into a concrete layout. By *meta layout* we denote an automatic process of generating abstract layouts.

Our contributions are a genetic representation of abstract layouts, metrics for convenient evaluation of *aesthetic criteria* [2], and operations for applying

T. Dwyer et al. (Eds.): Diagrams 2014, LNAI 8578, pp. 16–30, 2014.

an evolutionary algorithm for meta layout. Furthermore, we propose a simple method for adapting the weights of aesthetic criteria according to the user-selected layouts, supporting the approach of Biedl et al. mentioned above. We performed automated as well as user-based experiments in order to evaluate our proposed method. The results show that the method is well accepted by users and produces at least equally readable drawings compared to a manual configuration approach.

The remainder of this paper is organized as follows. In Sect. 2 we discuss related work on evolutionary graph layout and layout configuration. Sect. 3 introduces the necessary data structures and fitness function that enable the evolutionary process, which is described in Sect. 4. In Sect. 5 we report experimental results on the effectiveness and applicability of these methods. We conclude and outline prospective work in Sect. 6.

2 Related Work

Several authors have proposed evolutionary algorithms where the individuals are represented by lists of coordinates for the positions of the nodes of a graph [3,4,5,6,7,8,9]. Here, in contrast, we do not include any specific graph in our encoding of individuals, hence we can apply the result of our evolutionary algorithm to any graphs, even if they were not considered during the evolutionary process. Furthermore, we benefit from all features that are already supported by the existing algorithms, while previous approaches for evolutionary layout were usually restricted to draw edges as straight lines and did not consider additional features such as edge labels.

Other works have focused on integrating meta heuristics in existing layout methods. De Mendonça Neto and Eades proposed a system for automatic learning of parameters of a simulated annealing algorithm [10]. Utech et al. introduced a genetic representation that combines the layer assignment and node ordering steps of the layer-based drawing approach with an evolutionary algorithm [11]. Such a combination of multiple NP-hard steps is also applied by Neta et al. for the *topology-shape-metrics* approach [12]. They use an evolutionary algorithm to find planar embeddings (*topology* step) for which the other steps (*shape* and *metrics*) are able to create good layouts.

Bertolazzi et al. proposed a system for automatic selection of layout algorithms that best match the user's requirements [13]. The system is initialized by evaluating the available algorithms with respect to a set of aesthetic criteria using randomly generated graphs of different sizes. The user has to provide a ranking of the criteria according to her or his preference. When a layout request is made, the system determines the difference between the user's ranking and the evaluation results of each algorithm for graphs of similar size as the current input graph. The algorithms with the lowest difference are offered to the user.

Similarly, Niggemann and Stein proposed to build a database that maps vectors of structural graph features, e. g. the number of nodes and the number of connected components, to the most suitable layout algorithm with respect to some predefined combination of aesthetic criteria [14]. These data are gathered by applying the algorithms to a set of "typical" graphs. A suitable algorithm for a given input graph is chosen by measuring its structural features and comparing them with the entries present in the database. Both the approaches of Bertolazzi et al. and Niggemann and Stein are restricted to selecting layout algorithms. Here, in contrast, we seek to configure arbitrary parameters of algorithms in addition to their selection.

Archambault et al. combined graph clustering with layout algorithm selection in a *multi-level* approach [15]. The clustering process is tightly connected with the algorithm selection, since both aspects are based on topological features of the input graph. When a specific feature is found, e. g. a tree or a clique, it is extracted as a subgraph and processed with a layout algorithm that is especially suited for that feature. This kind of layout configuration depends on detailed knowledge of the behavior of the algorithms, which has to be encoded explicitly in the system, while the solution presented here can be applied to any algorithm independently of their behavior.

3 Genotypes and Phenotypes

The *genotype* of an individual is its genetic code, while the *phenotype* is the total of its observable characteristics. In biology a phenotype is formed from its genotype by growing in a suitable environment. We propose to use abstract layouts (configurations) as genotypes, and concrete layouts (drawings) as phenotypes. The "environment" for this kind of phenotypes is a graph. We generate the concrete layout $L(\lambda)$ that belongs to a given abstract layout λ by applying all parameters encoded in λ to the chosen layout algorithm A, which is also encoded in λ, and executing A on the graph given by the environment. This encoding of parameters and algorithm selection is done with a set of *genes*, which together form a *genome*. A gene consists of a *gene type* with an assigned value. The gene type has an identifier, a data type (integer, floating point, Boolean, or enumeration), optional lower and upper bounds, and an optional parameter controlling the standard deviation of Gaussian distributions.

We assign each layout algorithm to a *layout type* depending on the underlying approach implemented in the algorithm. The main layout types are *layer-based*, *force-based*, *circular*, *orthogonal*, *tree*, and *planar*. Each algorithm A has a set P_A of parameters that control the behavior of A. We consider the union $\mathcal{P} = \bigcup P_A$ of all parameters, which we call the set of *layout options*. Each genome contains a gene g_{T} for selecting the layout type, a gene g_{A} for selecting the layout algorithm, and one for each layout option in $\mathcal{P}$. It is also possible to use only a subset of these genes, as long as all generated genomes contain the same subset. Such a restriction can serve to focus on the selected layout options in the optimization process, while other options are kept constant.

Some genes of a genome are dependent of each other. The gene g_{A}, for instance, is constrained to a layout algorithm that belongs to the layout type selected in g_{T}. Furthermore, the layout algorithm A selected in g_{A} does not

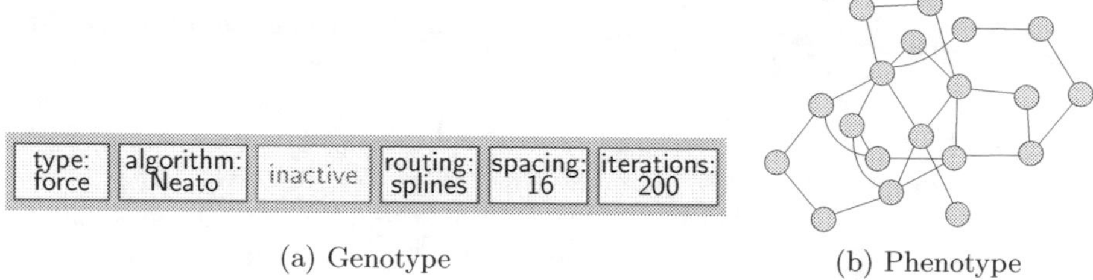

(a) Genotype (b) Phenotype

Fig. 1. (a) A genome with six genes. The layout type gene is set to force-based algorithms, the layout algorithm gene is set to a specific algorithm named "Neato", and three parameters of that algorithm are set with the remaining genes. One gene is inactive because the corresponding layout option is not supported by Neato. (b) A phenotype of the genome, represented by a layout generated by Neato for an arbitrary graph.

support all layout options in $\mathcal{P}$, therefore the options in $\mathcal{P} \setminus P_A$, i. e. those not supported by A, are marked as *inactive*. A genome with six genes and a possible phenotype are shown in Fig. 1.

Inactive genes of a genome X do not contribute to the characteristics of the phenotype of X, i. e. of its drawing, hence two genomes that differ only in their inactive genes may produce the same drawing. On the other hand, some layout algorithms are randomized and produce different drawings when executed twice with the same configuration. However, we assume that drawings that result from the same configuration tend to be similar with respect to our fitness function, hence this ambiguity is probably not noticeable in practice.

3.1 Fitness Function

Our genotypes have a completely different representation compared to previous evolutionary layout algorithms. The phenotypes, in contrast, are commonly represented by graph layouts, hence we can apply the same approach to fitness evaluation as previous solutions, that is the evaluation of aesthetic criteria [2].

Some authors used a linear combination of specific criteria as fitness function [16,4,5]. For instance, given a graph layout L, the number of edge crossings $\kappa(L)$, and the standard deviation of edge lengths $\delta(L)$, the optimization goal could be to minimize the cost function $f(L) = w_c\kappa(L) + w_d\delta(L)$, where suitable scaling factors w_c and w_d are usually determined experimentally. The problem of this approach is that the values resulting from $f(L)$ have no inherent meaning apart from the general assumption "the smaller $f(L)$, the better the layout L." As a consequence, the cost function can be used only as a relative measure, but not to determine the absolute quality of layouts.

An improved variant, proposed by several authors, is to normalize the criteria to the range between 0 and 1 [17,2,7,8,9]. However, this is still not sufficient to effectively measure absolute layout quality. For instance, Tettamanzi normalizes the edge crossings $\kappa(L)$ with the formula $\mu_c(L) = \frac{1}{\kappa(L)+1}$ [8]. For the complete graph K_5, which is not planar, even the best layouts yield a result of $\mu_c(L) = 50\%$, suggesting that the layout is only half as good as it could be. Purchase proposed to scale the number of crossings against an upper bound κ_{max} defined as the number that results when all pairs of edges that are not incident to the same node cross each other [2]. Her formula is $\mu_c(L) = 1 - \frac{\kappa(L)}{\kappa_{max}}$ if $\kappa_{max} > 0$ and $\mu_c(L) = 1$ otherwise. Purchase herself notes that this definition "is biased towards high values." For instance, the graph N14 used in her evaluations has 24 nodes, 36 edges, and $\kappa_{max} = 558$. All layouts with up to 56 crossings would result in $\mu_c(L) > 90\%$. When tested with a selection of 28 layout algorithms, all of them resulted in layouts with less that 56 crossings (the best had only 11 crossings), hence the formula of Purchase would assign a very high fitness to all these generated layouts.

We propose new normalization functions that aim at well-balanced distributions of values among typical results of layout algorithms. A *layout metric* is a function μ that maps graph layouts L to values $\mu(L) \in [0,1]$. Given layout metrics $\mu_1, \ldots, \mu_k$ with weights $w_1, \ldots, w_k \in [0,1]$, we compute the fitness of a graph layout L by

$$f(L) = \frac{1}{\sum_{i=1}^{k} w_k} \sum_{i=1}^{k} w_k \mu_k(L) \ . \tag{1}$$

In the following we describe some of the metrics we have used in conjunction with our proposed genotype representation and evolutionary algorithm. The goal of these metrics is to allow an intuitive assessment of the respective criteria, which means that the worst layouts shall have metric values near 0%, the best ones shall have values near 100%, and moderate ones shall score around 50%. The metrics should be parameterized such that this spectrum of values is exhausted for layouts that are generated by typical layout algorithms, allowing to clearly distinguish them from one another. We evaluated to which extent our proposed metrics satisfy these properties based on an experimental analysis. The results confirm that our formulae are very suited to the stated goals. However, we omit the details of these experiments due to space limitations; they are available in a technical report [18], which also contains formulae for more aesthetic criteria.

The basic idea behind each of our formulae is to define a certain *input split value* x_s such that if the value of the respective criterion equals x_s, the metric is set to a defined *output split value* μ^*. Values that differ from x_s are scaled towards 0 or 1, depending on the specific criterion. The advantage of this approach is that different formulae can be applied to the ranges below and above the split value, simplifying the design of metrics that meet the goals stated above. The approach involves several constants, which we determined experimentally.

Let $G = (V, E)$ be a directed graph with a layout L. Let $n = |V|$ and $m = |E|$.

Number of Crossings. Similarly to Purchase we define a virtual upper bound $\kappa_{max} = m(m-1)/2$ on the number of crossings [2]. We call that bound virtual because it is valid only for straight-line layouts, while layouts where edges have bend points can have arbitrarily many crossings. Based on the observation that crossings tend to be more likely when there are many edges and few nodes, we further define an input split value

$$\kappa_{\mathrm{s}} = \min\left\{\frac{m^3}{n^2},\ (1-\mu_{\mathrm{c}}^*)\kappa_{\max}\right\}\ . \tag{2}$$

μ_{c}^* is the corresponding output split value, for which we chose $\mu_{\mathrm{c}}^* = 10\%$. The exponents of m and n are chosen such that the split value becomes larger when the m/n ratio is high. We denote the number of crossings as $\kappa(L)$. Layouts with $\kappa(L) < \kappa_{\mathrm{s}}$ yield metric values above μ_{c}^*, while layouts with $\kappa(L) > \kappa_{\mathrm{s}}$ yield values below μ_{c}^*. This is realized with the formula

$$\mu_{\mathrm{c}}(L) = \begin{cases} 1 & \text{if } \kappa_{\max} = 0, \\ 0 & \text{if } \kappa(L) \geq \kappa_{\max} > 0, \\ 1 - \frac{\kappa(L)}{\kappa_{\mathrm{s}}}(1-\mu_{\mathrm{c}}^*) & \text{if } \kappa(L) \leq \kappa_{\mathrm{s}}, \\ \left(1 - \frac{\kappa(L)-\kappa_{\mathrm{s}}}{\kappa_{\max}-\kappa_{\mathrm{s}}}\right)\mu_{\mathrm{c}}^* & \text{otherwise.} \end{cases} \tag{3}$$

Area. Let $w(L)$ be the width and $h(L)$ be the height of the drawing L. The area required to draw a graph depends on the number of nodes and edges, hence we define a *relative area*

$$\alpha(L) = \frac{w(L)h(L)}{(n+m)^2} \tag{4}$$

that takes into account the number of elements in the graph. We square that number because we observed that many drawings of larger graphs require a disproportionately high area. We split the output values at two points $\mu_{\mathrm{a}}^* = 10\%$ and $\mu_{\mathrm{a}}^{**} = 95\%$, with corresponding input split values α_{s1} and α_{s2}. Values below μ_{a}^* are met when $\alpha(L) > \alpha_{\mathrm{s1}}$, values above μ_{a}^{**} are met when $\alpha(L) < \alpha_{\mathrm{s2}}$, and values in-between are scaled proportionally. The constants α_{s1} and α_{s2} have been determined experimentally as 1000 and 50, respectively. We define the area metric as

$$\mu_{\mathrm{a}}(L) = \begin{cases} \frac{\alpha_{\mathrm{s1}}}{\alpha(L)}\mu_{\mathrm{a}}^* & \text{if } \alpha(L) > \alpha_{\mathrm{s1}}, \\ 1 - \frac{\alpha(L)}{\alpha_{\mathrm{s2}}}(1-\mu_{\mathrm{a}}^{**}) & \text{if } \alpha(L) < \alpha_{\mathrm{s2}}, \\ \left(1 - \frac{\alpha(L)-\alpha_{\mathrm{s2}}}{\alpha_{\mathrm{s1}}-\alpha_{\mathrm{s2}}}\right)(\mu_{\mathrm{a}}^{**} - \mu_{\mathrm{a}}^*) + \mu_{\mathrm{a}}^* & \text{otherwise.} \end{cases} \tag{5}$$

Edge Length Uniformity. We measure this criterion with the standard deviation $\sigma_\lambda(L)$ of edge lengths and compare it against the average edge length $\bar{\lambda}(L)$, which we use as input split value. We define

$$\mu_{\mathrm{u}}(L) = \begin{cases} \frac{\bar{\lambda}(L)}{\sigma_\lambda(L)}\mu_{\mathrm{u}}^* & \text{if } \sigma_\lambda(L) \geq \bar{\lambda}(L), \\ 1 - \frac{\sigma_\lambda(L)}{\bar{\lambda}(L)}(1-\mu_{\mathrm{u}}^*) & \text{otherwise,} \end{cases} \tag{6}$$

where the output split value $\mu_{\mathrm{u}}^* = 20\%$ corresponds to the metric value that results when the standard deviation equals the average.

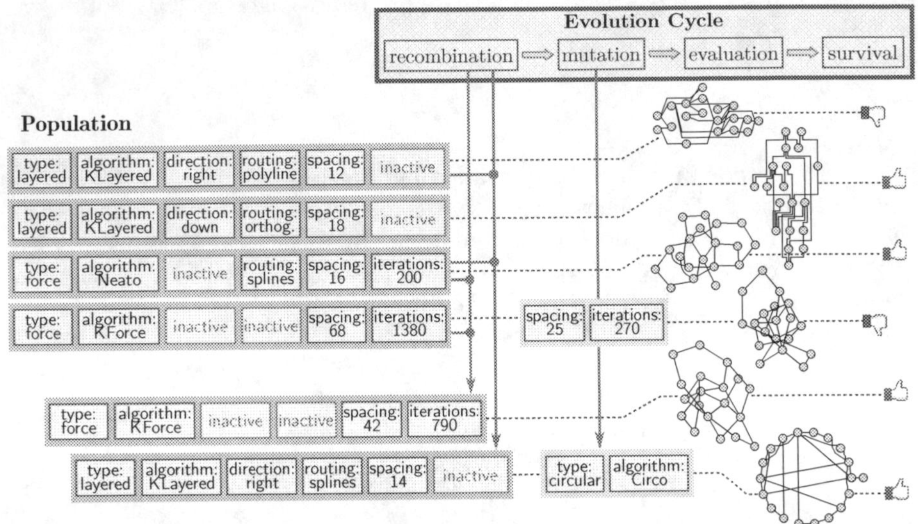

Fig. 2. Evolutionary layout example: starting with a population of four genomes, two new genomes are created through recombination, two genomes are mutated, and four of the resulting genomes survive after their evaluation

4 Evolutionary Process

The genetic encoding presented in Sect. 3 can serve as basis for numerous meta heuristics. In this section we discuss one possible heuristic with the goal of creating a starting point of further research, without claiming that this is the ultimate solution. We use an evolutionary algorithm, a popular method for searching large solution spaces.

A *population* is a set of genomes. An *evolution cycle* is a function that modifies a population with four steps, which are explained below. The evolutionary algorithm executes the evolution cycle repeatedly, checking some termination condition after each execution. Simple conditions for fully automatic optimization are to limit the number of iterations and to check whether the fitness of the best individual exceeds a certain threshold. Alternatively, the user can be involved by manually controlling when to execute the next evolution cycle and when to stop the process. The four steps of the evolution cycle are discussed in the following and exemplified in Fig. 2.

1. Recombination. New genomes are created by crossing random pairs of existing genomes. A crossing of two genomes is created by crossing all their genes. Two integer or floating point typed genes are crossed by computing their average value, while for other data types one of the two values is chosen randomly. Only a selection of the fittest individuals is considered for mating.

When the parent genomes have different values for the layout algorithm gene, the child is randomly assigned one of these algorithms. As a consequence, the *active / inactive* statuses of the other genes of the child must be adapted such that they match the chosen algorithm A: each gene g is made active if and only if A supports the layout option associated to g.

2. Mutation. Genomes have a certain probability of mutating. A mutation is done by randomly modifying its genes, where each gene g has an individual mutation probability p_g depending on its type. We assign the highest p_g values to genes with integer or floating point values, medium values to genes with Boolean or enumeration values, and the lowest values to the layout algorithm and layout type genes. Let g be a gene with value x. If the data type of g is integer or floating point, the new value x' is determined using a Gaussian distribution using x as its average and the standard deviation assigned to the gene type of g. If x' exceeds the upper or lower bound assigned to the gene type of g, it is corrected to a value between x and the respective bound. For genes with other types, which have no specific order, a new value is chosen based on a uniform distribution over the finite set of values, excluding the previous value. When the layout algorithm gene mutates, the *active / inactive* statuses of other genes must be updated as described for the recombination step.

3. Evaluation. A fitness value is assigned to each genome that does not have one yet (see Sect. 3.1), which involves executing the encoded layout algorithm in order to obtain a corresponding phenotype. The population is sorted using these fitness values.

4. Survival. Only the fittest individuals survive. Checking all genomes in order of descending fitness, we include each genome X in the set of survivors if and only if it meets the following requirements: (i) its fitness exceeds a certain minimum, (ii) the maximal number of survivors is not reached yet, and (iii) the distance of X to other individuals is sufficient. The latter requirement serves to support the diversity of the population. Comparing all pairs of individuals would require a quadratic number of distance evaluations, therefore we determine the distance only to some random samples X' from the current set of survivors. We determine the distance $d(X, X')$ of two individuals X, X' by computing the sum of the differences of the gene values. In order to meet the third requirement, $d(X, X') \geq d_{\min}$ must hold for a fixed minimal distance $d_{\min}$.

4.1 Choosing Metric Weights

The fitness function discussed in Sect. 3.1 uses layout metrics $\mu_1, \ldots, \mu_k$ and weights $w_1, \ldots, w_k \in [0, 1]$, where each w_i controls the influence of μ_i on the computed fitness. The question is how to choose suitable weights. Masui proposed to apply genetic programming to find a fitness function that best reflects the user's intention [19]. The computed functions are evolved as Lisp programs and are evaluated with layout examples, which have to rated as "good" or "bad" by the user. A similar approach is used by Barbosa and Barreto [3], with the main difference that the fitness function is evolved indirectly by modifying a set of weights with an evolutionary algorithm. Additionally, they apply another evolutionary algorithm to create concrete layouts of a given graph. Both algorithms are combined in a process called *co-evolution*: the results of the weights evolution are used for the fitness function of the layout evolution, while the fitness of the weights is determined based on user ratings of sample layouts.

We have experimented with two much simpler methods, both of which involve the user: (a) the user directly manipulates the metric weights with sliders allowing values between 0 and 1, and (b) the user selects good layouts from the current population and the metric weights are automatically adjusted according to the selection. This second method builds on the assumption that the considered layout metrics are able to compute meaningful estimates of the absolute quality of any given layout (see Sect. 3.1). The higher the result of a metric, the higher its weight shall be. Let $\bar{\mu}_1, \ldots, \bar{\mu}_k$ be the average values of the layout metrics $\mu_1, \ldots, \mu_k$ for the selected layouts. Furthermore, let $w_1, \ldots, w_k$ be the current metric weights. For each $i \in \{1, \ldots, k\}$ we determine a *target weight*

$$w_i^* = \begin{cases} 1 - \frac{1}{2}\left(\frac{1-\bar{\mu}_i}{1-\mu_{\mathrm{w}}^*}\right)^2 & \text{if } \bar{\mu}_i \geq \mu_{\mathrm{w}}^*, \\ \frac{1}{2}\left(\frac{\bar{\mu}_i}{\mu_{\mathrm{w}}^*}\right)^2 & \text{otherwise,} \end{cases} \tag{7}$$

where μ_{w}^* is a constant that determines which metric result is required to reach a target weight of 50%. We chose $\mu_{\mathrm{w}}^* = 70\%$, meaning that mediocre metric results are mapped to rather low target weights. The square functions in Equation 7 are used to push extreme results even more towards 0 or 1. The new weight of the layout metric μ_i is $w_i' = \frac{1}{2}(w_i + w_i^*)$, i. e. the mean of the old weight and the target weight.

4.2 User Interface

We have experimented with a user interface that includes both variants for modifying metric weights, shown in Fig. 3. The window visualizes populations by presenting up to 16 small drawings of the evaluation graph, which represent the fittest individuals of the current population. 13 metrics are shown on the side of the window. The user may use the controls in the window to

- view the computed values of the layout metrics for an individual,
- directly set the metric weights,
- select one or more favored individuals for indirect adjustment of weights,
- change the population by executing an evolution cycle ("Evolve" button),
- restart the evolution with a new initial population ("Restart" button), and
- finish the process and select the abstract layout encoded in a selected individual ("Apply" button).

The indirect method for choosing weights, which adapts them according to the user's selection of favored layouts, is in line with the *multidrawing* approach introduced by Biedl et al. [1]. The main concept of that approach is that the user can select one of multiple offered drawings without the need of defining her or his goals and preferences in the first place. The multidrawing system reacts

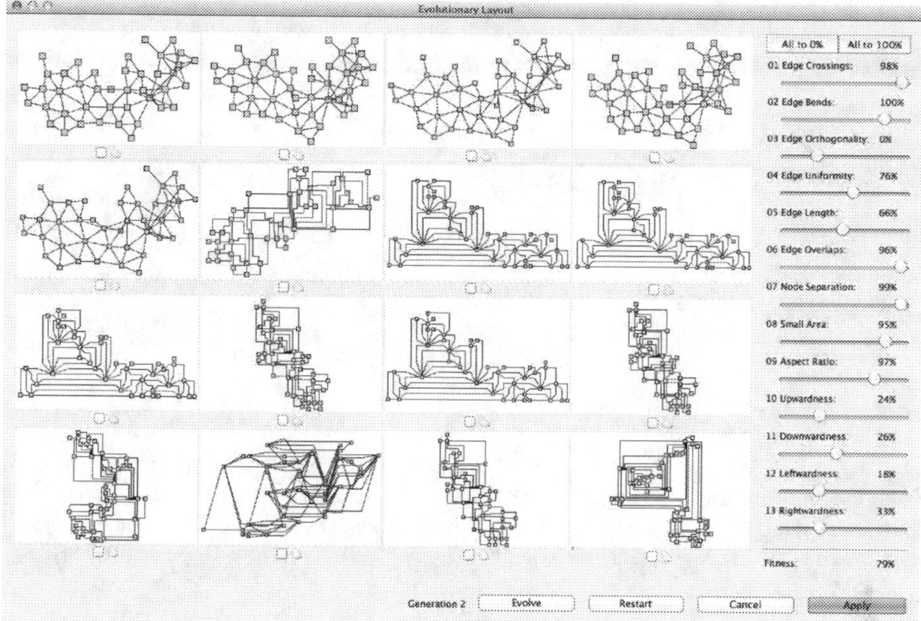

Fig. 3. User interface for evolutionary meta layout, showing drawings for 16 individuals of the current population. The check box below each proposed graph drawing is used to select favored layouts for automatic adaption of metric weights. The sliders on the right offer direct manipulation of the weights.

on the user's selection and generates new layouts that are similar to the selected ones. In our proposed method, this similarity is achieved by adjusting the fitness function such that the selected layouts are assigned a higher fitness, granting them better prospects in the competition against other layouts.

5 Evaluation

The methods presented in this paper have been implemented and evaluated in KIELER, an Eclipse-based open source project.[1] Our experiments included four layout algorithms provided by KIELER as well as five algorithms from the Graphviz library [20] and 22 algorithms from the OGDF library [21]. The total number of genes in each genome was 79.

5.1 Execution Time

We tested the performance of evolutionary meta layout on a set of 100 generated graphs with varying number of nodes $2 \le n \le 100$ and $e = 1.5\,n$ edges. The

[1] `http://www.informatik.uni-kiel.de/rtsys/kieler/`

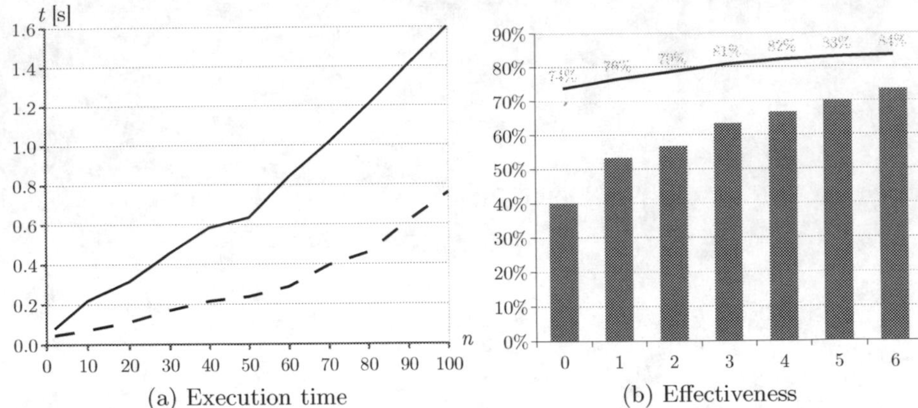

(a) Execution time (b) Effectiveness

Fig. 4. (a) Execution time t plotted by number of nodes n with single core execution (solid line) and multicore execution (dashed line). (b) Result of the edge uniformity experiment. The line on top shows the fitness values of the best genomes for iterations 0 to 6 (horizontal axis), while the bars show the fractions of genomes that are set to force-type algorithms.

tests have been executed with an Intel Xeon 2.5 GHz CPU. The population contained 16 genomes, the recombination operation bred 13 new genomes, and the mutation operation affected 60% of the whole population, thus 22.6 new genomes were created on average. This means that about 23 layout algorithm executions had to be performed for the evaluation operation of each evolution cycle, and each layout metric has been evaluated just as often. We measured the average execution time of one evolution cycle (i. e. a single iteration), which led to the results shown in Fig. 4a. The vast majority of time is spent in the evaluation step: on average 74% is taken by layout algorithm execution, and 20% is taken by metrics evaluation. The rather high execution time limits the number of evolution cycles that can be performed in an interactive environment. The consequence is that the evolutionary algorithm has to converge to an acceptable solution within few iterations. However, the evaluation step is very suitable for parallelization, since the evaluations are all independent. As seen in Fig. 4a, the total execution time can be reduced by half when run with multiple threads on a multicore machine (eight cores in this example).

5.2 Programmatic Experiments

We carried out three experiments in order to verify the effectiveness of the evolutionary approach. The experiments had different optimization goals: minimal number of edge crossings, maximal number of edges pointing left, and optimal uniformity of edge lengths. In each experiment the corresponding layout metric was given a weight of 100%, while most other metrics were deactivated (except some basic metrics avoiding graph elements overlapping each other). The optimization goals were chosen such that they can be mapped to certain kinds of layout algorithms, allowing to validate the results according to prior knowledge of their behavior. 30 randomly generated graphs were used as evaluation graphs. In the crossing minimization experiment, for 60% of the graphs a planarization-based algorithm was selected as the genome with highest fitness after three or less iterations. This confirms the intuitive expectation, since planarization methods are most effective in minimizing edge crossings. In the experiment that aimed at edges pointing left, for 90% of the graphs a layer-based algorithm was selected as the genome with highest fitness after three or less iterations. Additionally, the layout option that determines the main direction of edges had to be set to *left*, which was accomplished in 83% of the cases. In the edge uniformity experiment, a force-based algorithm was selected for 63% of the graphs after three iterations, and for 73% of the graphs after six iterations (see Fig. 4b). This result matches the expectation, too, because force-based methods aim at drawing all edges with uniform length. In all experiments it could be observed that the average rating of genomes was consistently increasing after each iteration, but this increase became smaller with each iteration. We conclude that our proposed evolutionary meta layout approach can effectively optimize given aesthetic criteria, and in most cases the kind of layout algorithm that is automatically selected is consistent with the intuition. A very relevant observation is that the process tends to converge very quickly, often yielding good solutions after few iterations, e. g. as illustrated in Fig. 4b. On the other hand, in some cases the computation is trapped in local optima, which could possibly be avoided by improving the parameters of the evolutionary computation.

5.3 User Study

We have conducted a user study to determine the practical usefulness of our approach. The study is based on a set of 8 graphs, inspired by real-world examples that were found on the web, with between 15 and 43 nodes and 18 to 90 edges. 25 persons participated in the study: four members of our research group, 17 computer science students, and four persons who were not involved in computer science. The research group members are experts in graph layout technology and are likely to have predetermined opinions about which layout configurations to use in certain contexts, while the other participants can be regarded as novices for that matter. We expected that novice users would benefit more strongly from the evolutionary meta layout approach compared to the experts.

For each graph, the participants were presented three tasks regarding connectivity, e. g. finding the shortest path between two given nodes. The participants then had to find a layout configuration which they regarded as useful for working on the tasks. The test instructions encouraged the participants to improve the layout configuration until they were sure they had found a well readable layout.

Four of the graphs were treated with the user interface presented in Sect. 4.2, named Evol in the following, which evolves a population of layout configurations and lets users pick configurations by their previews. They were free to use both the direct and the indirect method for modifying weights (Sect. 4.1). For the other four graphs, the participants were required to find layout configurations manually by choosing from a list of available layout algorithms and modifying parameters of the chosen algorithms. For each participant we determined randomly which graphs to treat with EVOL and which to configure with the manual method, called MANUAL in the following. After the participants had accepted a layout configuration for a graph, they worked on the respective tasks by inspecting the drawing that resulted from the configuration.

After all graphs were done, the participants were asked 6 questions about their subjective impression of the evolutionary approach. The overall response to these questions was very positive: on a scale from −2 (worst rating) to 2 (best rating), the average ratings were 1.0 for the quality of generated layouts, 0.8 for their variety, 1.2 for the time required for finding suitable layouts, 0.6 for the effectiveness of manually setting metric weights, and 1.5 for the effectiveness of adjusting metric weights by favoring individuals. Most notably, the indirect adjustment of metric weights was rated much higher than their direct manipulation. This indicates that most users prefer an intuitive interface based on layout proposals instead of manually setting parameters of the fitness function, since the latter requires to understand the meaning of all layout metrics.

The average rate of correct answers of non-expert users to the tasks was 77.4% for MANUAL and 79.8% for EVOL. The average time used to work on each task was lower by 7.5% with EVOL (131 seconds) compared to MANUAL (142 seconds). These differences are not statistically significant: the p-values resulting from a t-test on the difference of mean values are 29% for the correctness of answers and 23% for the working time. A more significant result ($p = 8.3\%$) is obtained when comparing the differences of EVOL and MANUAL working times between expert users and non-expert users. In contrast to the non-experts, expert users took more time to work on the tasks with EVOL (126 seconds) compared to MANUAL (107 seconds). Furthermore, the average rate of correct answers of expert users was equal for both methods. This confirms the assumption that the method proposed in this paper is more suitable in applications used by persons without expert knowledge on graph drawing.

Many participants commented that they clearly preferred EVOL over MANUAL. It could be observed that novice users were overwhelmed by the number of configuration parameters shown for the manual method. In many cases, they stopped trying to understand the effects of the parameters after some unsuccessful attempts to fine-tune the layout. Therefore the average time taken for finding a layout was lower for MANUAL (129 seconds) compared to EVOL (148 seconds). For the EVOL interface, on the other hand, similarly frustrating experiences were observed in few cases where the evolutionary algorithm apparently ran into local optima that did not satisfy the users' expectations. In these cases users were forced to restart the process with a new population.

The average number of applied evolution cycles was 3.1, which means that in most cases the participants found good solutions after very few iterations of the evolutionary algorithm. Furthermore, we measured the index of the layout chosen for working on the tasks on a scale from 0 to 15. The layout with index 0 has the highest fitness in the current population, while the layout with index 15 is the one with lowest fitness from the 16 fittest individuals. The average selected index was 2.3, a quite low value, suggesting that the computed fitness has a high correlation with the perceived quality of the layouts.

6 Conclusion

We introduced the notion of meta layout, which means creating an abstract layout by choosing and parameterizing a layout algorithm, which in turn generates a concrete layout of a graph. We presented a genetic representation of abstract layouts, layout metrics for building a fitness function, and an evolutionary algorithm for developing a population of abstract layouts. Furthermore, we proposed a simple method for the indirect adjustment of weights of layout metrics. Since the result of the evolutionary computation is not a concrete layout, but a layout configuration, it can be applied to any graph without repeating the process.

Our experiments partially confirmed the usefulness of the presented methods. Participants of the user study clearly preferred the evolutionary approach over the manual setting of parameters for layout algorithms, and they also liked to modify the fitness function indirectly rather than to adjust weights directly. The objective results about the effectivity working on tasks about graph connectivity were not statistically significant with respect to comparing our proposed method with the manual method. However, non-expert users clearly profited from the evolutionary method more than the experts did.

The evolutionary algorithm presented here is not the only heuristic for optimizing abstract layouts. Further work could evaluate other optimization heuristics that build on our genetic representation and compare them to the results of this paper. For instance, using a divide-and-conquer approach one could separately optimize each parameter of the layout algorithms, one after another.

References

1. Biedl, T.C., Marks, J., Ryall, K., Whitesides, S.H.: Graph multidrawing: Finding nice drawings without defining nice. In: Whitesides, S.H. (ed.) GD 1998. LNCS, vol. 1547, pp. 347–355. Springer, Heidelberg (1999)
2. Purchase, H.C.: Metrics for graph drawing aesthetics. Journal of Visual Languages and Computing 13(5), 501–516 (2002)
3. Barbosa, H.J.C., Barreto, A.M.S.: An interactive genetic algorithm with co-evolution of weights for multiobjective problems. In: Proceedings of the Genetic and Evolutionary Computation Conference (GECCO 2001), pp. 203–210 (2001)
4. Branke, J., Bucher, F., Schmeck, H.: Using genetic algorithms for drawing undirected graphs. In: Proceedings of the Third Nordic Workshop on Genetic Algorithms and their Applications, pp. 193–206 (1996)
5. Eloranta, T., Mäkinen, E.: TimGA: A genetic algorithm for drawing undirected graphs. Divulgaciones Matemáticas 9(2), 155–170 (2001)

6. Groves, L.J., Michalewicz, Z., Elia, P.V., Janikow, C.Z.: Genetic algorithms for drawing directed graphs. In: Proceedings of the 5th International Symposium on Methodologies for Intelligent Systems, pp. 268–276 (1990)
7. Rosete-Suarez, A., Ochoa-Rodriguez, A.: Genetic graph drawing. In: Nolan, P., Adey, R.A., Rzevski, G. (eds.) Applications of Artificial Intelligence in Engineering XIII, Software Studies, vol. 1. WIT Press / Computational Mechanics (1998)
8. Tettamanzi, A.G.: Drawing graphs with evolutionary algorithms. In: Parmee, I.C. (ed.) Adaptive Computing in Design and Manufacture, pp. 325–337. Springer, London (1998)
9. Vrajitoru, D.: Multiobjective genetic algorithm for a graph drawing problem. In: Proceedings of the Midwest Artificial Intelligence and Cognitive Science Conference, pp. 28–43 (2009)
10. de Mendonça Neto, C.F.X., Eades, P.D.: Learning aesthetics for visualization. In: Anais do XX Seminário Integrado de Software e Hardware, pp. 76–88 (1993)
11. Utech, J., Branke, J., Schmeck, H., Eades, P.: An evolutionary algorithm for drawing directed graphs. In: Proceedings of the International Conference on Imaging Science, Systems, and Technology (CISST 1998), pp. 154–160. CSREA Press (1998)
12. de Mendonça Neta, B.M., Araujo, G.H.D., Guimarães, F.G., Mesquita, R.C., Ekel, P.Y.: A fuzzy genetic algorithm for automatic orthogonal graph drawing. Applied Soft Computing 12(4), 1379–1389 (2012)
13. Bertolazzi, P., Di Battista, G., Liotta, G.: Parametric graph drawing. IEEE Transactions on Software Engineering 21(8), 662–673 (1995)
14. Niggemann, O., Stein, B.: A meta heuristic for graph drawing: learning the optimal graph-drawing method for clustered graphs. In: Proceedings of the Working Conference on Advanced Visual Interfaces (AVI 2000), pp. 286–289. ACM, New York (2000)
15. Archambault, D., Munzner, T., Auber, D.: Topolayout: Multilevel graph layout by topological features. IEEE Transactions on Visualization and Computer Graphics 13(2), 305–317 (2007)
16. Barreto, A.M.S., Barbosa, H.J.C.: Graph layout using a genetic algorithm. In: Proc. of the 6th Brazilian Symposium on Neural Networks, pp. 179–184 (2000)
17. Dunne, C., Shneiderman, B.: Improving graph drawing readability by incorporating readability metrics: A software tool for network analysts. Tech. Rep. HCIL-2009-13, University of Maryland (2009)
18. Spönemann, M., Duderstadt, B., von Hanxleden, R.: Evolutionary meta layout of graphs. Technical Report 1401, Christian-Albrechts-Universität zu Kiel, Department of Computer Science (January 2014) ISSN 2192-6247
19. Masui, T.: Evolutionary learning of graph layout constraints from examples. In: Proceedings of the 7th Annual ACM Symposium on User Interface Software and Technology (UIST 1994), pp. 103–108. ACM (1994)
20. Gansner, E.R., North, S.C.: An open graph visualization system and its applications to software engineering. Software—Practice and Experience 30(11), 1203–1234 (2000)
21. Chimani, M., Gutwenger, C., Jünger, M., Klau, G.W., Klein, K., Mutzel, P.: The Open Graph Drawing Framework (OGDF). In: Tamassia, R. (ed.) Handbook of Graph Drawing and Visualization, pp. 543–569. CRC Press (2013)

Seeing Around Corners: Fast Orthogonal Connector Routing

Kim Marriott[1], Peter J. Stuckey[2], and Michael Wybrow[1]

[1] Caulfield School of Information Technology,
Monash University, Caulfield, Victoria 3145, Australia
{Michael.Wybrow,Kim.Marriott}@monash.edu
[2] Department of Computing and Information Systems,
University of Melbourne, Victoria 3010, Australia
pstuckey@unimelb.edu.au

Abstract. Orthogonal connectors are used in drawings of many types of network diagrams, especially those representing electrical circuits. One approach for routing such connectors has been to compute an orthogonal visibility graph formed by intersecting vertical and horizontal lines projected from the corners of all obstacles and then use an A* search over this graph. However the search can be slow since many routes are in some sense topologically equivalent. We introduce obstacle-hugging routes which we conjecture provide a canonical representative for a set of topologically equivalent routes. We also introduce a new *1-bend visibility graph* that supports computation of these canonical routes. Essentially this contains a node for each obstacle corner and connector endpoint in the diagram and an edge between two nodes iff they can be connected using an orthogonal connector with one bend. We show that the use of a 1-bend visibility graph significantly improves the speed of orthogonal connector routing.

Keywords: orthogonal routing, visibility graphs, circuit diagrams.

1 Introduction

Most interactive diagram editors provide some form of automatic connector routing between shapes whose position is fixed by the user. Usually the editor computes an initial automatic route when the connector is created and updates this each time the connector end-points (or attached shapes) are moved. Orthogonal connectors, which consist of a sequence of horizontal and vertical line segments, are a particularly common kind of connector, used in ER and UML diagrams among others. Wybrow *et al.* [7] gave polynomial time algorithms for automatic object-avoiding orthogonal connector routing which are guaranteed to minimise length and number of bends.

The connector routing algorithm given in Wybrow *et al.* [7] uses a three stage process. The first stage computes an *orthogonal visibility graph* in which edges in the graph represent horizontal or vertical lines of visibility from the corners and connector ports of each obstacle. Connector routes are found using an A* search through the orthogonal visibility graph that finds a route that minimizes bends and overall connector length. Finally, the actual visual route is computed. This step orders and nudges apart

T. Dwyer et al. (Eds.): Diagrams 2014, LNAI 8578, pp. 31–37, 2014.

the connectors in shared segments so as to ensure that unnecessary crossings are not introduced, that crossings occur at the start or end of the shared segment and that connectors where possible run down the centre of alleys. Unfortunately, for larger dense graphs the approach can be quite slow, the dominating cost is the time taken to find the optimal route for each connector in the second stage.

One of the main reasons that orthogonal connector routing is slow is that there are many "topologically equivalent" routes of equal cost to each vertex, greatly increasing the search space size. Figure 1 illustrates the problem. However, for our purposes these routes are equivalent as computation of the actual visual route will lead to an identical final layout (using the dashed edge which goes midway between objects B and C). The main contributions of this paper are to:

- Identify a class of connector route we call *obstacle-hugging* that we conjecture provide a canonical representative for a set of topologically equivalent routes;
- Present a new kind of visibility graph for computing these routes which we call the *1-bend visibility graph* in which nodes are object vertices and a search direction and there is an edge between two nodes if they can be connected using an orthogonal connector with one bend;
- Provide theoretical and empirical proof that this new approach is significantly faster than the current approach.

Our new approach has similar characteristics to the standard visibility graph used in poly-line connector routing. If we have n objects then the orthogonal visibility graph has $O(n^2)$ nodes, $O(n^2)$ edges and an optimal route can be $O(n^2)$ in length. In contrast the 1-bend visibility graph has $O(n)$ nodes, $O(n^2)$ edges and any optimal (orthogonal) route is $O(n)$ in length. This is similar to poly-line connector routing where the *standard visibility graph* has the same asymptotic sizes [6]. It also bears some similarities to the rectangularization approach of Miriyala *et al.* [5], though rectangularization is heuristic and so unlike our approach is not guaranteed to find an optimal route.

Orthogonal connector routing has been extensively studied in computational geometry, in part because of its applications to circuit design. Lee *et al.* [3] provides an extensive survey of algorithms for orthogonal connector routing, while Lenguauer [4] provides an introduction to the algorithms used in circuit layout. The 1-bend visibility graph, is as far as we are aware, completely novel.

2 Obstacle Hugging Routes

For simplicity we assume obstacles are rectangles: more complex shapes can be approximated by their bounding rectangle and assume for the purposes of complexity analysis that the number of connector points on each object is a fixed constant.

We are interested in finding a poly-line route of horizontal and vertical segments for each connector. We specify such an orthogonal route as sequence of points $p_1, \ldots, p_m$ where p_1 is the start connector point, p_m the end connector point and $p_2, \ldots, p_{m-1}$ the bend points. Note that orthogonality means that either $x_{j+1} = x_j$ or $y_{j+1} = y_j$ where $p_i = (x_i, y_i)$. We require that the routes are *valid*: they do not pass through objects and only contain right-angle bends, i.e., alternate horizontal and vertical segments.

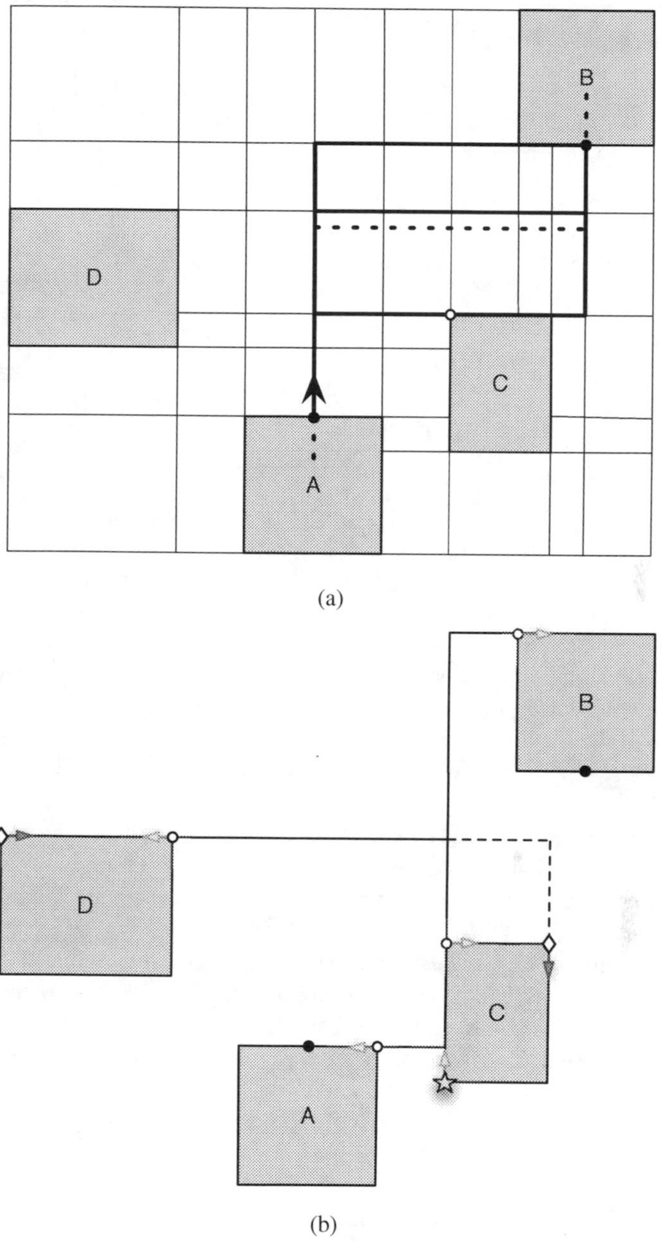

Fig. 1. (a) The orthogonal visibility graph and three topologically equivalent routes of equal cost (two bends plus overall length) between the centre of objects A and B. (b) Partial 1-bend visibility graph showing the visibility of nodes from the bottom left of C (star) going upwards to the circled nodes, with arrows indicating directions. The asymmetry is illustrated by the edge from the diamond node of D to the diamond node on C.

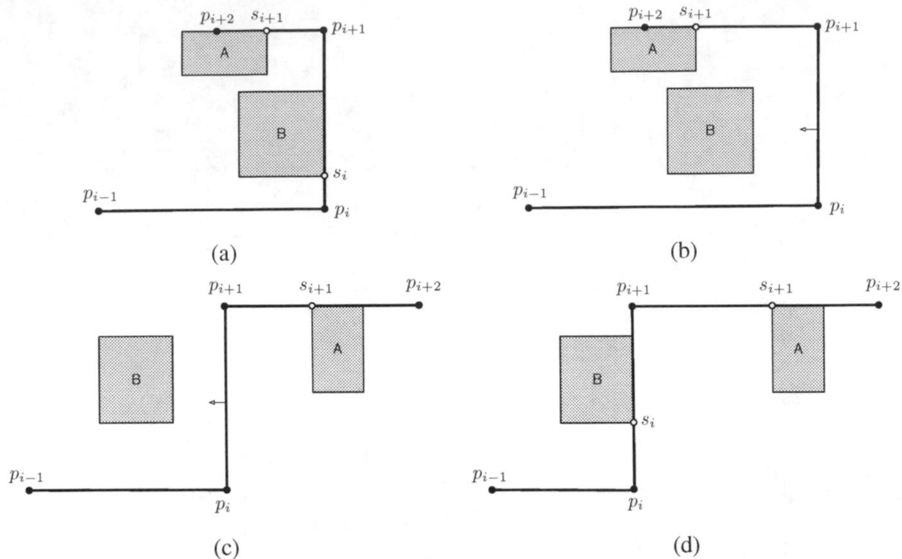

Fig. 2. Various cases in the construction of an optimal obstacle-hugging route

We wish to find routes that are short and which have few bends: We therefore assume our penalty function $\mathsf{p}(R)$ for measuring the quality of a particular route R is a monotonic function of the length of the path, $||R||$, and the number of bends (or equivalently segments) in R. A route R between two connector points is *optimal* if it is valid and it minimises $\mathsf{p}(R)$.

Two optimal orthogonal routes R_1 and R_2 from p_1 to p_m are *topologically equivalent* if no object is contained in the region between R_1 and R_2. Each of the routes shown in Figure 1 are topologically equivalent since there are no objects in the rectangular region between the routes.

One of the main reasons that existing orthogonal connector routing is slow is that the A* algorithm explores a large number of topologically equivalent routes. In order to reduce the search space we want to choose a single canonical route for each equivalence class. We conjecture that object-hugging routes provide such a canonical representative:

An orthogonal route $p_1, \ldots, p_m$ is *obstacle-hugging* if each segment $|p_j, p_{j+1}|$ for $j = 2, \ldots, m-2$ intersects the boundary of an object in the direction from p_{j-1} to p_j. This means the path bends around the far side of each object when following the route from its start.

The definition means that all intermediate (i.e., not the first or last) segments $|p_j, p_{j+1}|$ on an obstacle-hugging route have a supporting object o_j s.t. a vertex s_j of o_j lies on the segment. We call s_j the *support vertex* for the segment $|p_j, p_{j+1}|$.

In Figure 1 the bottom route between A and B is obstacle-hugging since the second segment intersects the top of the shape and the first segment goes towards the top. Neither the middle route or the top route along the bottom of B is obstacle-hugging.

Theorem 1. *Given an optimal orthogonal route $p_1,\ldots,p_m$ there is an optimal orthogonal route from p_1 to p_m topologically equivalent and which is obstacle-hugging.*

Proof. We sketch how, given the optimal route $R = p_1,\ldots p_m$, to construct an topologically equivalent route which has the same distance and number of bends and is obstacle-hugging.

The construction is iterative backwards for $i = m-1$ to 2. For each i it computes a support vertex s_i for segment $|p_i, p_{i+1}|$ such that $|s_i, p_{i+1}|$ has a nonempty overlap with the side of some object in the direction from p_{i-1} to p_i and possibly moves p_i and p_{i+1} while still maintaining the same topology and overall cost.

Assume that we have computed support vertex s_{i+1}, we show how to compute s_i. There are two cases to consider: p_{i-1}, p_i, p_{i+1} and p_{i+2} form a U-turn (Figure 2(a)) or p_{i-1}, p_i, p_{i+1} and p_{i+2} form a step (Figure 2(c)).

In the case of the U-turn, we must have an object B whose boundary lies along the segment $|p_i, p_{i+1}|$ as shown in Figure 2(a) because otherwise we could move the segment as shown in Figure 2(b) which would imply that R was not optimal. We therefore set s_i to be the vertex of object B closest to p_{i-1}. Clearly $|s_i, p_{i+1}|$ intersects the side of the object B in direction from p_{i-1} to p_i.

In the case of the step, we move the segment $|p_i, p_{i+1}|$ toward p_{i-1} by decreasing the length of segment $|p_{i-1}, p_i|$ and increasing the length of segment $|p_{i+1}, p_{i+2}|$ until it runs into some object B. This must happen before the length of segment $|p_{i-1}, p_i|$ decreases to 0 because otherwise we could have reduced the cost of R by removing this segment and the two bend points which would imply that R was not optimal. The construction is shown in Figure 2(c) and (d). We again set s_i to be the vertex of object B closest to p_{i-1}. And once more $|s_i, p_{i+1}|$ intersects the side of the object B in direction from p_{i-1} to p_i. □

Conjecture. *An optimal obstacle-hugging route is* canonical, *that is for each optimal route there is exactly one topologically equivalent optimal obstacle-hugging route.*

3 The 1-Bend Visibility Graph

While we could compute object-hugging routes using the orthogonal visibility graph by modifying the A* algorithm to prune non-object-hugging routes we can also compute them using a different kind of visibility graph. This follows from the observation that we can regard the support vertices $s_2,\ldots,s_{m-2}$ for a canonical optimal object-hugging route $p_1,\ldots,p_m$ as "split points" between a sequence of 1-bend visibility edges between object vertices, apart from the first and last segments which go to the start and end connector point respectively. Thus when we search for an optimal route we can search in a 1-bend visibility graph and build the object-hugging route as $p_1, s_2, s_3, \ldots, s_{m-2}, p_m$. This ensures we will only consider object hugging routes.

The 1-bend visibility graph is a directed graph where nodes are combination of connector points and corners of rectangles. The 1-bend visibility graph can be constructed as follows. Let I be the set of *interesting points* (x,y) in the diagram, i.e., the corners of the rectangular objects and the connector points.

1. Generate the *interesting horizontal segments* $H_I = \{|(l,y),(r,y)|$ where $(x,y) \in I$ and r is the biggest value where no object overlaps $|(x,y),(r,y)|$ and l is the least value where no object overlaps $|(l,y),(x,y)|$.
2. Generate the *interesting vertical segments* $V_I = \{|(x,b),(x,t)|$ where $(x,y) \in I$ and t is the greatest value no object overlaps $|(x,y),(x,t)|$ and b is the least value no object overlaps $|(x,b),(x,y)|$.
3. Compute the 1-bend visibility graph by intersecting all pairs of segments from H_I and V_I. Suppose we have intersecting segments $|(l,y),(r,y)| \in H_i$ and $|(x,b),(x,t)| \in V_i$ we add an edge from each vertex point (x,h) of an object o_1 on the horizontal segment where the segment $|(x,h),(x,y)|$ intersects the side of o_1 in the direction (x,h) to (x,y) to each vertex point (v,y) on an object o_2 in the direction (x,y) to (v,y) where the segment $|(x,y),(v,y)|$ only intersects the object at (v,y).
   Similarly we add an edge from each vertex point (v,y) of an object o_1 on the vertical segment in the direction (v,y) to (x,y) where $|(v,y),(x,y)|$ intersects the side of o_1 to each vertex point (x,h) on an object o_2 in the direction (x,y) to (x,h) where $|(x,y),(x,h)|$ only intersects the object at (x,h).
   The edges to and from connection points with directions are created similarly but there is no requirement for intersection/non-intersection with any object.

For example the directed edges from the bottom left corner (star) of C going upwards in the visibility graph of Figure 1(a) are shown in Figure 1(b) as circles with directions given by arrow. There are edges from the connection point on A in the up direction go to each of circled nodes except the one on A itself. Note that the graph is *not* symmetric! The only edge from the top side of shape D to shape C goes from the diamond node in the right direction, to the diamond node on C in the down direction.

Theorem 2. *The orthogonal visibility graph can be constructed in $O(n^2)$ time for a diagram with n objects using the above algorithm. It has $O(n)$ nodes and $O(n^2)$ edges.*

Proof. The interesting horizontal segments can be generated in $O(n \log n)$ time where n is the number of objects in the diagram by using a variant of the line-sweep algorithm from [1,2]. Similarly for the interesting vertical segments. The last step takes $O(n^2)$ time since there are $O(n)$ interesting horizontal and vertical segments. It follows from the construction that it has $O(n)$ nodes and $O(n^2)$ edges. □

Construction of a path from connector point p_1 to p_m starts from p_1 and constructs a path for each possible (feasible) direction, to reach p_m in any direction. The best path (according to the penalty function p) is returned for final visual route computation.

Consider the construction of the path from the connection point on A leaving vertically to the connection point on B entering vertically. The only nodes adjacent to the initial connection point are the circled nodes of B, C and D. Using A* search and the admissible heuristic described in [7], the node on C is preferable, and then we find an optimal route to the connection point on B (the bottom route in Figure 1). The remaining nodes are fathomed by the heuristic. Effectively we find the route with no search. Contrast this with the usual approach [7] where every node on the three paths in Figure 1(a) needs to be visited as well as others!

Table 1. Evaluation of different visibility graphs for several biological networks and circuit diagrams, named for their number of nodes (v) and edges (e). We give times (in milliseconds), number of edges in visibility graph, and the average number of steps in the search of each path.

	1-bend visibility graph					Orthogonal visibility graph				
	Total	Construction		Routing (avg)		Total	Construction		Routing (avg)	
Diagram	Time	Time	$\|E\|$	Time	Steps	Time	Time	$\|E\|$	Time	Steps
v185e225	197	41	35K	0.3	4	216	34	28K	0.4	34
v508e546	867	191	196K	0.5	3	1,964	185	136K	2.6	28
v4330e2755	180,831	14,070	25.9M	42	178	1,829,431	6,741	3.0M	645	14,859

4 Evaluation and Conclusion

Theoretically, searching for optimal obstacle-hugging connector routes over the 1-bend visibility graph should be considerably faster than finding optimal routes in the orthogonal visibility graph since the optimal route is $O(n)$ in length where n is the number of obstacles rather than the $O(n^2)$ length in the orthogonal visibility graph.

We compared the performance of a prototype implementation of the 1-bend visibility graph approach against the orthogonal visibility graph implementation in the **libavoid** C++ routing library. The test machine was a 2012 MacBook Pro with a 2.3 GHz Intel Core i7 processor and 16GB of RAM. As shown in Table 1, while 1-bend visibility graphs are larger and take longer to build, routing over them is significantly faster. For a very large diagram like our final example, use of a 1-bend visibility graph speeds up connector routing by a factor of ten!

We have presented a new kind of connector route and a new kind of visibility graph that significantly improves the speed of orthogonal connector routing. We plan to include the new approach in our widely used connector routing library **libavoid**.

Acknowledgments. We acknowledge the support of the ARC through Discovery Project Grants DP0987168 and DP110101390. We also thank the anonymous reviewers for their very helpful comments.

References

1. Dwyer, T., Marriott, K., Stuckey, P.J.: Fast node overlap removal. In: Healy, P., Nikolov, N.S. (eds.) GD 2005. LNCS, vol. 3843, pp. 153–164. Springer, Heidelberg (2006)
2. Dwyer, T., Marriott, K., Stuckey, P.J.: Fast node overlap removal—correction. In: Kaufmann, M., Wagner, D. (eds.) GD 2006. LNCS, vol. 4372, pp. 446–447. Springer, Heidelberg (2007)
3. Lee, D., Yang, C., Wong, C.: Rectilinear paths among rectilinear obstacles. Discrete Applied Mathematics 70(3), 185–216 (1996)
4. Lengauer, T.: Combinatorial Algorithms for Integrated Circuit Layout. John Wiley & Sons, Inc., New York (1990)
5. Miriyala, K., Hornick, S.W., Tamassia, R.: An incremental approach to aesthetic graph layout. In: Computer-Aided Software Engineering, pp. 297–308. IEEE (1993)
6. Wybrow, M., Marriott, K., Stuckey, P.J.: Incremental connector routing. In: Healy, P., Nikolov, N.S. (eds.) GD 2005. LNCS, vol. 3843, pp. 446–457. Springer, Heidelberg (2006)
7. Wybrow, M., Marriott, K., Stuckey, P.J.: Orthogonal connector routing. In: Eppstein, D., Gansner, E.R. (eds.) GD 2009. LNCS, vol. 5849, pp. 219–231. Springer, Heidelberg (2010)

Tennis Plots: Game, Set, and Match

Michael Burch and Daniel Weiskopf

Visualization Research Center, University of Stuttgart,
Allmandring 19, 70569 Stuttgart
{michael.burch,daniel.weiskopf}@visus.uni-stuttgart.de

Abstract. In this paper we introduce Tennis Plots as a novel diagram type to better understand the differently long time periods in tennis matches on different match structure granularities. We visually encode the dynamic tennis match by using a hierarchical concept similar to layered icicle representations used for visualizing information hierarchies. The time axis is represented vertically as multiple aligned scales to indicate the durations of games and points and to support comparison tasks. Color coding is used to indicate additional attributes attached to the data. The usefulness of Tennis Plots is illustrated in a case study investigating the tennis match of the women's Wimbledon final 1988 between Steffi Graf and Martina Navratilova lasting 1 hour, 19 minutes, and 31 seconds and being played over three sets (5:7, 6:2, 6:1). Interaction techniques are described in the case study in order to explore the data for insights.

Keywords: time-varying data, sports data, hierarchical data.

1 Introduction

Tennis matches are time-dependent sports events which can be subdivided into several playing phases. Finding insights in such time-varying data is difficult by just inspecting the data manually in a text file or by just watching the match, which can sometimes last a couple of hours. Statisticians or visualizers often analyze the data by aggregating the whole match or single sets into some statistical numbers which do not allow one to find insights in the evolution of the match. But exploring time-dependent sports data is interesting for sportsmen, managers, the mass media, as well as for the spectator in order to get informed by a simple all-showing diagram.

In this paper we introduce Tennis Plots: an interactive visualization technique that visually encodes the playing and pausing phases of a tennis match in a multiple aligned scales representation similar to layered icicle plots. Such a visualization first gives an overview of the complete match and simultaneously shows different phases and the match structure. A static diagram for dynamic data has many benefits due to mental map preservation, reduction of cognitive efforts, good performance of comparison tasks, and the application of interaction techniques [1].

T. Dwyer et al. (Eds.): Diagrams 2014, LNAI 8578, pp. 38–44, 2014.

In this visualization design we map the scored points as well as the time spent to score them and the pauses as differently thick and color coded rectangles depending on the durations. We visually map the time dimension to the vertical axis that is used to visualize the durations of the single events such as the time until a point is scored and the pauses in-between. The horizontal axis is also used as time axis showing the overall temporal structure of the tennis match in the form of sets. Interaction techniques are integrated targeting a better exploration of the time-varying data.

2 Related Work

Analyzing time-varying data can be challenging, depending on the type of data, the characteristics of the time axis, and the applied visual metaphor [2]. In particular, tennis matches are based on a time axis which can be split into a sequence of intervals which is also characteristic for calendar-based representations [3]. Klaasen and Magnus [4] propose a method for forecasting tennis match winners. They do not base their analysis on the beginning of the match but in particular during the running match. In the work of Terroba et al. [5] a tennis model based on Markov decision processes is presented for describing the dynamic interaction between tennis players. The goal of the work is to extract optimal strategies. Also a transformation of tennis videos into time-varying discrete data is proposed.

Visualization techniques are rarely applied strategies but are a good means to uncover visual patterns which can give insights into interesting phenomena hidden in time-varying data. For example, Burch and Weiskopf [6] illustrate and classify time-varying patterns by a visualization technique investigating dynamic water level data which are of quantitative nature. In particular, for tennis matches, the TennisViewer [7] by Jin and Banks has been proposed. Although the hierarchical structure of the tennis match is visible combined with a color coding on treemap [8] [9] boxes to indicate which of the two players scored which point, the comparison of multiple time axes is difficult. In our work we add explicit timelines to the hierarchical organization of the tennis match in order to derive time-varying insights on different levels of temporal granularity.

Jin and Banks [7] use competition trees to organize the tennis match. They also follow the idea of encoding the hierarchical structure of the match into layered representations similar to layered icicle plots [10] used for hierarchy visualization. However, in their work, the explicit temporal aspect in the match is not displayed, i.e. we see which point was scored before which other, but there is no absolute time axis integrated telling the viewer about the actual spent time when either playing or pausing in the match. Tu and Shen [11] also apply the concept of treemaps to illustrate time-varying data by adding explicit timelines to the treemap boxes, but the timeline is directly integrated into each treemap box making interval comparison tasks hard to solve visually.

We, instead, rely on a time-to-space mapping of the time-varying data, which has several benefits compared to animated diagrams. Comparisons, which are important tasks in visualizations, are done visually which would be problematic with animation. The concept which we follow is denoted as mapping to identical aligned scales [12]. This data mapping strategy is ranked very high for comparing quantitative data values. In our work we follow the Visual Information Seeking Mantra [13]: Overview first, zoom and filter, then details-on-demand.

3 Visualization Technique

In this paper we describe a visualization technique for exploring a tennis match on different levels of structural granularity, i.e. either on match, set, game, or score levels.

3.1 Data Model

We model a tennis match as an ordered finite sequence S_I of $n \in \mathbb{N}$ intervals $S_I := \{I_1, \ldots, I_n\}$. Each single interval $I_i \in S_I$ is attached by two timestamps t_{s_i} and t_{e_i}, expressing start and end time points for that interval I_i. The division into intervals depends on the events occurring in a tennis match, i.e. different playing and pausing phases, which again depend on the performances of the single players.

By this data model we subdivide the time axis into non-overlapping intervals which completely cover the time axis, i.e. $t_{e_i} = t_{s_{i+1}}\ \forall 1 \leq i \leq n-1$ and $t_{s_i} \leq t_{e_i}\ \forall 1 \leq i \leq n$. It may be noted that if we allow $t_{s_i} = t_{e_i}$ also time points instead of time intervals can be modeled. The duration of an interval I_i is defined as $t_{d_i} := t_{e_i} - t_{s_i}$.

Moreover, each interval I_i has additional data attributes which can be modeled as an ordered list $L_i := \{a_1, \ldots, a_m\}$. The a_i's can for example be used to attach the current score in the match, additional events such as faults, double faults, net faults, aces, pauses, breaks, or the player who is to serve in this game.

The attached score data is taken to derive structural timeline subdivisions, i.e. a tennis match consists of sets, games, and the single scores. This allows us to generate a hierarchical organization of the match while still preserving the chronological order between the intervals S_I.

Moreover, this data model can be used to compute derived data values based on several filter criteria, for example the average duration of pauses taken by a specific player, the maximum length until a point is scored, or the sum of durations for single sets or games, i.e. derived data on different structural granularities.

3.2 Design Decisions

To design a suitable visualization technique for such time-varying data with attached data attributes having an inherent hierarchical structure we first illustrate how the sequence of intervals S_I is visually encoded.

Figure 1 (a) shows the vertical stacking of time intervals $S_I^{'} \subseteq S_I$. Each interval $I_i \in S_I^{'}$ is mapped to a rectangle where the width is fixed and the

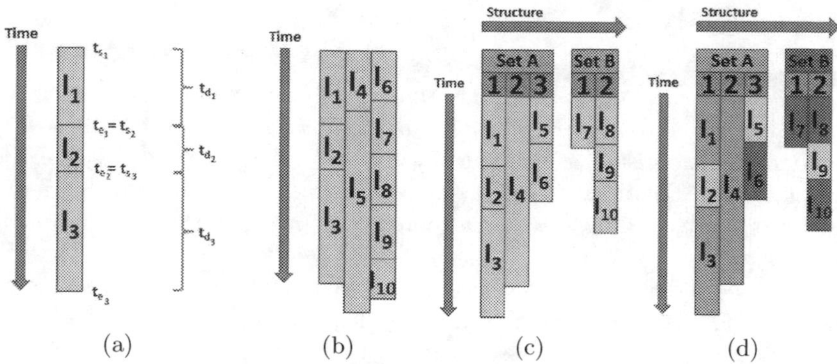

Fig. 1. Generating a Tennis Plot: (a) Stacking of time intervals. (b) Multiple sequences of time intervals. (c) Combination of time axes and the match structure. (d) Color coding to visualize categorical data.

height encodes the duration t_{d_i} of I_i. This visualization design allows the user to visually explore a number of intervals by inspecting the sum of heights of all intervals stacked on top of each other.

If we have to deal with m sets of intervals, i.e. $\mathbb{S} := \{S_{I_1}, \ldots, S_{I_m}\}$ where each $S_{I_k} \subseteq S_I$ we need a representation that supports comparison tasks among all interval sets. Figure 1 (b) illustrates how we design such a visual representation. The single S_{i_k} are mapped in chronological order by their start time points from left to right in order to achieve a readable diagram. The sum of durations for each interval sequence can be inspected while additionally allowing comparisons among all of them.

The additional structure of the tennis match, i.e. the organization into sets, games, and scores is illustrated in Figure 1 (c). We use the vertical direction for representing time and the horizontal direction to represent the additional match structure, which has also a temporal nature similar to a calendar-based representation [3], i.e. the division into time intervals by preserving the inherent match structure.

Color coding is used as the final design step to map additional attributes to the plot. Typically, this additional data is of categorical nature, e.g. the single players, pausing and playing phases, or net faults and faults (see Figure 1 (d)).

Further information can easily be added to such a plot aligned to each timeline. Such data attachments are for example derived values such as average set durations, game duration, average pause durations, or average time for waiting until a player will serve for the next point (for each player separately) and the like.

We support several interaction techniques to browse, navigate, and explore the data. Some of the most important ones are event, time, structure, and text filters. Moreover, we support time normalization, logarithmic scaling, zooming, color codings, highlighting, and details-on-demand. Also additional derived values such as average, maximum, minimum and the like are displayable by statistical plots.

4 Case Study

We illustrate the usefulness of Tennis Plots by first transforming time-varying tennis match data into a specific data format combined with additional events. In the second step we visualize this data by the novel diagram and finally explore it by interaction techniques to find interesting insights which are hard to be found by either watching the tennis match or browsing through the textual time-varying data.

In our case study, we inspect the women's Wimbledon final of 1988 between Steffi Graf and Martina Navratilova, which lasted 1 hour, 19 minutes, and 31 seconds. The match was played over three sets, resulting in 5:7, 6:2, and 6:1.

There are several visual patterns which can be detected when inspecting the plot. Those can be classified by either looking at single time axes, comparing a certain number of time axes, and by investigating the additional match structure.

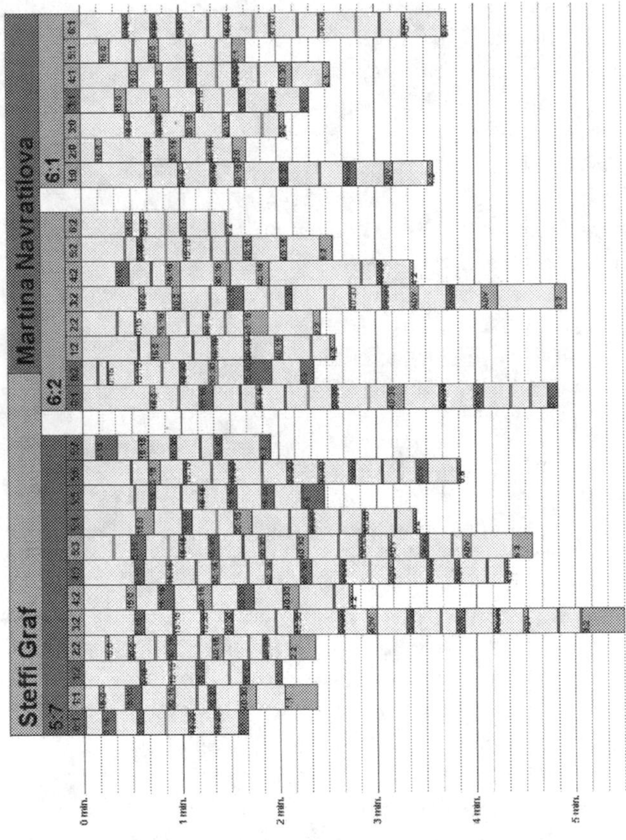

Fig. 2. Tennis Plot of the women's Wimbledon final in 1988 between Steffi Graf and Martina Navratilova. Time axes are integrated and the single point durations are shown side-by-side.

When looking at single time axes in Figure 2, we directly see that the pausing phases are much longer than the playing phases during a tennis match. This can be done for each time axis separately. Apart from doing this inspection visually the tool can also show additional derived values for average durations of both categories of phases which is not shown in this figure. Looking at those additional statistics we can confirm our first impression obtained by reading the Tennis Plot. Moreover, if we have a look at single time axes we see the sequence of points whereas the color coding and the labeling information helps to rapidly understand which player scored at what time and also in which sequential order.

Comparing several time axes is beneficial to explore the different durations taken until one game is over. Here, we can for example see that the 3:2 game in the opening set took much longer than all other games (Figure 2). The 6:2 game in the second set took the least time in the match, which can be observed by looking at the timeline integrated into the plot. For this game we also directly see that Steffi Graf (blue color) won by scoring four points in a row and all of them in a very short time. To make these observations time-based comparison tasks have to be conducted first and then the single time axes have to be visually explored.

When taking the match structure into account, i.e. the subdivision into points and sets, we can furthermore see that the first set took the most time of all sets. This insight is obtained by interactively stacking all intervals for each set separately on top of each other. Moreover, Martina Navratilova won the opening set but she lost the second and the third one. Looking at the time axes in Figure 2 we can see that Steffi Graf wins more and more points the more the match is coming to an end. Asking for details-on-demand again, we get the information that the number of won points changes from 38:46 (first set) to 31:18 (second set), and finally to 28:14 (third set), i.e. in the last set Steffi Graf won twice as many points as Martina Navratilova. Here we can detect in a more efficient way that the number of Steffi Graf's scored points is increasing towards the end of the match, i.e. in the second and third set.

5 Conclusion and Future Work

In this paper we introduced Tennis Plots as an interactive diagram supporting a viewer in inspecting time-varying data acquired from tennis matches. Instead of showing the data as statistical numbers the diagram focuses more an depicting the data on an interval basis. Playing and pausing phases as well as additional events and attributes can be visually inspected together with the match hierarchy, i.e. the subdivision into points, games, and sets. Interaction techniques are integrated to browse, navigate, and explore the data on different levels of time and structural granularities.

Although this approach is able to solve some exploration tasks, it might be extended by additional features and visual components. For example, a direct combination to the actual video of the match might be of interest to help users to directly browse to a specific point in time in the match in which they can get better impressions of the real data and what actually happened there. Such detail information cannot be provided by the Tennis Plots. Moreover, other sports data might be of interest to be visually analyzed such as volleyball, basketball, and the like. Apart from visualizing sports data on an interval basis also more general time-varying data with an additional hierarchical structure might be interesting to be represented. For more general scenarios we are aware of the fact that additional problems will occur such as finding a suitable subdivision into intervals. Finally, a user study should be conducted investigating the performances of participants when analyzing data by these plots.

References

1. Tversky, B., Bauer Morrison, J., Bétrancourt, M.: Animation: Can it Facilitate? International Journal of Human-Computer Studies 57(4), 247–262 (2002)
2. Aigner, W., Miksch, S., Schumann, H., Tominski, C.: Visualization of Time-Oriented Data. Springer (2011)
3. van Wijk, J.J., van Selow, E.R.: Cluster and Calendar Based Visualization of Time Series Data. In: Proceedings of Infovis, pp. 4–9 (1999)
4. Klaassen, F.J.G.M., Magnus, J.R.: Forecasting the Winner of a Tennis Match. European Journal of Operational Research 148(2), 257–267 (2003)
5. Terroba Acha, A., Kosters, W.A., Varona, J., Manresa-Yee, C.: Finding Optimal Strategies in Tennis from Video Sequences. IJPRAI 27(6) (2013)
6. Burch, M., Weiskopf, D.: Visualizing Dynamic Quantitative Data in Hierarchies – TimeEdgeTrees: Attaching Dynamic Weights to Tree Edges. In: Proceedings of International Conference on Information Visualization Theory and Applications, pp. 177–186 (2011)
7. Jin, L., Banks, D.C.: TennisViewer: A Browser for Competition Trees. IEEE Computer Graphics and Applications 17(4), 63–65 (1997)
8. Johnson, B., Shneiderman, B.: Tree Maps: A Space-Filling Approach to the Visualization of Hierarchical Information Structures. In: Proceedings of IEEE Visualization, pp. 284–291 (1991)
9. Shneiderman, B.: Tree Visualization with Tree-Maps: 2-D Space-Filling Approach. ACM Transactions on Graphics 11(1), 92–99 (1992)
10. Kruskal, J., Landwehr, J.: Icicle Plots: Better Displays for Hierarchical Clustering. The American Statistician 37(2), 162–168 (1983)
11. Tu, Y., Shen, H.-W.: Visualizing Changes of Hierarchical Data Using Treemaps. IEEE Transactions on Visualization and Computer Graphics 13(6), 1286–1293 (2007)
12. Cleveland, W.S., McGill, R.: Graphical Perception: Theory, Experimentation, and Application to the Development of Graphical Methods. Journal of the American Statistical Association 79(387), 531–554 (1984)
13. Shneiderman, B.: The Eyes Have It: A Task by Data Type Taxonomy for Information Visualizations. In: Proceedings of the IEEE Symposium on Visual Languages, pp. 336–343 (1996)

Coloured Modelling Spider Diagrams

Paolo Bottoni[1], Andrew Fish[2], and Alexander Heußner[3]

[1] Dipartimento di Informatica - "Sapienza" University of Rome, Italy
bottoni@di.uniroma1.it
[2] School of Computing, Engineering and Mathematics - University of Brighton, UK
Andrew.Fish@brighton.ac.uk
[3] University of Bamberg, Germany
alexander.heussner@uni-bamberg.de

Abstract. While Euler Diagrams (EDs) represent sets and their relationships, Coloured Euler Diagrams (CEDs [1]) additionally group sets into families, and sequences of CEDs enable the presentation of their dynamic evolution. Spider Diagrams (SDs) extend EDs, permitting the additional expression of elements, relationships between elements, and set membership, whilst Modelling Spider Diagrams (MSDs [2]) are used to specify the admissible states and evolutions of instances of types, enabling the verification of the conformance of configurations of instances with specifications. Transformations of MSDs generate evolutions of configurations in conformity with the specification of admissible sequences. We integrate CEDs and MSDs, proposing Coloured Modelling Spider Diagrams (CMSDs), in which underlying curves represent properties of a family of sets, whether this be state-based information or generic attributes of the domain elements and colours distinguish different families of curves. Examples of CMSDs from a visual case study of a car parking model are presented.

Keywords: spider diagrams, coloured Euler diagrams, visual modelling languages.

1 Motivation, Examples and Discussion

What do we gain when going from Coloured Euler Diagrams to Coloured Spider Diagrams? We obtain the ability to simultaneously express multiple complex properties of elements (spiders): since colouring of curves can induces colouring of zones, a colouring of curves can impose an identification of the colouring properties on the elements (spiders). Independent zone colourings are also possible to express properties orthogonal to those induce from curve colourings.

What do we gain when going from SD to MSD and (generalised)-MSD? In MSDs we have a type-instance diagram distinction and an associated theory of *policies*, expressed via sequences of type-SDs. While MSDs permit only state based information, in (generalised)-MSDs we gain the ability to represent additional attribute-based information of elements. Since sequences are vital to express policy-conforming system evolution, we obtain the benefit of an easy sequential comparison of diagrams, using colour matching. Quick identification of families of sets is also a benefit: e.g. one can distinguish between state and non-state attribute curves via a distinguished form of colouring highlighting the importance of state.

T. Dwyer et al. (Eds.): Diagrams 2014, LNAI 8578, pp. 45–47, 2014.

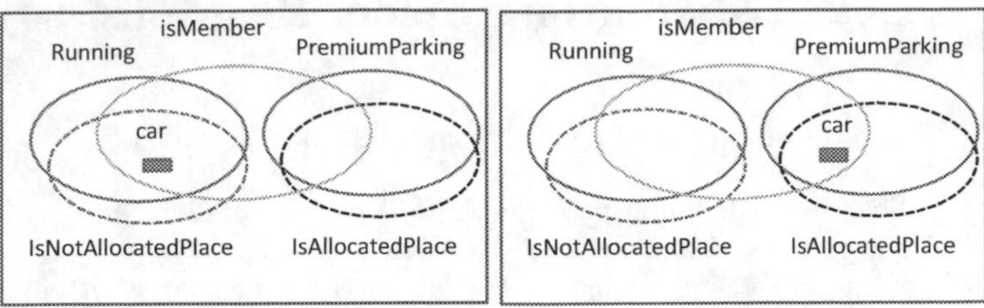

(a) Evolution in time: from state `Running` to state `Premium Parking`.

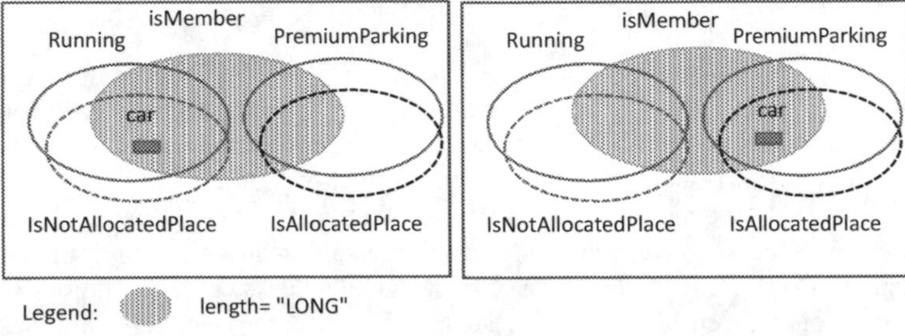

(b) Using colouring to represent additional length information.

Fig. 1. Car parking policy specifications. (a) depicts a CMSD transition, e.g., as part of a policy specification in a car parking domain, where a car can be in a state of either being `Running` or being `Premium Parking`. The `isMember` curve is from the related domain in which the car owner is a member of a scheme which permits the use of premium parking spaces. The dashed curves indicate states of allocation of parking places to a particular car, taken from a different domain model and so presented using a different form of colouring. The diagrams demonstrate an evolution in time (reading diagrams from left to right): when the car moves from `Running` to the `Premium Parking` state, it is successfully allocated the parking place (depicted via the move from `isNotAllocatedPlace` to `isAllocatedPlace`) and additionally fulfilled the prerequisite of being a premium parking member (via `isMember`). (b) shows the same sequence, with additional information, expressed using colouring in the form of internal hashing; the legend at the bottom indicates that the internal hashing corresponds to an additional constraint on the length of the car (`length="LONG"`), and one would have suitable constraints in place so that the length of the allocated place of parking is suitable for the length of the car.

Coloured Modeling Spider Diagrams (CMSDs) generalise the notion of MSDs (thus permitting arbitrary attributes to be expressed via curves, instead of just state information), as well as incorporate colouring as in CEDs (thus assigning a vector of colouring functions to the curves of a diagram). One can adopt certain graphical conventions of usage which may provide long-term advantages for readers due to the developed familiarity with the conventions. For example, one can highlight the importance of state properties within the domain by mapping any state-based curve colourings to curves with solid line boundaries, whilst non-state based curve colourings are mapped to non-solid line boundaries. In terms of the graphical presentation, a colouring of curves can be used to perceptually identify a family of curves within the same diagram (e.g. at the concrete level one may use a set of curves to depict a single set) or across a sequence of diagrams (e.g. tracing the evolution of a set over time). Then, within a single

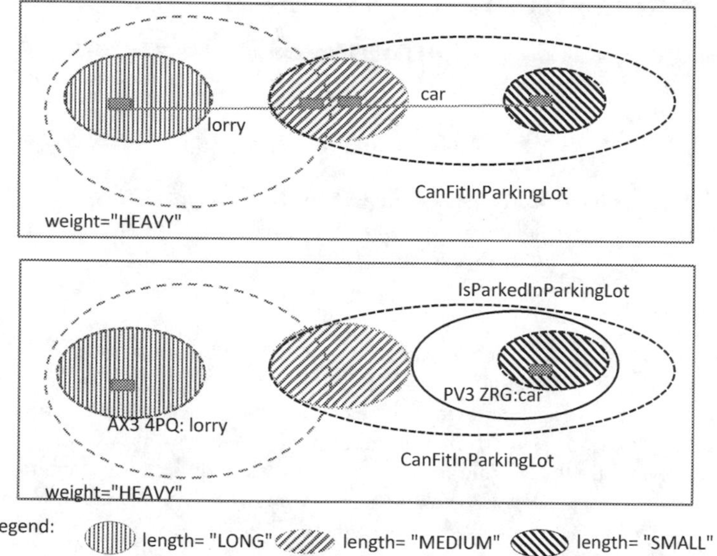

Fig. 2. Conformance of instance-CMSD (bottom) to type-CMSD (top). The dual coding of the length attribute (as label and colour) is also depicted. The top diagram indicates that instances of the car type can be of length small or medium, instances of the lorry type can be of length medium or large and are of heavy weight, but lorries of long length do not fit into the parking lot (but lorries of medium length and cars do fit). The bottom diagram indicates that there is a particular heavy and long lorry, with registration AX3 4PQ, which does not fit in the parking lot, plus a particular small car, with registration PV3 ZRG, which is in fact parked in the parking lot.

diagram, two regions which are within sets of curves with distinct colourings will indicate different sets of attributes for spiders (i.e. elements) placed within those regions. When considering sequences of diagrams, one must be slightly more careful since each individual diagram will express a set of attributes corresponding to the curves present in that diagram, which may differ from those presented in another diagram. We can extend the use of the labelling of curves to express precise semantic information, effectively permitting labels to be specialised to indicate an attribute (or set of attributes, say) with variables, allowing an arbitrary number of labels (e.g. if real-values labels are used). For instance, within the parking application domain a curve label "length=2.5" could be used to express the property that all spiders placed within the interior of the curve represent cars of length precisely 2.5 metres.

References

1. Bottoni, P., Fish, A.: Coloured Euler diagrams: a tool for visualizing dynamic systems and structured information. In: Goel, A.K., Jamnik, M., Narayanan, N.H. (eds.) Diagrams 2010. LNCS, vol. 6170, pp. 39–53. Springer, Heidelberg (2010)
2. Bottoni, P., Fish, A.: Extending spider diagrams for policy definition. Journal of Visual Languages & Computing 24(3), 169–191 (2013)

Graphical Representations of Context-Free Languages

Benedek Nagy

Eastern Mediterranean University, Famagusta, North Cyprus, Mersin-10, Turkey
Faculty of Informatics, University of Debrecen, Debrecen, Hungary
nbenedek@inf.unideb.hu

Abstract. Regular languages can be represented by finite automata and by railroad diagrams. These two visual forms can be converted to each other. Context-free languages can also be described by (finite sets of) railroad diagrams. Based on the analogy we develop a new type of automata, the fractal automata: they accept the context-free languages. Relations between pushdown automata and fractal automata are also established.

Keywords: fractal automata, syntax diagrams, pushdown automata.

1 Introduction, Preliminaries

The wide-spread use of the *regular languages* (REG) is based on the fact that *finite automata* (FA) accept them and it is easy to deal with them in a visual way. However there are several places where REG are not enough, more complex languages, e.g., *context-free languages* (CF) are needed. In this paper we consider REG and CF ([1]) from a special point of view: we analyze some of their diagrammatic representations. There are various tools that represent (describe) CF (and programming) languages. Such tool is the *railroad diagram system* (RDS). They were used for syntactic description of the Pascal language in [2]. An RDS consists of a finite number of *railroad diagrams* (RD). RD are defined iteratively, they contain arrows, terminals and nonterminals (as vertices). The used operations, the concatenation, alternatives, option and iteration are equivalent to the regular operations [3]. Actually, each nonterminal (there is a finite number of them) is defined by an RD in an RDS. Each RD describes full paths from its entry point (starting arrow) to its end point (finishing arrow). When a nonterminal is reached in a path, then a word described by the RD defining that nonterminal must be substituted there. In this way recursion can be used, e.g., using the nonterminal itself in the description. There is a main diagram; it defines the *described language*.

2 Representations of Regular Languages

REG can be described by *regular expressions* and accepted by FA. FA are usually defined in a graphical way by their graphs: states are vertices (circles), transitions

T. Dwyer et al. (Eds.): Diagrams 2014, LNAI 8578, pp. 48–50, 2014.

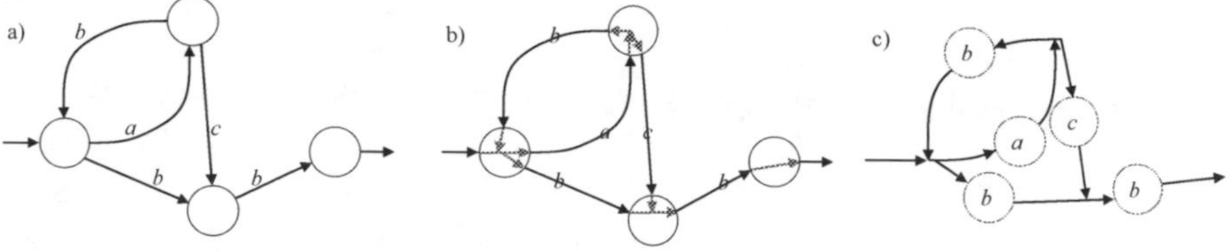

Fig. 1. Representations of a regular language: a. FA c. RD, and b. an intermediate representation by the transformation between them.

are labeled arrows (edges), initial and accepting states are marked by an in and by out arrows. A run of an automaton on an input can be interpreted as a path in the graph. Regular expressions are represented by special RDS [3]: RDS with 1 RD without nonterminals describe REG. Now, we have two types of visual descriptions of REG (see Figure 1a and c). Actually, RD can be seen as a kind of dual graphs of FA. The (labeled) edges of FA play the role of labeled nodes, while instead of the states of FA in RD there are some arrows, i.e., edges instead of the vertices of the graph of FA. We can go from FA to RD and back by graph-transformation algorithms (a mid-representation step is also shown in Figure 1b). If all RD of an RDS can be ordered in such a way that in every RD only those nonterminals can be used that were already defined (not allowing to use the nonterminal being currently defined by this RD), then the described language is REG. In this case substituting the RD of nonterminals instead of their appearances, an RD without nonterminals is obtained as the main diagram.

3 The Context-Free Case and the Fractal Automata

CF are generated by *context-free grammars* (CFG), and they are accepted by *pushdown automata* (PDA). It can also be proven in a constructive way that exactly the class of CF is described by RDS, and thus, a CFG can be represented by a finite set of RD. By using the same type of transformation as we used in Section 2 we arrive to the concept of *fractal automata* (FRA), that is, a new diagrammatic representation of CF: Analogously to the regular case, now, one can obtain a finite set of FA linked by recursion (that is called the *finite representation* of FRA). If instead of the nonterminals their diagrams are substituted, then the obtained automaton is the *infinite representation* of the FRA for the given CF. Since the recursion can be arbitrarily deep in these diagrams, the transformation and substitution process 'results' an infinite – but somehow regular, self-similar – system. FRA are very similar to FA but has an infinite number of states (in non regular case). There are 3 types of transitions in FRA: 'local' transitions of any of the FA obtained from RD; transitions into recursion (allowing to step to the initial state of an embedded FA at the cases where nonterminal is reached in RD) and transitions out of a recursion (stepping from the final state of an embedded FA to one level out: simulating the continuation of the flow in an RD after a nonterminal). Even the set of states is infinite, there

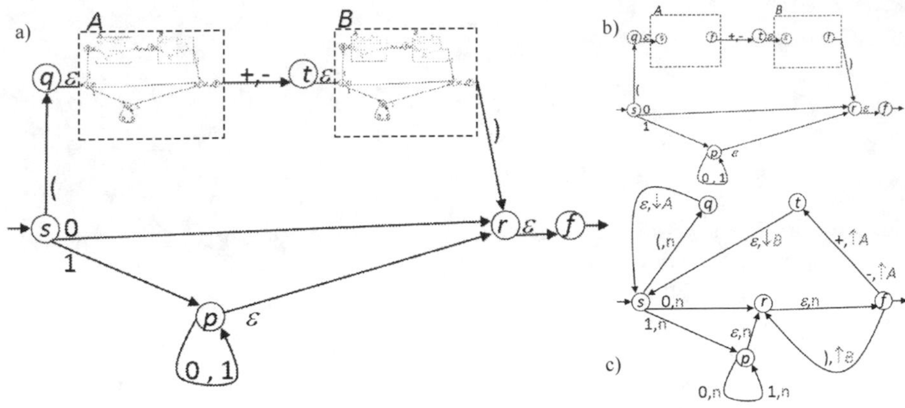

Fig. 2. a. an FRA, b. its finite representation and c. an equivalent PDA. ε denotes the empty word.

is a convenient way to address (name) the states of the infinite representation of FRA. An FRA is shown in Figure 2a that accepts the CF of the correct bracketed expressions using two binary operators $+, -$ and binary (nonnegative) integers. The self-similar feature of the automaton can easily be observed. The family of FRA accepts exactly the family of CF. When more than one RD are used in an RDS to describe the language, the recursions (the parts that we can zoom in as in fractals) can be varied. FRA are useful tools to prove some properties of CF, e.g., closure under regular operations and under recursion (i.e., pumping). The representation of an FRA can be the finite set of automata obtained from the set of RD, connecting them according to the recursions (based on the nonterminals). See Figure 2b. The chain of the recursion may be traced by the help of a pushdown stack (as it is done in computers as well in recursive function calls). Based on this idea an equivalent PDA can be constructed with 3 types of transitions: push ($\downarrow$) operation: a new element is pushed to the stack and the next state is independent of the earlier stack contents; pop ($\uparrow$) operation: the top element is popped out from the stack and the next state depends on that symbol as well; no change (n): in this transition the stack is not used, the next state does not depend on the stack contents, and it has not been changed during this transition. A link between PDA and the (automatic) pushdown stack used in programming at recursive function calls can easily been established. Figure 2c shows the PDA that simulates the FRA on 2a.

References

1. Rozenberg, G., Salomaa, A. (eds.): Handbook of Formal Languages. Springer (1997)
2. Wirth, N.: The Programming Language Pascal (July 1973)
3. Nagy, B.: Programnyelvek elemeinek szintaktikus leírása normál formában (Syntactic Description of Programming Languages in a Normal Form, in Hungarian). In: Conference on Informatics in Higher Education 2005, Debrecen, 6+1 pages (2005)

An Example HyperVenn Proof

Dave Barker-Plummer[1], Nik Swoboda[2], and Michael D. Murray[1]

[1] CSLI/Stanford University, Stanford, California, 94305, USA
{dbp,mikem05}@stanford.edu
[2] Universidad Politécnica de Madrid,
Campus de Montegancedo S/N, Boadilla del Monte, Madrid, 28660, Spain
nswoboda@fi.upm.es

Abstract. HyperVenn is a heterogeneous logic resulting from combining the individual homogeneous logics for Euler/Venn diagrams and blocks world diagrams. We provide an example proof from the system along with a brief discussion of some of its inference rules.

Keywords: heterogeneous logic, Euler/Venn diagrams, blocks world diagrams.

1 Introduction

We introduce a new heterogeneous logic called HyperVenn. A heterogeneous logic is one in which representations from different systems are used together to convey information about a single problem. Typically, as in Hyperproof [3], Diamond [4] and Diabelli [6], at least one of the representation systems participating in the logic is sentential. HyperVenn, by contrast, involves two diagrammatic representations: blocks world diagrams, drawn from the Hyperproof system [3], and Euler/Venn diagrams [5]. Our interest in developing the system is to investigate the challenges that such a heterogeneous logic presents.

A formal specification of the HyperVenn logic is forthcoming [2]; here we present an example proof to illustrate some of the key ideas in the logic

The HyperVenn logic is implemented as a component of our Openbox framework [1]. As we had preexisting implementations of each of the homogeneous systems, HyperVenn was implemented solely by providing the heterogeneous rules of inference (all non-logical concerns are handled by the framework).

2 An Example Proof

Fig. 1 shows the premises of our example proof. The first premise is a blocks world diagram, these diagrams are used to represent situations involving blocks on a checkerboard. This diagram includes each of the three possible shapes that blocks may have: tetrahedra, cube and dodecahedra. Blocks also have one of three alternative sizes: small, medium or large. The cylinder named b represents a block whose size and shape is not known. The cylinder with a badge represents a dodecahedron of unknown size. The medium-sized paper bag to the right

T. Dwyer et al. (Eds.): Diagrams 2014, LNAI 8578, pp. 51–53, 2014.

represents a medium size block of unknown shape, and the cylinder displayed as being off the checkerboard corresponds a block of unknown location.

The second premise is an Euler/Venn diagram. The regions enclosed by the curves labeled Medium, Dodec and LeftOfA represent the corresponding sets of objects with those properties. The shading of a region is used to indicate that the corresponding set is empty; for example we can see that there is no object which is left of the object named a and which is not medium size. When used alone or in sequences, constants or existentially quantified variables can be placed in regions to convey that the object represented by the constant is a member of the set represented by the region.

Fig. 1. The Premises

Fig. 2. Main Proof Structure and One Case Diagram

Fig. 2 shows our example proof. The first two steps contain the premise diagrams. The left hand side of the figure shows the structure of the proof. The initial steps represent a proof by cases. The block named b must be of one of three sizes, and there is one case for each possible size (the one in which b is large is shown on the right of Fig. 2). These cases are shown to be exhaustive using an inference rule of the homogeneous blocks world logic at step 8 of the proof. Of these three cases, two of them are contradictory, since the Euler/Venn diagram indicates that every block to the left of a is of medium size. This contradiction is detected using the Close Venn rule of the heterogeneous logic. The homogeneous blocks world logic contains an inference rule, Merge, which allows the joining of information in the diagrams in the open subproofs to be inferred from an exhaustive set of cases, and so we infer that the block b is of medium size in the main proof, at step 9.

The same reasoning applies to the other block of unknown size which is to the left of a. The heterogeneous logic contains a rule, APPLY VENN TO BLOCKS, with which we can infer this in one step. By citing the Euler/Venn premise diagram and the blocks world diagrams just inferred, we can infer a new diagram in which this other block is also of medium size. APPLY VENN TO BLOCKS examines the differences between the desired diagram and the premise diagram of the same type, and verifies that had these differences been modified in alternative ways, then each of them would have contradicted the Euler/Venn premise.

Fig. 3. The Conclusion Diagrams

A similar rule, APPLY BLOCKS TO VENN, is used in the last step of this proof. Since the blocks world diagram now represents the existence of a medium dodecahedron that is to the left of a, we can infer a new Euler/Venn diagram in which there is an existential variable in the region representing the conjunction of these three properties. Again, any alternative extension of the Euler/Venn diagram, i.e., shading that region, would contradict the information displayed in the blocks world diagram. Fig. 3 shows the content of the last two steps.

3 Conclusion

Through designing and studying the purely diagrammatic and yet heterogeneous HyperVenn reasoning system we hope to gain further insight into diagrammatic reasoning in general.

References

1. Barker-Plummer, D., Etchemendy, J., Murray, M., Pease, E., Swoboda, N.: Learning to use the openbox: A framework for the implementation of heterogeneous reasoning. In: Cox, P., Plimmer, B., Rodgers, P. (eds.) Diagrams 2012. LNCS, vol. 7352, p. 3. Springer, Heidelberg (2012)
2. Barker-Plummer, D., Swoboda, N., Murray, M.D.: Hypervenn: A diagrammatic heterogeneous logic (in preparation)
3. Barwise, J., Etchemendy, J.: Hyperproof. CSLI Publications (1995)
4. Jamnik, M.: Mathematical Reasoning with Diagrams - From Intuition to Automation. CSLI, Stanford (2001)
5. Swoboda, N., Allwein, G.: Using dag transformations to verify euler/venn homogeneous and euler/venn fol heterogeneous rules of inference. Software and Systems Modeling 3(2), 136–149 (2004), `http://dx.doi.org/10.1007/s10270-003-0044-8`
6. Urbas, M., Jamnik, M.: Diabelli: A heterogeneous proof system. In: Gramlich, B., Miller, D., Sattler, U. (eds.) IJCAR 2012. LNCS, vol. 7364, pp. 559–566. Springer, Heidelberg (2012)

Visualizing Concepts with Euler Diagrams

Jim Burton, Gem Stapleton, John Howse, and Peter Chapman

University of Brighton, UK
www.ontologyengineering.org
{j.burton,g.e.stapleton,john.howse,p.b.chapman}@brighton.ac.uk

Abstract. We present a new ontology visualization tool, that uses Euler diagrams to represent concepts. The use of Euler diagrams as a visualisation tool allows the visual syntax to be well matched to its meaning.

Keywords: semantic web, visualisation, Euler diagram.

1 Introduction

An ontology comprises a set of statements (called axioms) that capture properties of individuals, concepts and roles. Individuals represent particular elements of the modelled domain, with concepts and roles corresponding to classes and binary relations, respectively. The primary (formal) notations for ontology modelling are symbolic, such as description logics (DLs) or OWL. To aid with accessibility and understandability, standard ontology editors often provide visualisation support. For example, Protégé[1] includes a plugin visualisation package, OWLViz, that shows derived hierarchical relationships but does not show complete information about the known relationships between the concepts.

This paper presents a new ontology visualization tool, *ConceptViz*, that uses Euler diagrams to represent concepts. These diagrams have the advantage that their topological properties reflect the semantic information that they convey; the structural correspondence between topology and semantics called *iconicity* by Peirce [3]. For instance, the containment of one curve by another reflects a subsumption relationships (i.e. set inclusion) between concepts. The asymmetric, transitive nature of curve containment reflects the asymmetric, transitive nature of (proper) subsumption. These properties motivate the choice of Euler diagrams as an effective medium for ontology visualisation. Furthermore, we combine *asserted* and *inferred* ontology information to directly visualise facts that users would otherwise need to consult several different sources to ascertain.

2 The ConceptViz Tool

ConceptViz has been developed as a plugin for Protégé. Figure 1 shows a screenshot from the *Pizza* ontology[2], which defines *Spiciness* to be equivalent to the

[1] `http://protege.stanford.edu`, accessed December 2013.

[2] Protégé OWL tutorial: `http://owl.cs.manchester.ac.uk/tutorials`, accessed December 2013.

T. Dwyer et al. (Eds.): Diagrams 2014, LNAI 8578, pp. 54–56, 2014.

union of three pairwise disjoint concepts – the disjointness is conveyed by the placement of the circles, while the shading tells us "nothing is spicy unless it is hot, medium or mild". For contrast, the OWLViz view of the *Spiciness* concept, displayed in figure 2, does not show the information that *Mild*, *Medium* and *Hot* are pairwise disjoint, or that they form a partition of *Spiciness*. Several parts of the standard Protégé interface must be inspected to gather the information provided by figure 1. We see the increased co-location of information as one of the strengths of ConceptViz.

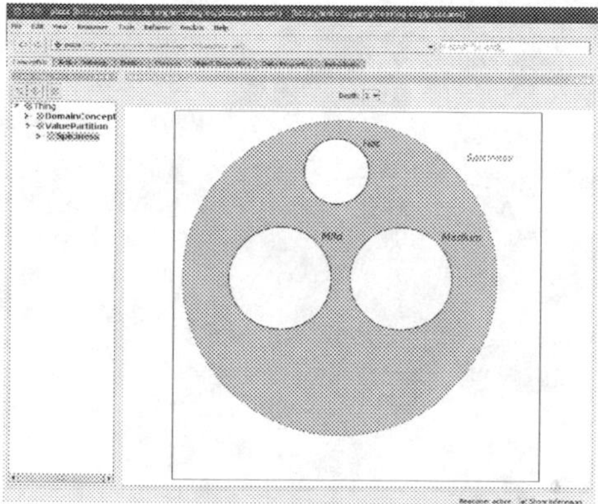

Fig. 1. ConceptViz view in Protégé

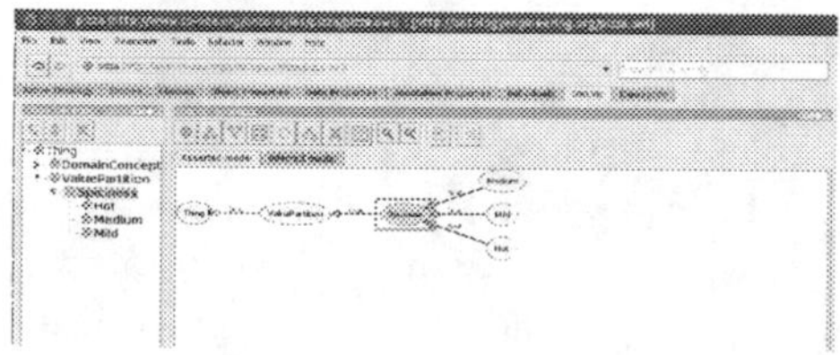

Fig. 2. OWLViz view in Protégé (whitespace cropped)

In order to produce the Euler diagram in figure 1, ConceptViz extracts information about the concepts given in the associated OWL file via the Protégé OWL API. Protégé also incorporates automated reasoners which provide the user with information that can be inferred from that which the user has directly asserted. After initiating a reasoner, Protégé users are able to inspect the inferred information, helping them to understand their ontology more fully and to identify any inconsistencies. ConceptViz interacts with reasoners to discover inferred information about equivalence, unions and so on. Figure 1, for example, includes inferred information and was produced by ConceptViz after a reasoner, started by the user, made inferred information available.

ConceptViz generates an *abstract description* of the to-be-drawn Euler diagram. The abstract description is passed to the *iCircles* library [4], which automatically produces the drawn diagram. Figure 3 shows a diagram generated by the tool alongside its abstract syntax.

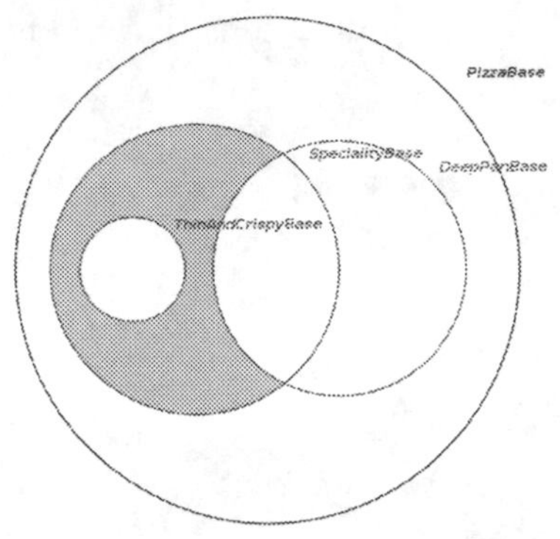

Fig. 3. A diagram generated by ConceptViz

1. labels: {*PizzaBase*, *SpecialityBase*, *ThinAndCrispyBase*, *DeepPanBase*}
2. curves: $\{c_p, c_s, c_t, c_d\}$
3. regions (called zones): $\{\emptyset, \{c_p\}, \{c_p, c_s\}, \{c_p, c_d\}, \{c_p, c_s, c_t\}, \{c_p, c_s, c_d\}\}$
4. shaded zones: $\{\{c_p, c_s\}\}$.

Fig. 4. Abstract syntax

Work is under way to extend ConceptViz to support a notation, *concept diagrams* [1], which is expressive enough to visualise entire ontologies. Concept diagrams extend Euler diagrams with syntax to represent individuals (by solid dots) and roles (by arrows); their use will allow us to provide a more compact and expressive ontology visualisation than currently exists. This work is also informed by the ontology visualisation literature. For instance, a common shortcoming of existing ontology visualisation tools is the tendency to clutter [2]; concept diagrams have been designed with an emphasis on clutter reduction where possible. Finally, we intend to equip ConceptViz with editing features, enabling our goal of round-trip visual ontology engineering.

References

1. Howse, J., Stapleton, G., Taylor, K., Chapman, P.: Visualizing ontologies: A case study. In: Aroyo, L., Welty, C., Alani, H., Taylor, J., Bernstein, A., Kagal, L., Noy, N., Blomqvist, E. (eds.) ISWC 2011, Part I. LNCS, vol. 7031, pp. 257–272. Springer, Heidelberg (2011)
2. Katifori, A., Halatsis, C., Lepouras, G., Vassilakis, C., Giannopoulou, E.: Ontology visualization methods - a survey. ACM Comput. Surv. 39(4) (November 2007)
3. Peirce, C.S.: Collected Papers, vol. 4. Harvard University Press (1933)
4. Stapleton, G., Flower, J., Rodgers, P., Howse, J.: Automatically drawing euler diagrams with circles. Journal of Visual Languages and Computing 23(3), 163–193 (2012)

Argument Mapping for Mathematics in Proofscape

Steve Kieffer

Caulfield School of Information Technology,
Monash University, Caulfield, Victoria 3145, Australia
Steve.Kieffer@monash.edu

Abstract. The Proofscape argument mapping system for mathematical proofs is introduced. Proofscape supports argument mapping for informal proofs of the kind used by working mathematicians, and its purpose is to aid in the comprehension of existing proofs in the mathematical literature. It supports the provision of further clarification for large inference steps, which is available on demand when a proof is explored interactively through the Proofscape browser, and theory-wide exploration is possible by expanding and collapsing cited lemmas and theorems interactively. We examine how an argument map makes the structure of a proof immediately clear, and facilitates switching attention between the detailed level and the big picture. Proofscape is at http://proofscape.org.

Keywords: argument mapping, informal proofs, theory exploration.

1 Introduction

Proofs in the mathematical literature use words like "therefore" and "hence" to introduce assertions, but such language fails to indicate from precisely which prior statements the new one is meant to follow. In other words, the *inferential structure* of the proof is not made explicit. Moreover, proofs may explain too little, leaving the reader to fill in difficult gaps, or explain too much, dissipating attention and obscuring the overview. For these reasons prose proofs can be difficult to understand.

In this paper I present Proofscape, an *argument mapping system* designed specifically for mathematics. Each assertion, assumption, and definition made in a literature proof is placed on a node in a Proofscape diagram, and the nodes are linked together with arrows indicating the inferential structure of the proof unambiguously. In fact the diagrams are not mere graphs but *DASHgraphs*, which I define in Sec. 3, and examples of which are seen in Fig. 2.

The overview provided by the diagram counteracts the problem of too much detail, or obscuration of the general plan, in prose proofs. As for the opposite problem of not enough detail, or difficult inferences, this is counteracted by supplementary clarifications, written by and for users of the system. These clarifications can be interactively opened and closed, meaning that extra nodes and edges are added to or taken away from the diagram. Theory-wide interactive

T. Dwyer et al. (Eds.): Diagrams 2014, LNAI 8578, pp. 57–63, 2014.

exploration is also supported by allowing users to open and close cited lemmas and theorems.

Proofs are represented internally using a simple system I have designed for transcribing what happens in informal proofs, but proper discussion of that system requires a separate paper contrasting it to related proposals e.g. Wiedijk's [1]. Nor is there room here to discuss the layout algorithms that are used. On the contrary, the focus in the present paper is on *the design of the visual encoding* (Sec. 3), and on the demonstration of *how Proofscape diagrams support study of proofs* (Sec. 4). I review related work in Sec. 2. The main contributions of this paper are the visual notation, and the working Proofscape system at `http://proofscape.org`.

2 Related Work

The problems of obscure inferential structure and lack of modularity – i.e. the ability to expand and collapse levels of detail – in prose proofs were confronted by Lamport [2], who proposed presenting proofs in the form of indented outlines whose levels could be opened and closed, and in which claims cited previous lines by number. Lamport envisioned implementation in hypertext, which was later realised by Cairns and Gow [3], but this system did not use the diagrammatic nodes-and-arrows style of argument mapping systems. An earlier related proposal by Leron [4] used nodes and arrows to diagram not the steps of a proof but the relations between the modular levels into which it was to be decomposed.

The mapping of "what follows from what" is essentially the aim of all argument mapping systems (see e.g. Harrell's survey in [5]), but the structure and visual encoding of the diagram may depend on the intended subject matter. In particular, most existing argument mapping systems seem to be intended for use with subjects like law, philosophy, business, or politics (e.g. Gordon and Walton [6] or van Gelder [7]), and accordingly rely on the Toulmin model [8] or variants thereof. The Toulmin model is adapted for the questionable nature of claims regarding the real world, accommodating things like rebuttals, and citations of factual evidence, but none of this is appropriate for mathematics, where claims are either true or false, never debatable, and factual information about the real world is irrelevant. Moreover, systems like these omit many features that *are* needed for mathematical arguments, which I discuss in Sec. 3.

Some existing software does generate diagrams of mathematical proofs, but these seem almost universally to be proofs that have been encoded in some formal logical system. Here existing software is of two kinds: (1) interfaces for semi-automated theorem provers or formal proof assistants (e.g. Siekmann et. al. [9]); or (2) educational software intended to teach the inference rules of basic formal logics (e.g. Watabe and Miyazaki [10]). Systems of the former kind may present proofs at a high level, but still in a formal system, and their visual representations often represent progress in the construction of a proof, not a complete literature proof as it stands. Meanwhile the latter sort of system is concerned with the low-level atomic steps sanctioned by a rudimentary formal logic.

The *e-Proofs* system [11] guides students through informal textbook-style proofs, but does so using pre-generated instructional videos. In one of its three modalities the system shows a proof written out as a paragraph of text, and in successive frames draws boxes around various parts of the proof and arrows linking these, but these disappear in the next frame, and a graph is not built up (nor would the graph be laid out well if it did appear, its arrangement dictated by the line-breaks in the prose). The system is not meant for theory-wide exploration, each video being dedicated to a single proof, nor can the user interactively explore and hide supplementary clarifications for inference steps.

Proofscape thus fills a need for an argument mapping system designed specifically for informal mathematical proofs, with interactive access to further clarification, and theory-wide exploration.

3 Visual Encoding

In Proofscape, a proof is represented by a directed, acyclic, styled, hierarchical graph, or *DASHgraph*. The directed edges of the graph may represent both *deduction* and *flow* (see below). The graph is acyclic because deduction never involves circular logic. It is said to be styled because additional information is carried by the shape and stroke style (plain, bold, or dashed) of node boundaries, as well as the stroke style of edges. Finally, a DASHgraph is hierarchical in that some of its nodes may be nested inside of others, in order to represent subproofs and to contain cited theorems and their proofs.

The nodes carry the content of a proof, such as definitions, assumptions, and conclusions. Edges drawn with solid stroke are called *deduction arrows*, and indicate which assertions follow from which others; specifically, when the transcriber believes that $A_1, A_2, ..., A_n$ are the nodes which are, in an informal sense, "used" in inferring node B, then there is to be a deduction arrow from each A_i to B. Meanwhile edges drawn with dashed stroke are called *flow arrows*, and indicate the order in which the author of the proof would like you to explore the graph. This helps to ensure that the argument is built up in a sensible sequence, e.g. so that definitions are encountered before they are used.

There are five *basic* node types, two *compound* types, and several "special nodes" of which just two are considered here (see Fig. 1). The bulk of the content of a proof is carried by the first three basic node types: *Intro* nodes, *Assumption* nodes, and *Assertion* nodes.

Intro nodes are for the introduction of objects and symbols in the proof. For example, the label could state "Let K be a number field," or "Let G be the Galois group of E over K." Their boundary is rectangular and drawn with a bold stroke.

Assumption nodes are for the introduction of the premises of a theorem, and for additional assumptions made during the course of a proof, such as at the beginning of a proof by contradiction, or proof by cases. Thus the labels state suppositions like "Suppose G is Abelian," or "Suppose φ is surjective." The boundary of an assumption node is rectangular with a dashed stroke.

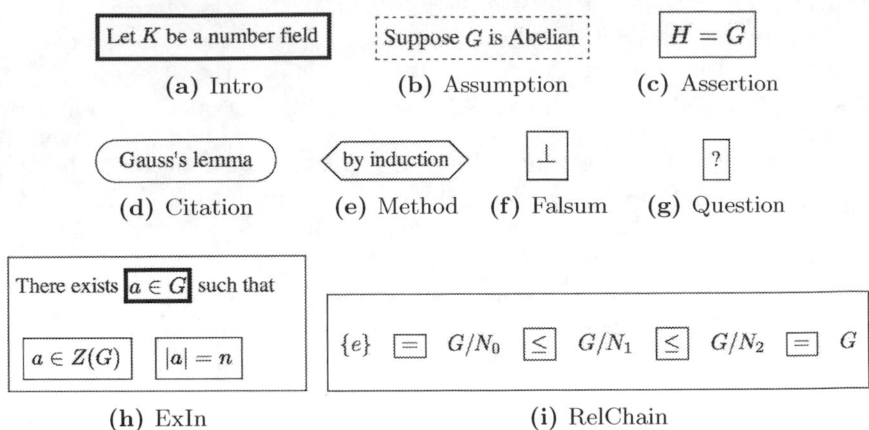

Fig. 1. Basic node types: (a),(b),(c),(d),(e); Special nodes: (f),(g); Compound node types: (h),(i)

There is a difference between the premises of a theorem and any additional assumptions introduced during the course of a proof, in that the latter will be discharged by the end of the proof whereas the former will not; in order to make this difference obvious to the reader, Assumption nodes belonging to top-level deductions are drawn both dashed and bold.

Assertion nodes are the most common type of node, carrying the assertions of a proof. Since it is up to the deduction arrows in the DASHgraph to indicate which assertions follow from which others, all logical language such as "therefore," "hence," "it follows that," etc. is omitted from the node labels. For example a label would state simply "$H = G$," never "therefore $H = G$." Assertion nodes have a rectangular boundary with plain stroke.

The node styles described so far have been chosen deliberately. It was felt that a dashed boundary is a natural metaphor for the contingent nature of assumptions, so is suitable for Assumption nodes in general, while the addition of a bold stroke suggests the somewhat more permanent nature of the premises of top-level deductions. The simple bold boundary of Intro nodes reflects the introduction of objects and symbols by fiat. Meanwhile Assertion nodes are the most common, and so should have a plain style, while the special styles of Assumption and Intro nodes help the reader to locate them quickly.

The final two basic node types are *Citation* nodes, which can be used *internally* to cite other results in the Proofscape library, and *externally* to cite any result in the literature; and *Method* nodes, carrying phrases like, "by induction," or, "substituting c for x," which clarify *how* an inference is made. Citation and Method nodes are given "stadium" and hexagonal shapes, respectively, which were chosen because they can easily stretch horizontally. The boundaries of Citation nodes are drawn dashed and solid to indicate internal and external citations respectively.

The first compound node type is the *Existential Introduction* or *ExIn* node, which handles formulas like, "There exists $a \in G$ such that $a \in Z(G)$ and $|a| = n$," in which an object is both stated to exist and introduced for further use. It features internal Intro and Assertion nodes for the components of the statement. The other compound node type is the *Relation Chain* or *RelChain* node which handles expressions of the form, "$A_1 = A_2 = \cdots = A_n$," by putting each infix relation $=$, $\leq$, $\geq$, etc. on its own internal Assertion node. Arrows to and from internal nodes may cut across hierarchy levels in the DASHgraph.

Among special nodes is the *Falsum* node, which is a special Assertion node featuring only the falsum symbol $\bot$, and which is to be used at the end of any proof by contradiction; and the *Question* node, featuring only a question mark, for use when the transcriber is not sure how an inference is to be made, thus to mark a point to return for further study.

4 Use Case and Conclusions

We conclude with an illustration of the use of Proofscape. Suppose you wanted to understand Mihăilescu's proof of the Catalan Conjecture [12]. You begin by locating the work in the Proofscape library and opening the main result, "Theorem 5," in the Proofscape browser. Initially only the theorem statement is shown. You click to open the proof. See Fig. 2a. The first thing you notice is that it is a proof by contradiction, with a Falsum node pointing to the final conclusion. You see the contradiction assumption in the node with dashed boundary in the upper left. Tracing backward along the deduction arrows that point to the Falsum node you see that the contradiction is between an assertion that $p > q$ which appears to follow relatively easily from the assumption, and another, $q > p$, which seems to require the bulk of the work in the proof.

From the external Citation node at the top right you see exactly which step in the argument relies on Baker's methods for linear forms in logarithms. Meanwhile the internal Citation nodes refer to three results from the present work, on which the main line of reasoning leading to $q > p$ depends. Suppose you decide to delve into Proposition 1 first, by selecting it in the sidebar. The Citation node is then replaced by the full statement of Prop. 1. After reading the statement you click again to open its proof. See Fig. 2b.

You might begin by tracing backward from the conclusion to see where it came from, and see that we get $q > p$ from $q - 2 \geq p - 1$. Expressions like $q - 2$ or $p - 1$ tend to arise because they represent some meaningful quantity, and you continue tracing backward to see what these might have been. On the far left you see that $q - 2$ is an upper bound on the number of zeroes of a certain polynomial F, whose definition is easily located in the Intro node above this. On the other hand there is a longer line of reasoning that shows that this polynomial F has at least $p - 1$ zeroes. The polynomial F is thus a *pivot* in the sense of Leron [4], a constructed object on whose properties the proof hinges.

Next you may wish to see where the one premise of the Proposition is used. You easily locate it at the top right, in the node with the bold and dashed

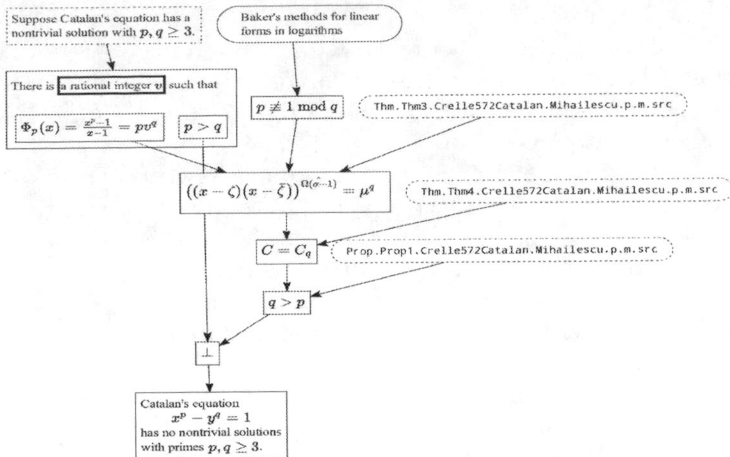

(a) Mihăilescu's Thm. 5 and its proof are open.

ζ is a qth power

$C_q = C$

$\hat{C}$ is primary

For all c in P, $1 + \zeta^c$ is q-primary.

$\exists$ $\nu \in \mathbb{Z}[\zeta]$ such that

$1 + \zeta^c \equiv \nu^q \bmod q^2 \cdot \mathbb{Z}[\zeta]$

Let r satisfy $qr \equiv 1 \bmod p$.

Consider the polynomial $F(X) \in \mathbb{F}_q[X]$ obtained by reducing the coefficients of $\frac{(1+X)^q - (1+X^q)}{qX}$ mod q.

$\nu \equiv 1 + \zeta^{cr} \bmod q \cdot \mathbb{Z}[\zeta]$

$\exists$ $\gamma \in \mathbb{Z}[\zeta]$ such that

$\nu = 1 + \zeta^{cr} + q\gamma$

$C = C_q$

$\mathbb{L}$ is a field

$F(X) = \sum_{k=1}^{q-1} (-1)^{k-1} \cdot k^{-1} \cdot X^{k-1}$

F has at most $q-2$ zeroes in $\mathbb{L}$.

Let Ω be a maximal ideal of $\mathbb{Z}[\zeta]$ lying above q.

Raise both sides to the qth power.

Let $\mathbb{L} = \mathbb{Z}[\zeta]/\Omega \subset \bar{\mathbb{F}}_q$

$1 + \zeta^c \equiv (1 + \zeta^{cr})^q \bmod q^2 \cdot \mathbb{Z}[\zeta]$

char $\mathbb{L} = q$

Apply the substitution $\zeta \mapsto \zeta^q$.

For all $c \in P$, let $X_c = \zeta^c \bmod \Omega \in \mathbb{L}$.

$1 + \zeta^{cq} \equiv (1 + \zeta^c)^q \bmod q^2 \cdot \mathbb{Z}[\zeta]$

p, q are distinct primes

$\zeta^c - \zeta^d = w_p$

For all $c \in P$, we have $F(X_c) = 0$.

The powers of ζ are all distinct modulo Ω.

F has at least $p-1$ zeroes in $\mathbb{L}$.

$q - 2 \geq p - 1$

$q > p$

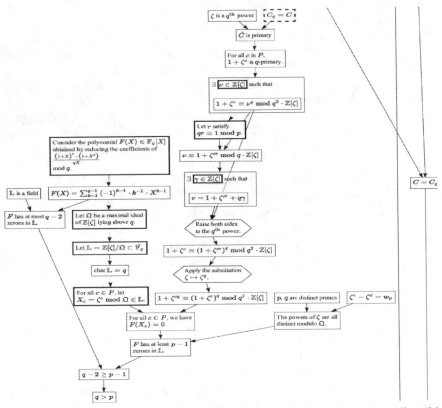

(b) Prop. 1 and its proof have been opened. Only one node of Thm. 5 is still visible on the far right, while the others are above and below the large box for Prop. 1.

Fig. 2. Using Proofscape

boundary, and can see that it contributes to a long line of reasoning which finally helps to get the $p-1$ zeroes of F.

You have learned all this about the structure of the argument without yet having spent any time working through difficult inferences. At this point you have a choice: If you want to understand the proof of Proposition 1 thoroughly you should begin to study each of the inferences one by one. As you go you may find user-supplied clarifications. On the other hand, if you are satisfied with this overview then you may close the proof by clicking again in the sidebar, and return to your study of the main theorem.

These are some of the ways in which Proofscape can help to show the structure of a proof, and to organise the work involved in studying it. The user quickly learns how the parts of the proof fit together, where the assumptions are used, and where the bulk of the work is apt to lie. The forked lines of reasoning leading off from any pivot objects are easily found, and the purpose of those objects thus made clear. The user can open cited results and clarifications, and close them when finished. The diagram serves as a map to chart the user's progress in confirming each individual inference, and then allows the user to quickly return from the details to the big picture and the overall strategy of the proof.

References

1. Wiedijk, F.: Formal proof sketches. In: Berardi, S., Coppo, M., Damiani, F. (eds.) TYPES 2003. LNCS, vol. 3085, pp. 378–393. Springer, Heidelberg (2004)
2. Lamport, L.: How to write a proof. The American Mathematical Monthly 102(7), 600–608 (1995)
3. Cairns, P., Gow, J.: A theoretical analysis of hierarchical proofs. In: Asperti, A., Buchberger, B., Davenport, J.H. (eds.) MKM 2003. LNCS, vol. 2594, pp. 175–187. Springer, Heidelberg (2003)
4. Leron, U.: Structuring mathematical proofs. The American Mathematical Monthly 90(3), 174–185 (1983)
5. Harrell, M.: Using argument diagramming software in the classroom. Teaching Philosophy 28(2), 163–177 (2005)
6. Gordon, T.F., Walton, D.: The Carneades argumentation framework. Frontiers in Artificial Intelligence and Applications 144, 195 (2006)
7. Van Gelder, T.: Argument mapping with reason!able. The American Philosophical Association Newsletter on Philosophy and Computers 2(1), 85–90 (2002)
8. Toulmin, S.: The Uses of Argument. Cambridge University Press, Cambridge (1969)
9. Siekmann, J., et al.: LΩUI: Lovely Ωmega User Interface. Formal Aspects of Computing 11, 326–342 (1999)
10. Watabe, T., Miyazaki, Y.: Visualization of logical structure in mathematical proofs for learners. In: Lee, R. (ed.) Computer and Information Science 2012. SCI, vol. 429, pp. 197–208. Springer, Heidelberg (2012)
11. Alcock, L.: e-proofs: Online resources to aid understanding of mathematical proofs. MSOR Connections 9(4), 7–10 (2009)
12. Mihăilescu, P.: Primary cyclotomic units and a proof of Catalan's conjecture. J. Reine Angew. Math. 572, 167–195 (2004)

Towards a General Diagrammatic Literacy: An Approach to Thinking Critically about Diagrams

Brad Jackel

Australian Council for Educational Research (ACER), Camberwell, Australia, 3124
Brad.Jackel@acer.edu.au

Abstract. Despite our increasing reliance on diagrams as a form of communication there is little guidance for educators on how to teach students to think critically about diagrams in general. Consequently, while we teach students to read specific kinds of diagrams within specific contexts, we lack a coherent approach to thinking critically about diagrams *per se*. This paper presents a theoretical framework for a general diagrammatic literacy, based on conceptualizing diagrams in terms of function rather than form. Approaching diagrams functionally generates a framework for thinking critically about diagrams (in general) that is simple, robust and exhaustive. In addition to this functional approach, the role of context and language to the internal definition of any given diagram is emphasized.

Keywords: diagrammatic literacy, understanding diagrams, teaching diagrams, visual literacy, visual critical thinking, teaching critical thinking, teaching visual literacy, visual representation.

1 Defining Diagrams by Functional Categories

Our reliance on visually communicating information has never been greater. Therefore a coherent theory of diagrammatic literacy is essential[1]. Unfortunately that theory, inasmuch as it exists at all, is badly under-developed[2]. As students we are exposed to various types of diagrams in specific contexts. Yet we do not teach a general diagrammatic literacy *per se*. We may well simply assume that this does not need to be taught: diagrams are what we use to teach other things[3]. This is a dangerous assumption, given the importance of diagrams in communication and decision making. Diagrams, even of hard data, are rarely neutral, are almost always used as a 'rhetorical device for presenting a particular analysis'[4]. How can we learn to analyze this 'visual rhetoric'? How can we encourage a general diagrammatic literacy that will allow for consideration of diagrams in a consistently critical way? This paper is an attempt to begin answering those questions.

The variety of forms diagrams can take is so heterogeneous that any attempt to classify them by form is futile. This creates significant difficulty in attempting to think about diagrams in general: 'research on ... multimodal integration is currently in a premature state due to abundant possible variations of the external representations'[5]. My aim here is not primarily to define the boundaries of what constitutes a

T. Dwyer et al. (Eds.): Diagrams 2014, LNAI 8578, pp. 64–70, 2014.

diagram, but to do so in a way which enables a critical approach that is teachable to undergraduates throughout the curriculum. What classification can we come up with that covers all diagrams, without it being so loose as to include a Vermeer?

All diagrams are visual representations of something, but not all visual representations are diagrams. Diagrams, therefore, are a specific subset of visual representation. Given this, a way forward can be found by defining the boundaries of that subset, defining diagrams by the kinds of things they represent, rather than by the form that representation takes. That is, we can classify diagrams by function, not form.

I posit that *all* diagrams, in terms of function, visually represent at least one of the following: 1. Data; 2. System and/or Process; 3. Conceptual Relationship(s). This can be represented as:

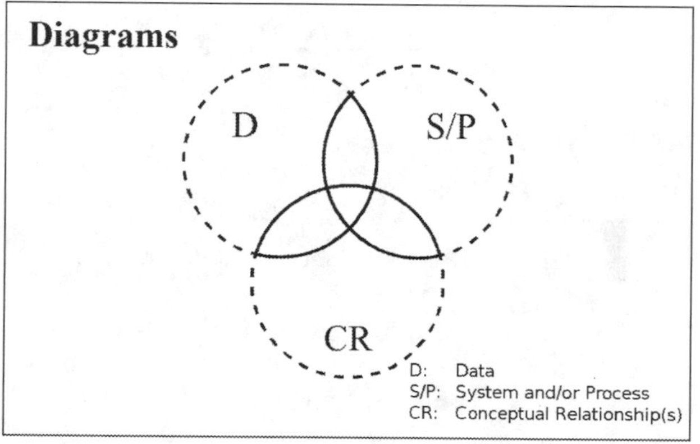

Fig. 1. A functional approach to diagrams

This framework locates the point at which a visual representation becomes a diagram: a stylized picture of the continent of Australia is not a diagram any more than a photograph of a tree is. If, however, we draw objects onto that picture representing boundaries, contours, and so forth, we have a diagram: a visual representation of both data and a number of systems (political, spatial, social), commonly referred to as a 'map'. Questions on intelligence tests often present a number of diagrammatic objects with one 'missing': this is diagrammatic: the candidate must discern some kind of system and/or process, then choose a diagrammatic object that will complete it.

In practice, the vast majority of diagrams visually represent more than one of these categories at a time: the dotted lines in Fig. 1 attempt to draw attention to the prevalence of this categorical overlap. Not only are most diagrams located within one of the overlapping areas in Fig. 1, that overlap is frequently the very reason we use diagrams in the first place: a stock chart (representation of data) shows at a glance some kind of process (e.g. the explosion of Google's value over a given period). A family tree visually represents all three categories at once.

My aim with this functional definition is not primarily to classify diagrams, but rather to encourage the conscious interrogation of the various ways a given diagram functions. This can be done by formally asking three questions derived from this functional understanding of what diagrams visually represent:

- *Is this diagram a representation of data?*
- *Is this diagram a representation of system/process?*
- *Is this diagram a representation of conceptual relationship(s)?*

Suppose we take four measurements of a variable over time and represent that data visually on a chart, with time being the horizontal axis as in Fig 2a:

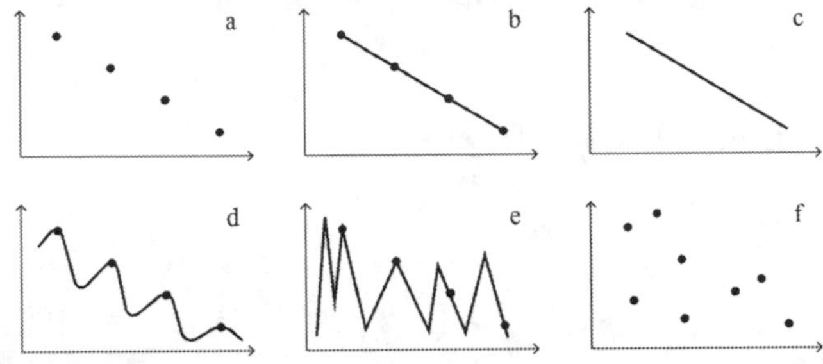

Fig. 2. Same data, different visual rhetoric

We could also represent that data as in Fig. 2b. Both of these diagrams are clearly visual representations of data. However we need to learn to formally ask the question: do these diagrams function as visual representations of process? In both cases the answer is yes, however Fig. 2a weakly asserts process while Fig. 2b strongly asserts it by 'joining the dots' for us. With Fig. 2c the rhetorical assertion of process is complete, any connection to the four points of data erased. Yet we typically think of this kind of diagram as a self-evident representation of quantitative fact. Asking these questions also focuses attention on the manner in which the visual rhetoric of the diagram asserts process (a solid line). Paradoxically, the more overt the assertion the less we see an assertion: with Fig. 2a we are aware we are projecting process onto the four points of data. With Fig. 2b and 2c we see the self-evident 'fact' of the process itself.

This kind of 'fact' is more problematic than it may appear. Consider the same set of data as pictured in Fig. 2d and 2e. Joining the dots with a straight line is a huge interpretation that has been made for us in Fig. 2b and 2c: we see the process and don't perceive an interpretation or contention. The more rhetorical a diagram is the less we notice that rhetoric. In fact the decision to join the dots in any way at all is a substantial act of reification, asserting some identifiable process that binds the measurements together, that exists in some real sense even when our measurement of it does not. In fact the situation could be that of Fig. 2f.

Reification of process is merely one example of diagrams functioning within the overlap of representational domains in Fig. 1. The examples of Fig. 2 may appear contrived, yet it is frequently witnessed. In the context of education, for example, a graphical representation of a decline in numeracy skills, based on three measurements over nine years, presents a process we can 'see'. Not only will that process presumably continue downwards indefinitely unless we take action, it typically acts as de-facto proof for a given contention about both its cause and its solution within the context of public debate.

Understanding diagrams in terms of these overlapping functional categories generates a set of simple questions that we can ask in order to guide critical thinking about diagrams, allowing us to critically examine their rhetorical impact.

2 Context and Language in Diagrams

As defined in any dictionary diagrams are almost exclusively visual objects. Yet the vast majority of diagrams are explicitly contained within some context[6] that limits what the diagram means. The presence of a context which constructs the 'correct' signification of a given diagram is non-trivial, as is the fact that this context is almost invariably verbal, that is, of words. Even in the case of 'pure' geometrical diagrams, it is notable that the 'interaction between diagrams and language appears to have been what gave rise to the first invention of proof in Greek geometry'[7].

It is in relating the diagram to its context that the possible meanings of a diagram are heavily restricted in a way that Van Gogh's *Church at Auvers* is not. Consider the following, taken from a letter from Joseph Conrad to H.G. Wells in 1903 [8]:

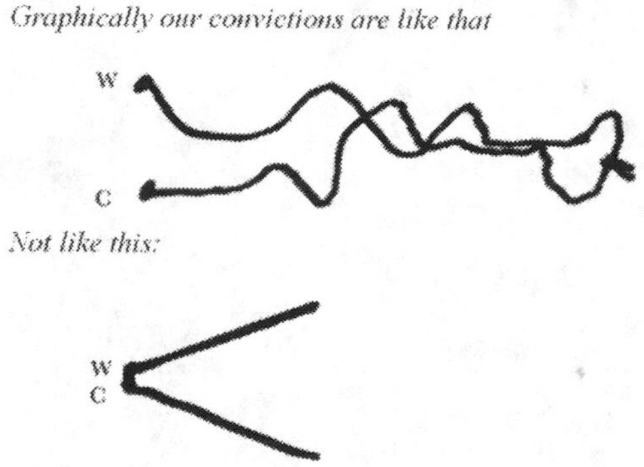

Fig. 3. Joseph Conrad to H.G. Wells

These images are clearly diagrams. W and C represent Wells and Conrad. The lines represent their convictions and the diagrams tell us something quite specific about the nature of their relationship. But it is the words that allows the lines to function as diagrams. Without that context the images could coherently refer to almost anything: the only reason we would call them diagrams at all is due to implicit context: they are similar in form to diagrammatic conventions we are familiar with. Diagrams, to function as diagrams, must be constrained by a context that limits their interpretation[1]. Moreover, it is clearly observable that the diagrams we habitually use

[1] Even the wordless instructions in a flat-pack from IKEA depend on context. We are familiar with diagrams, flat-pack furniture, Allen keys; the flat-pack of components themselves give context. There is so much context an entire civilisation is implied.

and need to critically consider are constrained by a contextual definition that is, in practice, nearly always verbal: while useful in many ways, the either/or dichotomy of distinctions between sentential and diagrammatic representations[9] can be misleading. It is the explicit or implicit context of a diagram, usually sentential, that shapes our interpretation of it. Therefore there is one overriding question we must keep in mind when considering any diagram while we formally consider the three functional questions listed above: *How is the interpretation of the diagram constructed by the context and language that defines it?*

3 Applying the Approach: Florence Nightingale

The following diagram was published by Florence Nightingale[10] in order to represent visually the statistics of the Royal Commission she headed into the medical care of British soldiers:

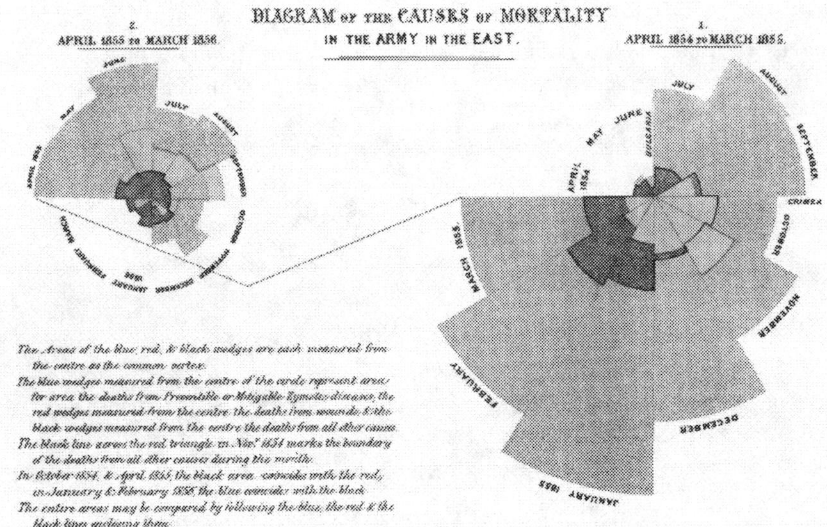

Fig. 4. Florence Nightingale's mortality diagram

Firstly, how is the interpretation of this diagram constructed by the context and language that defines it? The key to the diagram explains the mathematical relationship between the data and the diagram, but the most significant component of it is the way it describes what each color represents[2]: Blue is defined as 'Deaths from Preventable or Mitigable Zymotic Diseases'; Red as 'Deaths from wounds' and Black as 'Deaths from all other causes'. The phrase 'deaths from wounds' is morally neutral. The phrase 'deaths from all other causes' is morally neutral. The phrase 'deaths from

2 Blue is the large medium-grey area on the outer of each spiral. A color version can be found at http://upload.wikimedia.org/wikipedia/commons/1/17/Nightingale-mortality.jpg

Preventable or *Mitigable* Zymotic Diseases' is not morally neutral: they are not merely deaths, they are 'preventable' deaths. Defining the blue area as 'preventable' death immediately begs a very big question: why weren't they prevented? As presented to Queen Victoria it was as much a fiercely intelligent exercise in marketing and politics as a simple representation of data.

An astonishingly high rate of mortality for wounded soldiers thanks to zymotic (a process involving fermentation) diseases all the way from gangrene through to dysentery was, at the time, simply an accepted part of war. It can be difficult to appreciate now just how normalized this was for soldiers in military hospitals even while hygiene related mortality was no longer a systemic problem in civilian hospitals, a discrepancy many of Nightingale's diagrams drew attention to. The argument that most of those deaths were preventable (i.e., the fault of the British) was thus contentious, even scandalous. However, diagrams effectively form a 'puzzle' which we have to solve[11]. We see all the blue, search for a definition and find 'preventable death.' We now *see* the blue as 'preventable death'. That is, the diagram causes us to see 'preventable death' as a fact rather than consider that alleged preventability as a contention. Once we have analyzed the way the language structures our reading we can then ask the questions that form the basis of this approach to diagrammatic literacy:

1. Is it a visual representation of data? Yes. But formally asking that question encourages us to think about the data and the choices made in its presentation. What if Nightingale had individually represented the mortality statistics for each disease (gangrene, dysentery etc.) rather than grouping them together as 'preventable'? What are the rhetorical effects of the manner in which she grouped the data?
2. Is it a visual representation of a system/process? Yes. Indeed the power of this diagram derives from how it functions as a representation of a system of medical care that is far more effective at killing soldiers than the enemy.
3. Is it a visual representation of some kind of conceptual relationship(s)? Yes. The diagram sets up a conceptual relationship between two forms of mortality: one acceptable, the other not. This was a contentious notion at a time that conceived of combat mortality as a single category. Because the diagram caused contemporary viewers to *see* intolerable death rather than consider the proposition that some intolerable category might exist, it stopped being a radical idea. It became the truth.

4 Conclusion

The method of thinking critically about diagrams outlined above allows us to identify and explicitly consider the visual rhetoric of a diagram and what it is doing. It is relatively easy to teach, as it simply involves conceiving of diagrams as visual objects that function to represent at least one of three overlapping domains and that are constrained by some kind of (usually verbal) context that must be explicitly considered. The three questions that derive from understanding diagrams in terms of elementary functional categories allow us to quickly hone in on the visual rhetoric of any given diagram, to understand and make explicit how a diagram works on an audience.

My aim here is not quite to define what diagrams are. A definition tells us what a thing is. Rather I have attempted to develop a way of approaching any given diagram in order to guide critical thought about what that diagram *does*. As the torrents of information we now routinely deal with forces us to rely ever more on communicating that information 'at a glance', thinking critically about diagrams is a skill we must begin to formally teach.

References

1. Novick, L.R.: The Importance of Both Diagrammatic Conventions and Domain-Specific Knowledge for Diagram Literacy in Science: The Hierarchy as an Illustrative Case. In: Barker-Plummer, D., Cox, R., Swoboda, N. (eds.) Diagrams 2006. LNCS (LNAI), vol. 4045, pp. 1–11. Springer, Heidelberg (2006)
2. Skaife, M., Rogers, Y.: External cognition, interactivity and graphical representations. In: Proceedings of the IEE Colloquium: Thinking with Diagrams, pp. 8/1–8/6 (1996)
3. Blackwell, A.: Introduction Thinking with Diagrams. In: Blackwell, A.F. (ed.) Thinking with Diagrams, pp. 1–3. Kluwer Academic Publishers (2001)
4. Elzer, S., Carberry, S., Demir, S.: Communicative Signals as the Key to Automated Understanding of Simple Bar Charts. In: Barker-Plummer, D., Cox, R., Swoboda, N. (eds.) Diagrams 2006. LNCS (LNAI), vol. 4045, pp. 25–39. Springer, Heidelberg (2006)
5. Acarturk, C., Habel, C., Cagiltay, K.: Multimodal Comprehension of Graphics with Textual Annotations: The Role of Graphical Means Relating Annotations and Graph Lines. In: Stapleton, G., Howse, J., Lee, J. (eds.) Diagrams 2008. LNCS (LNAI), vol. 5223, pp. 335–343. Springer, Heidelberg (2008)
6. Feeney, A., Hola, A.K.W., Liversedge, S.P., Findlay, J.M., Metcalf, R.: How People Extract Information from Graphs: Evidence from a Sentence-Graph Verification Paradigm. In: Anderson, M., Cheng, P., Haarslev, V. (eds.) Diagrams 2000. LNCS (LNAI), vol. 1889, pp. 149–161. Springer, Heidelberg (2000)
7. Stenning, K.: Distinctions with Differences: Comparing Criteria for Distinguishing Diagrammatic from Sentential Systems. In: Anderson, M., Cheng, P., Haarslev, V. (eds.) Diagrams 2000. LNCS (LNAI), vol. 1889, pp. 132–148. Springer, Heidelberg (2000)
8. Conrad, J.: Letter to HG Wells, 19/09/1903. In: Karl, F., Davies, L. (eds.) The Collected Letters of Joseph Conrad, p. 62. Cambridge University Press (1988)
9. Larkin, J.H., Simon, H.A.: Why a diagram is (sometimes) worth ten thousand words. Cognitive Science 11, 65–100 (1987)
10. Nightingale, F.: Diagram of the causes of mortality in the army in the East. Notes on Matters Affecting the Health, Efficiency and Hospital Administration of the British Army. Harrison and Sons (1858)
11. Niño Zambrano, R., Engelhardt, Y.: Diagrams for the Masses: Raising Public Awareness – From Neurath to Gapminder and Google Earth. In: Stapleton, G., Howse, J., Lee, J. (eds.) Diagrams 2008. LNCS (LNAI), vol. 5223, pp. 282–292. Springer, Heidelberg (2008)

Item Differential in Computer Based and Paper Based Versions of a High Stakes Tertiary Entrance Test: Diagrams and the Problem of Annotation

Brad Jackel

Australian Council for Educational Research (ACER), Camberwell, Australia, 3124
Brad.Jackel@acer.edu.au

Abstract. This paper presents the results from a tertiary entrance test that was delivered to two groups of candidates, one as a paper based test and the other as a computer based test. Item level differential reveals a pattern that appears related to item type: questions based on diagrammatic stimulus show a pattern of increased difficulty when delivered on computer. Differential in performance was not present in other sections of the test and it would appear unlikely to be explained by demographic differences between the groups. It is suggested this differential is due to the inability of the candidates to freely annotate on the stimulus when delivered on computer screen. More work needs to be done on considering the role of annotation as a problem solving strategy in high-stakes testing, in particular with certain kinds of stimulus, such as diagrams.

Keywords: Diagrams, computer based assessment, paper based assessment, high-stakes tests, standardized test, annotation, diagrammatic assessment, NAPLAN.

1 Introduction

The literature on Computer Based Assessment (CBA) is extensive and broadly optimistic: the general picture is that 'computer-based assessment offers enormous scope for innovations in testing and assessment'[1] and that 'electronic delivery will bring major benefits'[2]. The literature concerning performance differential between CBA and Paper Based Assessment (PBA) is also extensive, if somewhat mixed and inconclusive, as is often noted[3–5]. Where differential has been found between PBA and CBA versions of tests these 'mode effects' are typically explained by demographic differences among the candidates (such as varying degrees of computer literacy), or problems with the CBA layout (such as stimulus material that requires excessive scrolling). Leeson's meta-review of mode effect studies labels these categories 'Participant Characteristics' and 'User Interface'[6].

This paper is written from my own point of view as a research fellow with ACER where my primary role is to write and construct high-stakes cognitive skills tests. From this point of view there are some identifiable reasons that the literature regarding the psychometric equivalence of CBA and PBA is so mixed and inconclusive.

T. Dwyer et al. (Eds.): Diagrams 2014, LNAI 8578, pp. 71–77, 2014.

One of them is that studies on PBA and CBA equivalence typically focus on aggregated scores that have been decoupled from individual item responses and item type (multiple choice questions, including all four options and any subsidiary stimulus material provided, are referred to as 'items'). When we only consider the aggregated results of an entire test the differentials between the aggregated scores are often within the realm of 'not statistically significant'. This approach can hide significant patterns in item differential that reveal mode effects unrelated to demographics or problems with CBA layout.

What this paper attempts to shed light on is whether or not there is an identifiable performance differential between CBA and PBA that is related to item type, not demographics or layout.

2 Method

The results below are from one of ACER's tertiary entrance tests, which I will refer to as the TEST. The TEST is used by various universities in Australia for entry to a variety of courses. Due to the fact that some institutions wish the TEST to be delivered on paper and others on computer, it is delivered in both media, with identical stimulus and items delivered in the same order in both versions. In the CBA version the stimulus material is presented on the left of the screen with each question appearing to the right, and the candidate is free to go forwards and backwards within the TEST. The CBA candidates were applying to one group of institutions, the PBA candidates to another, and in this sense there is no disadvantage to the candidates regardless of which version they take as their entry scores are only being compared with those who took the TEST in the same mode. A total of 1652 candidates took the TEST. 1134 of these took the computer based version with 518 taking the paper based version.

There are three sections to the TEST: Critical Reasoning (making decisions on the basis of information, logically analyzing scientific, technical, business and certain forms of socio-cultural stimulus), Verbal and Plausible Reasoning (interpretation of stimulus material in a socio-cultural context, reasoning typically needed in the arts, humanities and social sciences) and Quantitative Reasoning (mathematical and scientific information and problem solving). Item validation is based on analysis generated by ACER's Rasch model ProQuest software.

This approach has some advantages. Firstly, the sample sizes are large enough for the psychometric data to be valid. Secondly, the candidates are attempting to enter university: they are performing at their best. Thirdly, the questions have been developed and revised by an experienced team of test developers and have been trialed for validity prior to their inclusion as scoring items. However, this approach does have some drawbacks, primarily that the questions are secure: presenting sample questions or sample workbooks showing candidate annotations is not possible.

3 Results

The following table presents the Quantitative Reasoning section of the TEST, ordered by the increase in difficulty from PBA to CBA:

Table 1. Comparison of CBA and PBA difficulty.

Item	% Correct (PBA)	% Correct (CBA)	Increase in CBA difficulty (differential)	Type
44	39.2	27.9	11.3	Diagram/Calculation
13	36.9	26.2	10.7	Diagram/Calculation
59	26.3	16	10.3	Diagram/Calculation
43	49.6	39.6	10	Diagram/Calculation
3	86.3	76.4	9.9	Diagram/Calculation
58	74.5	64.9	9.6	Diagram/Calculation
60	27.6	18.2	9.4	Diagram/Calculation
15	73.2	64.1	9.1	Table/Calculation
14	52.3	44.2	8.1	Table/Calculation
4	77.2	69.4	7.8	Diagram/Calculation
35	46.5	39.1	7.4	Pattern Recognition Diagram
76	38.8	32.1	6.7	Chart Reading
34	46.3	40.2	6.1	Pattern Recognition Diagram
72	52.1	47	5.1	Table/Calculation
74	51	46.1	4.9	Table/Calculation
78	44	39.6	4.4	Chart Reading
90	33.2	29	4.2	Non Fiction Prose/Calculation
12	63.5	59.7	3.8	Diagram/Calculation
52	70.8	67.3	3.5	Chart Reading
73	50.8	47.3	3.5	Table/Calculation
89	29.2	25.8	3.4	Non Fiction Prose/Calculation
42	25.3	22.3	3	Non Fiction Prose/Calculation
51	78.4	75.4	3	Chart Reading
50	44.8	42.9	1.9	Chart Reading
77	47.9	47.1	0.8	Chart Reading
33	53.7	53.3	0.4	Pattern Recognition Diagram
41	65.4	65.3	0.1	Non Fiction Prose/Calculation
49	82	82.9	-0.9	Chart Reading
79	27	29.2	-2.2	Non Fiction Prose/Calculation
75	51.4	56.6	-5.2	Table/Calculation

Nearly all of these items were harder on computer than paper. There are very few where the difference is negligible, and only one item was easier with CBA delivery in any degree approaching significance. On average, this group of questions was 5% harder on computer screen than paper. Most importantly the results show a pattern of differential related to item type. In particular, questions based on certain operations on diagrams stack the top of the scale in terms of the degree to which they are harder with computer delivery than paper delivery.

The differentials in the Quantitative Reasoning section of the TEST do not appear to be dependent on the difficulty of the question. As there was no average differential between the paper and computer based Critical Reasoning sections and almost none with the Verbal and Plausible Reasoning sections, the hypothesis that the mode effect is a result of demographic differences between those two groups of candidates is implausible. Certainly, if the difference in difficulty between the two modes of delivery was solely due to differing ability levels of mathematical and scientific ability between the candidates who took the CBA and PBA versions of the TEST, then we would expect to see at least some level of differential reflected in the Critical Reasoning section. There was none. If the differential was simply based on differing levels of mathematical ability between the CBA and PBA groups, then I would also expect a significant differential for calculation items based on passages of non-fiction. This cannot be seen. Finally, if the differential was based on differing levels of 'visual literacy' between the two groups then I would expect to see that reflected in diagrammatic stimulus with questions based on simply reading charts. Neither can the pattern of differential be explained by the CBA layout: items 58, 59 and 60 required the candidate to scroll through the stimulus material. While this may account for at least some of the increase in difficulty for those items when delivered by computer, none of the other questions required scrolling. Therefore the differential would appear related to the type of stimulus *and* the kinds of questions depending on it rather than demographic differences between candidates or problems with CBA layout.

The average differential between the CBA and PBA delivery of the Critical Reasoning section of the TEST was effectively nil. While some items were harder with CBA delivery, this was balanced by some items that were easier with CBA delivery and there was a large group in the middle where the differential between the two modes of delivery was negligible. There was no discernible pattern of differential that can be related to item type. The Verbal and Plausible Reasoning section of the TEST showed a similar lack of pattern. These questions, considered as a group, were very slightly harder overall with CBA delivery than PBA delivery: the percentage of candidates who selected the correct answer was an average of 1.2% higher when answering on paper than on a computer screen. While some of these questions were harder with CBA delivery, this was largely balanced by items that were easier with CBA delivery. Again there was no discernible pattern in the differential with relation to the item type.

4 Discussion: The Problem of Annotation

Wherever calculation is involved, whether that be numerical or spatial or a hybrid of the two, we like to scribble, particularly when considering diagrams and tables. Scribbling and drawing is an inherent component of certain kinds of cognitive processing:

the reason that we still use whiteboards in schools and universities it is that when we are communicating or thinking about certain kinds of problems a whiteboard is a better technology than a MacBook. Annotation seems particularly important with certain kinds of thinking: 'making visualizations is integral to scientific thinking'[7]. Studies of annotation patterns in text books show that textbooks on 'Organic Chemistry and Calculus had more penciled-in marginalia … than many of the other textbooks'[8] and that these annotations were used to both solve problems in context and record interpretive activity: 'the act of annotation is a very powerful active reading and learning mechanism'[9]. Studies of the way architects use sketches show that:

> One role of a sketch is to check the completeness and internal consistency of an idea, especially a spatial idea … They are a kind of external representation serving as a cognitive tool to augment memory and information processing … relieving the dual burden of holding the content and also simultaneously operating on it.[10]

Removing that tool from candidates who are attempting to solve quantitative and spatial problems is non-trivial. With certain kinds of stimulus and certain kinds of questions the mode of delivery may not change the nature of the problem to be solved, but it certainly does change the manner in which the candidate can solve it.

In addition to this, with politically sensitive state or nation-wide benchmarking tests such as the Australia's *National Assessment Program – Literacy and Numeracy* (NAPLAN) there would serious political and educational consequences were the results to show a drop in the order of 5% for the measured numeracy of an entire country's students after a move to CBA. In its paper version the numeracy sections of NAPLAN are heavily, almost exclusively, reliant on visual and diagrammatic stimulus[11]. As tests such as NAPLAN begin to transition to CBA the challenges posed by the CBA delivery of diagrammatic stimulus will, I fear, begin to put real downward pressure on its selection for use within tests.

The CBA candidates for the TEST were provided with scrap paper. However, with complex diagrams, tables, calculations, there remains a huge difference between scribbling on the object itself and scribbling on a separate piece of paper on a separate visual plane. For one thing, it is not possible to look at what you are writing or drawing at the same time as the table or diagram itself, which makes thinking about it more difficult. For another it makes revision harder as the candidate no longer has a record of their logic written on the object they are grappling with. We scribble *on* problems we are trying to solve, as anyone who has spent any time looking at used test booklets can attest. The literature on CBA assumes that interactivity is an obvious advantage of CBA[12–14]. Yet in one very important respect PBA is far more interactive than CBA: we cannot freely draw on a computer screen[1].

---

[1] Unless, of course, we are talking about stylus equipped tablets. CBA with that technology could potentially solve the problem of annotation. However when we refer to CBA we are currently referring to technology which is relatively ubiquitous in education: computers with screens and mice, not stylus equipped tablets.

5 Conclusion

This pattern of item type differential has been noted before. Large samples, psychometric validity, and researchers who pay attention to patterns of item differential generate the same results and conclusions:

> Looking across grades, the results seemed to suggest a pattern of item-by-mode effect for mathematics items involving graphing and geometric manipulations. In these cases, a common strategy employed by students to draw on the item to solve the problem was not readily available in the online mode.[15]

Given adequate sample sizes, psychometric item validation and attention to item type differential (as opposed to aggregated score differential), the evidence suggests there is a problem with the CBA delivery of certain kinds of items that rely on diagrams, geometry or tables. It is not related to demographics or layout. It is fundamental to current technology: we can't write or draw or annotate on computer screens in the unrestricted manner we rely on when interacting with this class of problem on paper.

While this aspect of CBA has received little attention in the literature on it, the problem of annotation is well known within the literature on e-reading technologies:

> If we want people to read on the screen, we're going to have to provide them with the facilities to annotate. The standard keyboard-mouse interaction paradigm is not sufficient for the kind of unselfconscious interaction that we observe in practice.[16]

I have been somewhat limited in the context of this paper by my inability to present examples of annotations due to the secure nature of the TEST: the descriptors I have used in the table of data give *some* indication of the kinds of stimulus material and problem solving required, however, I would note that not all 'Diagram/Calculation' problems have me reaching instinctively for a pen. What I can say, from a test-developer's point of view, is that what I see when I look at results like those presented is a pattern of differential that is related to the kind of stimulus presented and the kinds of questions asked about it, rather than demographic differences or problems with layout. I would expect the difficulty of individual items to naturally fluctuate up and down a little between two groups. I would also expect those individual fluctuations to effectively cancel each other out, as they did in the other two sections of the TEST. But when we have a large number of items with differentials in the order of 10%, all in the same direction, then something is causing it.

Certainly the results suggest that more work needs to be done to investigate the impact of removing a candidate's potential to use free-form annotations as a problem-solving strategy when sitting tests delivered by computer. Assessment is the engine that drives education. Amidst the somewhat enthusiastic rush to CBA we need to carefully consider what we are doing when we deny candidates the ability to annotate, the ability to use their own diagrammatic reasoning when solving certain kinds of problems.

References

1. McDonald, A.: The impact of individual differences on the equivalence of computer-based and paper-and-pencil educational assessments. Computers & Education 39, 299–312 (2002)
2. McGaw, B.: A test to suit the 21st century (2013), http://www.theaustralian.com.au/archive/national-affairs/a-test-to-suit-the-21st-century/story-fnd17met-1226654600162
3. van der Kleij, F.M., Eggen, T.J.H.M., Timmers, C.F., Veldkamp, B.P.: Effects of feedback in a computer-based assessment for learning. Computers & Education 58, 263–272 (2012)
4. Noyes, J.M., Garland, K.J.: Computer- vs. paper-based tasks: Are they equivalent? Ergonomics 51, 1352–1375 (2008)
5. Way, W.D., Lin, C.-H., Kong, J.: Maintaining score equivalence as tests transition online: Issues, approaches and trends. Annual meeting of the National Council on Measurement in Education, New York, NY (2008)
6. Leeson, H.V.: The Mode Effect: A Literature Review of Human and Technological Issues in Computerized Testing. International Journal of Testing 6, 1–24 (2006)
7. Ainsworth, S., Prain, V., Tytler, R.: Drawing to learn in science. Representations 3, 5 (2011)
8. Marshall, C.C.: Annotation: from paper books to the digital library. In: Proceedings of the Second ACM International Conference on Digital Libraries, pp. 131–140. ACM (1997)
9. Plimmer, B., Apperley, M.: Making paperless work. In: Proceedings of the 8th ACM SIGCHI New Zealand Chapter's International Conference on Computer-Human Interaction: Design Centered HCI, pp. 1–8. ACM (2007)
10. Tversky, B.: What do sketches say about thinking. In: 2002 AAAI Spring Symposium, Sketch Understanding Workshop, Stanford University, AAAI Technical Report SS-02-08. pp. 148–151 (2002)
11. The Tests | NAP, http://www.nap.edu.au/naplan/the-tests.html
12. Greiff, S., Wüstenberg, S., Holt, D.V., Goldhammer, F., Funke, J.: Computer-based assessment of Complex Problem Solving: concept, implementation, and application. Educational Technology Research and Development 61, 407–421 (2013)
13. Saadé, R., Morin, D., Thomas, J.: Critical thinking in E-learning environments. Computers in Human Behavior 28, 1608–1617 (2012)
14. Terzis, V., Moridis, C.N., Economides, A.A.: How student's personality traits affect Computer Based Assessment Acceptance: Integrating {BFI} with {CBAAM}. Computers in Human Behavior 28, 1985–1996 (2012)
15. Keng, L., McClarty, K.L., Davis, L.L.: Item-Level Comparative Analysis of Online and Paper Administrations of the Texas Assessment of Knowledge and Skills. Applied Measurement in Education 21, 207–226 (2008)
16. Marshall, C.C.: Reading and interactivity in the digital library: Creating an experience that transcends paper. In: Proceedings of CLIR/Kanazawa Institute of Technology Roundtable, pp. 1–20 (2005)

Students' Spontaneous Use of Diagrams in Written Communication: Understanding Variations According to Purpose and Cognitive Cost Entailed

Emmanuel Manalo[1] and Yuri Uesaka[2]

[1] Faculty of Science and Engineering, Waseda University, Tokyo, Japan
[2] Graduate School of Education, The University of Tokyo, Tokyo, Japan
emmanuel.manalo@gmail.com, y_uesaka@p.u-tokyo.ac.jp

Abstract. This study examined the amounts of information that students represented in diagrams compared to text when taking notes (self-directed communication) and when constructing explanations for others (others-directed communication). The participants were 98 Japanese university students who read one of two passages (differing in imageability) in Japanese (L1) and in English (L2). While reading, they could take notes, and were subsequently requested to produce an explanation of the passage using L1 or L2. The students represented more information in diagrams in notes they took from the passage of higher imageability in L1. However, in their explanation of that same passage for others – still using L1 – they represented more information in text. This finding suggests perceptual differences about the functions of diagrams in self- and others-directed communication. Results also confirmed that passage imageability and students' language proficiency affect cognitive processing cost, which in turn influences the extent to which diagrams are used.

Keywords: spontaneous diagram production, strategy use perceptions and beliefs, written communication, cognitive processing cost, text and diagrammatic representations.

1 Introduction

Educational research has demonstrated that, when used appropriately, diagrams are efficacious tools for enhancing communicative effectiveness [1–4]. The efficacy that results from their use comes about because diagrammatic representations have a capacity for grouping related information together and supporting many perceptual inferences, thereby making them more computationally efficient compared to text or sentential representations [5]. Because processing of communicative information occurs in working memory [6, 7], some of these studies have also shown, among other things, that processing capacity limitation is a crucial consideration when designing such information – especially for instructional purposes [8, 9]. Mayer and Moreno [9] explained, for example, that the cognitive resources available in a learner's information processing system are severely limited, making it crucial that multimedia instructional design aims at reducing or eliminating any unnecessary cognitive load.

T. Dwyer et al. (Eds.): Diagrams 2014, LNAI 8578, pp. 78–92, 2014.

Apart from the benefits that can be derived from integrating diagrams in prepared materials for learning and instructional purposes, researchers have also emphasized the importance of cultivating students' abilities in creating and using diagrams [2, 10]. Ainsworth, Prain, and Tytler [2], for example, pointed out that learning how to construct diagrammatic representations is important as it facilitates the development of deeper understanding, reasoning, knowledge organization and integration, as well as capabilities in communication. They noted that, through the generation of visual representations, people can clarify their ideas for others, enable broader dissemination of those ideas, and promote greater opportunities for exchange of ideas with peers.

1.1 The Processes Involved in Producing Diagrams

Despite general agreement that cultivating students' abilities to use multiple representations – including diagrams – is highly desirable in educational settings, there are relatively fewer studies that have considered and examined the processes involved in students' *production* of diagrammatic representations. Where diagram construction and use in problem solving is concerned, Cox [11] noted variations in the extent to which people externalize their reasoning, from those who construct comprehensive diagrams to those who seemingly require no need to construct such representations. He suggested that individual differences (e.g., in possession of prior knowledge) as well as demands of the task could influence predispositions toward such construction. Uesaka, Manalo, and Ichikawa [10] identified some obstacles to students' diagram use in math word problem solving: their results showed that when students lacked adequate confidence and perceived difficulties in using diagrams, and if they did not view diagrams as part of their own repertoire of learning strategies, they were unlikely to spontaneously use them. There are, however, instructional approaches that appear to promote spontaneity in such use: Uesaka and Manalo [12] reported that placing students in a peer instruction condition made interactive communication necessary, which often led to students using diagrammatic representations when they found it too difficult to explain with words alone. As a consequence of personally experiencing the benefits of using diagrams during those explanation sessions, the students subsequently evidenced greater spontaneous use of diagrams when working on their own in solving other math word problems. In a follow-up study, Uesaka, Manalo, and Ichikawa [13] confirmed the importance of appreciating the value of diagram use: they empirically demonstrated that such appreciation, together with possession of adequate skills for the construction of the required diagram, were key requirements for students to more spontaneously *produce and use* diagrams when working on solving math word problems.

Where diagram production and use in written communication is concerned, again only a few studies have considered and examined the processes involved. One such study, by Manalo and Uesaka [14], reported evidence indicating that the production of text and diagrammatic representations shares the same cognitive processing resources in working memory. In their study, Manalo and Uesaka found that diagram use (when writing explanations of what had been learned) was related to Japanese university students' English (their second language or L2) competence – but only when the students were writing their explanations in English, and not when they were writing in Japanese (their native language or L1). More specifically, the less competent the students were in English – and thus the more cognitive resources they presumably needed to write in that language – the fewer diagrams they produced, suggesting a cross modal (i.e., verbal to visual processing) depletion of cognitive resources available for such production. This finding offered a viable explanation for previously reported observations that students – even at the university level – do not adequately use diagrams when constructing written explanations of what they have learned [15]. In other words, because cognitive resources for the generation of text and diagrammatic representations are shared in working memory, when the generation of text uses up most of the available resources, what remains may not be enough for the generation of any diagrams. There is also evidence from the Manalo and Uesaka study [14] suggesting that, apart from individual factors like L2 competence, task-related factors like imageability of the information to be explained (i.e., how easy/difficult it is to imagine) could affect demand on cognitive processing resources. In other words, information of lower imageability would demand greater cognitive processing resources for the construction of an appropriate diagram (which is likely to be more abstract in nature), and thus whatever resources remain in working memory (following efforts at text representation of that information) would likely be inadequate for such construction.

However, in the Manalo and Uesaka study [14] described above, only the production or non-production of diagrams in explanations *for others* was investigated. They did not examine the relative amounts of information that students represented, or whether differences might be manifested, when students produce explanations or notes *for their own use*. In the present study, therefore, these details and variations were examined to better understand how and why diagrams are produced – in relation to text – when constructing written communication.

1.2 Overview of the Present Study

The present study's main aim was to contribute to the development of understanding about the mechanisms and processes involved in students' spontaneous production and use of diagrams. More specifically, it sought to further elucidate how and why students might employ diagrams when constructing written communication, including the extent to which purpose of communication – whether it is for others or for the self – might influence such use of diagrams. The study also examined whether the influence of cognitive cost (i.e., the amount of mental effort and resources required to execute tasks) associated with individual and task-related factors (e.g., L2 proficiency, imageability) would be detectable also in the amounts of information students represent in the written communication they construct.

In contrast to writing explanations for others, taking notes from information provided is usually self-directed: notes provide a record *for the self* of the information that has been presented, and they are usually selective – focusing on what students anticipate would be important for them to know [16, 17]. Previous authors have pointed out that the notes students make can include not just words or text, but also diagrams or the use of diagrammatic formats [17, 18]. In fact, some of these authors have argued that non-linear note taking, including the use of graphs and concept maps, is far more effective as it fosters more careful selection, organization, and reordering of information [19–21]. Piolat, Olive, and Kellogg [17] also stressed that note taking is a complex task that necessitates the use of high amounts of cognitive processing resources in working memory. They referred to research indicating that factors like linguistic skill can influence the relative amounts of effort that students need to deploy for note taking purposes: for example, taking notes in an L2 that one has lower proficiency in would require much greater cognitive effort compared to taking notes in the L1.

Although ample research has been carried out on the topic of student note taking, the present authors were not aware of any studies that have examined the spontaneous production and use of diagrams in note taking. Such an examination, however, could prove useful, particularly if comparisons are made with what students do when constructing written explanations *for others*, as it could help clarify the influence of purpose and/or audience on the forms of representations students employ when communicating. Understanding those kinds of influences could, in turn, assist in the formulation of strategies for more effective cultivation of students' abilities in constructing and using diagrammatic representations.

The present study examined students' inclusion and representation of units of information (contained in a passage they had been given to read) in notes and written explanations they produced. It investigated how much of that information they represented in text and in diagrams. Compared to simply determining whether students used diagrams or not, examining amounts of information representation in diagrams could potentially provide a better gauge of the priority decisions that students make and the effects of capacity limitations in cognitive processing.

Several related hypotheses were tested. The first was that students would represent more information in diagrammatic format than in text format when taking notes. Higher information representation in diagrammatic format was predicted in note taking because diagrams would more likely afford easier reorganization and summarization of the information provided. However, it was also hypothesized that the relative amounts would be affected by the language that students were required to use (L1 or L2) and features of the passage they were supplied (high or low imageability). Thus, when using an L2 and when taking notes from a passage of low imageability, the first hypothesis above was expected *not* to apply: L2 use could significantly deplete resources in working memory, and the resource demand of constructing a more abstract diagram for the hard-to-imagine information was expected to exceed that available in working memory. Hence, the higher representation of information in diagrams compared to text was predicted *only when students were using their L1 in taking notes from a passage of high imageability*.

Other hypotheses tested were that, in general, students would represent more information (i.e., in both text and diagrams) when using L1 (compared to L2), and when taking notes (compared to writing explanations). These predictions were based on assumptions about the relative cognitive costs entailed in performance under those conditions: lower cost could allow more cognitive resources to be used for constructing representations of more information in the written communications being produced. Finally, it was also hypothesized that, overall, students would represent more information in text compared to diagrams. This prediction was made based on an assumption that text representation would generally be less cognitively costly: as the stimulus passage to be provided to the students would be text-based with no accompanying diagrams, it would be less cognitively costly to produce notes and explanations also in text, compared to having to transform the text-based information from the passage to a diagrammatic format. The only anticipated exception was when taking notes using L1 from an easy-to-imagine passage, as explained above.

2 Method

2.1 Participants

The participants were 98 Japanese undergraduate university students in science and engineering disciplines (mean age = 19.86 years, SD = .90 year; females = 14). The students participated voluntarily in the study and received no monetary compensation for their participation.

2.2 Materials

The participants were given a booklet that contained the experimental materials, and a separate sheet for note taking purposes. There were four versions of the booklet, which were distributed randomly to students in approximately equal numbers. The booklets contained either a passage about how music is played from a CD [adapted from 22] or a passage about how the human blood circulation system works [adapted from 23]. These two passages were deemed as differing in their imageability based on the kinds of diagram considered most appropriate to use in explaining their content. Although both passages concern processes, for the CD passage, a flow chart of the different component parts involved would be appropriate: most of those parts are difficult to imagine (e.g., microscopic indentations on the surface of a CD, photocell detector, binary number decoder) and the key purpose of the diagram would be to show the sequence involved in generating, coding, and converting data that eventually become music. In contrast, for the circulation passage, a drawing or sketch of the parts of the body involved, together with indicators for directions of blood flow, would be considered appropriate: these (e.g., heart, lungs, blood vessels) are easier to imagine and the key purpose of the diagram would be to show the directions of blood flow through these parts of the system. The diagram that is appropriate for the CD passage (i.e., the flow chart) is more abstract compared to the one that is appropriate for the circulation passage (i.e., the sketch or drawing of the organs). Uesaka and Manalo [24] had earlier explained that constructing a more abstract diagram requires more transformational steps, and therefore involves greater cognitive processing cost.

The passages used in this study contained only words; no diagrams or illustrations were included. Modifications were made to the original versions of these passages [22, 23] so that the two passages were equivalent in length, and contained approximately the same number of discreet information units or segments that convey distinct, meaningful information. To clarify what those units pertained to, the following example sentence was considered as containing four of those units as indicated by the segmentation slashes used here: The list of numbers representing the music is 'burned' on a CD / using a laser beam / that etches bumps (called "pits") / into the shiny surface of the CD. In this example, the first unit pertains to what is done, the second indicates what is used, the third states how it is done, and the fourth refers to the location. In each booklet, the passage (CD or circulation) was provided first in the Japanese language, and then in the English language.

Apart from differences in the passage content, the booklets also differed in the language that the participants were asked to use in a subsequent tasks that required them to explain what they had learned: approximately half of the booklets asked for an explanation in Japanese (the L1) and the other half in English (the L2). Thus, the four versions of the booklet were: (1) CD passage requiring L1 explanation, (2) CD passage requiring L2 explanation, (3) circulation passage requiring L1 explanation, and (4) circulation passage requiring L2 explanation. In all other respects (i.e., the instructions given, the questions asked, etc.), the four versions were identical.

2.3 Procedure

Two of the factors manipulated in this study (i.e., kind of passage, and language of explanation required) were between-subject variables. As noted above, manipulation of these factors was reflected in the four versions of the booklet used. The four participant groups corresponding to these booklets, with numbers of participants in brackets, were as follows: CD-L1 (27), CD-L2 (25), Circulation-L1 (22), and Circulation-L2 (24). A third factor that was manipulated, the kinds of written communication participants had to produce (i.e., notes, explanations) was a within-subject variable.

Data collection was carried out at the end of one of the students' regular, scheduled classes. After distribution of the booklets to those who were willing to participate, the students were provided verbal instructions about what to do. Equivalent instructions were provided in the booklets in written form. The students were informed that they would be reading a passage – in Japanese, then in English – and that later they would be asked questions about it, including explaining its content in Japanese *or* in English. They were informed that the language they would need to use would be given later.

The students were asked to read the passage they were allotted in Japanese (8 minutes) and then in English (8 minutes). The passages were provided in both L1 and L2 to avoid the possibilities of misunderstandings about the content, and subsequent inability to explain the content due to unfamiliarity with the appropriate language (e.g., technical terms and phrases) to use. During the 8-minute reading periods allowed, the students were informed that they could use the extra sheet of paper they were provided to take notes as they wished, and that they could consult the notes they made during the entire experiment. (Note that in general the students used Japanese to take notes from the L1 passage and English to take notes from the L2 passage, although in a few cases the English notes were supplemented with a few explanatory words and/or phrases in Japanese.) The students were also informed, however, that they were not allowed to return to, or re-read, the original Japanese and English versions of the passage once the time allocated for reading those had expired.

Next, the students were given 2 minutes to answer five questions relating to their perceptions about the passage they had read. Due to space constraints, their responses to those questions are not considered and reported in this paper.

After this, the students were provided space on the following page to provide an explanation of the passage they had read. For this, they were asked to imagine that their audience was a fellow student who did not know anything about the topic. At this point, the students could see in their booklet whether the language they had to use was Japanese or English. The students were given 10 minutes to complete this task.

Following this, the students were given 4 minutes to answer four passage comprehension questions on the next page of the booklet. Finally, on the last page of the booklet, the students were asked to providc some demographic details (e.g., their age, gender, year at university).

2.4 Analysis

The factors examined in this study were: the passage administered (CD or circulation), the language required for communication tasks given (L1 or L2), the communication type (notes for self or explanations for others), and representation type used (text or diagrams). For the purposes of this study, a diagram was defined as any representations produced by the participants, other than representations in the form of words, sentences, or numbers on their own. For example, drawings and charts counted as diagrams, as did arrows and similar symbols when these were used to link three or more concepts or ideas.

As previously indicated, the present study focused particularly on the *amount of information* the students represented in text or diagrams in the written communication they produced. It did not simply determine whether the students used diagrams or not. Instead, *how much* of the content of the passage the students represented in text and in diagrams was analyzed.

The CD and circulation passages each contained three idea components (e.g., how music is stored on a CD as numbers, or the general purpose and structure of the circulation system): combined, these three components contained a total of 15 idea units (or IUs) in each passage. These idea components and IUs are shown in Appendices 1 and 2. The text and diagrammatic representations in the notes and explanations that the students produced were analyzed for presence of these IUs. The proportions (out of the total) of IUs contained were used in the statistical analyses conducted.

3 Results

For brevity, the factors examined will from hereon be referred to as: passage (CD or circulation), language (L1 or L2), communication-type (notes or explanations), and representation-type (text or diagrams).

An analysis of variance (ANOVA) revealed significant main effects for language, $F(1, 94) = 9.91$, $p = .002$; communication-type, $F(1, 94) = 190.70$, $p < .001$; and representation-type, $F(1, 94) = 134.86$, $p < .001$. These results indicated that a higher proportion of IUs were represented when the students were using L1 ($M = .48$, $SD = .11$) compared to L2 ($M = .41$, $SD = .10$), and when writing notes ($M = .42$, $SD = .12$) compared to explanations ($M = .25$, $SD = .10$). Additionally, the students used text ($M = .47$, $SD = .16$) more than diagrams ($M = .20$, $SD = .13$) in representing IUs. However, no significant effect due to passage was found, $F(1, 94) = .05$, $p = .824$, with the proportions of IUs represented from the CD passage ($M = .45$, $SD = .12$) and the circulation passage ($M = .45$, $SD = .11$) being virtually the same.

Significant interaction effects were also found between passage and representation-type, $F(1, 94) = 21.97$, $p < .001$; language and communication-type, $F(1, 94) = 16.22$, $p < .001$; and communication-type and representation-type, $F(1, 94) = 35.74$, $p < .001$; as well as between language, communication-type, and representation-type, $F(1, 94) = 7.72$, $p = .007$. Analysis of simple main effects revealed that, where the CD passage was concerned, students represented a significantly higher proportion of IUs in text compared to diagrams in both the notes and the explanations they produced ($p < .001$ for both). In both notes and explanations, the students also represented a higher proportion of IUs when using L1 compared to L2 ($p < .001$ for both). These differences can be seen in Figure 1.

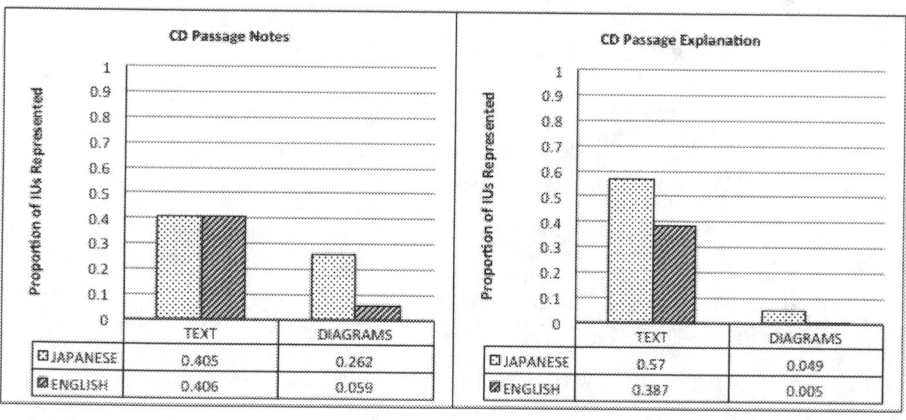

Fig. 1. Mean proportions of idea units represented in text and diagrams, in the notes and explanations that students produced, in Japanese and in English, after reading the CD passage

Where the circulation passage was concerned, overall, the students also represented a higher proportion of IUs in text compared to diagrams, and when writing in L1 compared to L2 ($p < .001$ for both). However, as can be seen in Figure 2, when using L1 and taking notes from the circulation passage, the students actually represented *a higher proportion of IUs in diagrams* compared to text ($p < .001$). Note that this higher diagrammatic representation of IUs was no longer present when the students produced their explanations: then, they represented more IUs in text compared to diagrams ($p < .001$).

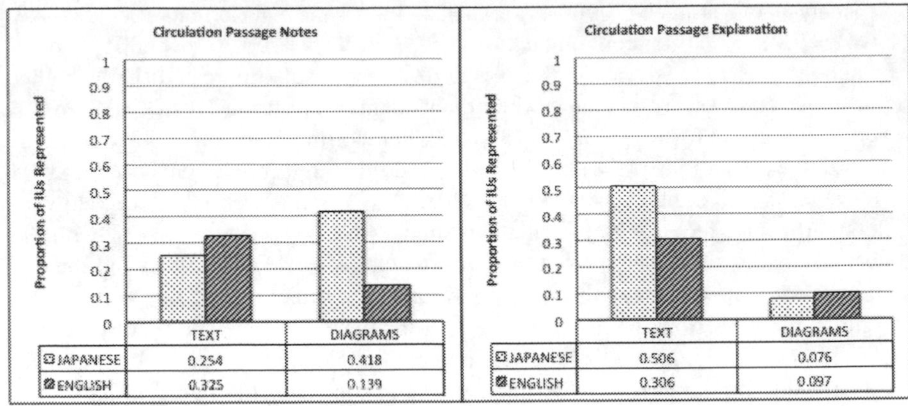

Fig. 2. Mean proportions of idea units represented in text and diagrams, in the notes and explanations that students produced, in Japanese and in English, after reading the circulation passage

This finding about the relative uses of text and diagrams to represent information when students were constructing notes compared to explanations is exemplified in the example shown in Figure 3.

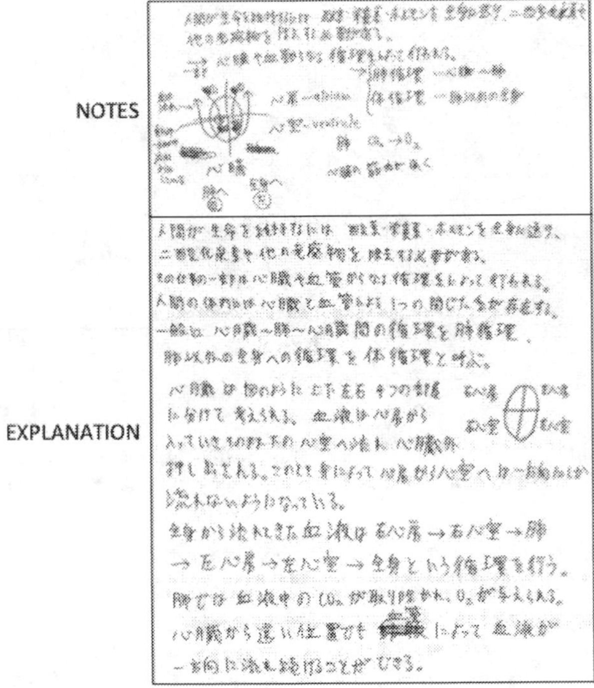

Fig. 3. Circulation passage notes and explanation that one student produced in Japanese, depicting the relatively higher diagrammatic representation of information in the notes that was not present in the explanation

4 Discussion

The hypotheses and predictions tested in this study were confirmed by the results. More information was represented in diagrammatic format – compared to text format – when the students were taking notes using L1 from the passage of higher imageability. However, overall, the students represented more information in text than in diagrams. They also represented more information when writing in their L1 compared to L2, and when taking notes compared to writing explanations for others.

The results confirm previous findings [14] about the influence of individual factors (like language proficiency) and task features (like imageability) on students' use of diagrammatic representations, suggesting not only resource limitations in working memory as being crucial in diagram production, but also that the processing resources available are shared in the production of text and diagrams. Hence, when the production of one form of representation (e.g., text) uses up most of the available resources, what remains may not be adequate for the production of the other form of representation (e.g., diagrams). In the present study, these influences on representational use were confirmed with the relative *proportions of information* students represented in text and in diagrams, rather than simply being based on the use or lack of use of diagrams as had been reported in previous studies [14, 15].

4.1 Different Purposes Served in Different Tasks

In addition to confirming previous findings, the results of the present study provide new insights into how students might view the use of text and diagrams differently when writing notes compared to writing explanations for others. More diagrams, for example, were used in notes, and one possible explanation for this is that diagrams might have been perceived as serving more useful functions then – such as assisting in the summarization of main points and connection of ideas. These purpose-related issues that might affect decisions about the use of text and diagrams in note taking and explaining to others are summarized in Table 1.

This table shows the proposed similarities and differences in the influences on, and functions of, text and diagrammatic representations in notes as compared to explanations. Clarity – that is *not reliant* on non-externalized knowledge possessed by the self – is crucial in the construction of explanations for others, so as to avoid the potential for any misunderstanding. In contrast, such apparent clarity may not be essential in constructing notes, as the note taker is aware of the knowledge that he or she possesses and can likely later retrieve when cued by what he or she does put in the notes: hence, leaving those "non-essential" details out when taking notes for the self could be deemed acceptable.

While the availability of cognitive processing resources is crucial if diagrams are to be used in either notes or explanations, diagrams may be viewed as serving more useful functions in the construction of notes – functions such as clarifying and visually depicting relationships and structures to make them more accessible and memorable to the self. In contrast, diagrams may not be viewed as serving as many useful functions in explanations as, depending on their concreteness and the amount of information included with them, they are likely to require at least some interpretation from the (other) person viewing them. Usually, they do not exactly spell out what is meant, and therefore leave more room for potential miscommunication. As such, diagrams in explanations may be viewed largely as supplements to text – to help clarify or illustrate, rather than being the key vehicle for conveying what needs to be explained.

Table 1. The purposes of text and diagrams in notes and explanations.

Considerations	Taking notes	Writing an explanation
For who?	For self	For another person
Purposes, qualities?	Summarize main points Clarify structural connections Assist memory Details may not be essential (can fill in for self)	Explain important points Facilitate understanding Must avoid lack of clarity and potential misunderstanding
Use text? (what kinds?) Why?	Yes (words and phrases, bulleted points) To capture key points and details	Yes (sentences, bulleted points) To explicitly explain important points and key details
Use diagrams? (what kinds?) Why?	Yes – if enough cognitive resources remain (flow charts of ideas, structural schemes, illustrations) To remind about important connections, and how things might look like	Possibly – if enough cognitive resources remain (illustrations, flow charts of ideas) To help make the text clearer (as a supplement); to show how things might look like

This proposed difference in perceived functions of diagrams in notes as compared to explanations is supported by the findings depicted in Figure 2: when taking notes in the L1 on the circulation passage, the students represented more information in diagrammatic format. This suggests they viewed diagrams as being useful in taking notes for their own selves. Those same notes, however, were available to them to use when they were constructing their explanation for others, but they represented more information in text format instead. As the students were using L1, writing about a passage of high imageability, and had their own notes available to consult, they were likely to have had adequate cognitive resources available to use in depicting more information in diagrammatic format in the explanation they were producing *if they had wanted to*. But, by and large, they chose not to: a decision that is most likely attributable to inadequate appreciation of the value of diagrams in such communicative contexts. The example notes and explanation shown in Figure 3 is from only one student, but it is typical of what many students produced when dealing with the circulation passage: plenty of diagrammatic representations in the notes, the majority of which 'disappear' from the explanations they produce for others.

4.2 Implications for Practice

The findings of this study suggest that students are not taking adequate advantage of the many potential benefits that diagram use could bring to written communication – as authors like Ainsworth et al. [2] have explained. More specifically, the results indicate that students rely a lot more heavily on text than on diagrams in representing information to explain to others. In part, this may be due to inadequate skill or confidence in producing the diagrams that may be considered "correct" or "appropriate" in clearly explaining to others. As already suggested, this may also arise from a perceptual or belief problem: students may simply believe that diagrams cannot serve as many useful functions in written communication for others as they might in the construction of notes for the self, or in comparison to the functions that text could serve. Both these possible explanations suggest a need to develop student knowledge and skills in the use of diagrams for effective communication with others.

In research in the area of math word problem solving, Uesaka et al. [10] have reported evidence that linked lack of spontaneity in diagram use to beliefs that diagrams are teachers' tools (rather than tools for one's own use) and to lack of confidence and anticipation of difficulties in diagram use. In subsequent studies, Uesaka et al. [12, 13] also demonstrated that enhancing student perceptions of the value of using diagrams in problem solving, as well as providing students with training in diagram construction skills, improved their spontaneity in using diagrams when attempting to solve math word problems. A similar approach may be indicated here: to address student perceptions about the value of diagrams in written communication intended for others, combined with enhancement of their skills in constructing the appropriate diagrams to use.

Apart from skills development in the use of diagrams when writing explanations, the results of this study suggest that skills development in diagram construction in general is much needed. Students need to acquire sufficient competence in diagrammatically representing more than what something might look like or where something might be located. They need to be able to construct more abstract diagrams to appropriately represent more abstract concepts, such as, processes, change, and relationships – all of which would be key knowledge components in much of what they have to learn through formal and informal education, and which they may have to explain to others in various situations like test taking, report production, and seminar presentation.

4.3 Conclusion

The findings of this study indicate that the use of diagrams in written communication is influenced, not only by the availability of cognitive processing resources in working memory, but also by student perceptions about the purposes that diagrams serve in different forms of written communication. While it was encouraging to find that students appeared to appreciate the usefulness of diagrams when taking notes for their own selves, it was rather concerning to find that they used much fewer diagrams – if any – in constructing explanations for others. Clearly there is a need to develop students' appreciation of the value of diagram use in effectively communicating with others. It would be important in future research to develop appropriate instructional programs and/or interventions to address these kinds of belief and performance issues among students, and to evaluate their effectiveness particularly in promoting greater spontaneous use of diagrams in both self-directed as well as others-directed written communication.

Acknowledgments. This research was supported by a grant-in-aid (23330207) received from the Japan Society for the Promotion of Science. The authors would like to thank Masako Tanaka for her help in data collection and processing.

References

1. Ainsworth, S.: DeFT: A Conceptual Framework for Considering Learning with Multiple Representations. Learning and Instruction 16, 183–198 (2006)
2. Ainsworth, S., Prain, V., Tytler, R.: Drawing to Learn in Science. Science 333, 1096–1097 (2011)
3. Mayer, R.E.: Multimedia Learning. Cambridge University Press, New York (2001)
4. Mayer, R.E.: The Promise of Multimedia Learning: Using the Same Instructional Design Methods Across Different Media. Learning and Instruction 13, 125–139 (2003)
5. Larkin, J.H., Simon, H.A.: Why a Diagram is (Sometimes) Worth Ten Thousand Words. Cognitive Science 11, 65–99 (1987)
6. Baddeley, A.D.: Working Memory. Oxford University Press, Oxford (1986)
7. Baddeley, A.D.: Human Memory. Allyn & Bacon, Boston (1998)
8. Mayer, R.E., Moreno, R.: A Split-Attention Effect in Multimedia Learning: Evidence for Dual Processing Systems in Working Memory. Journal of Educational Psychology 90, 312–320 (1998)
9. Mayer, R.E., Moreno, R.: Nine Ways to Reduce Cognitive Load in Multimedia Learning. Educational Psychologist 38, 43–52 (2003)
10. Uesaka, Y., Manalo, E., Ichikawa, S.: What Kinds of Perceptions and Daily Learning Behaviors Promote Students' Use of Diagrams in Mathematics Problem Solving? Learning and Instruction 17, 322–335 (2007)
11. Cox, R.: Representation Construction, Externalised Cognition and Individual Differences. Learning and Instruction 9, 343–363
12. Uesaka, Y., Manalo, E.: Peer Instruction as a Way of Promoting Spontaneous Use of Diagrams When Solving Math Word Problems. In: McNamara, D.S., Trafton, J.G. (eds.) Proceedings of the 29th Annual Cognitive Science Society, pp. 677–682. Cognitive Science Society, Austin (2007)
13. Uesaka, Y., Manalo, E., Ichikawa, S.: The Effects of Perception of Efficacy and Diagram Construction Skills on Students' Spontaneous Use of Diagrams When Solving Math Word Problems. In: Goel, A.K., Jamnik, M., Narayanan, N.H. (eds.) Diagrams 2010. LNCS (LNAI), vol. 6170, pp. 197–211. Springer, Heidelberg (2010)
14. Manalo, E., Uesaka, Y.: Elucidating the Mechanism of Spontaneous Diagram Use in Explanations: How Cognitive Processing of Text and Diagrammatic Representations Are Influenced by Individual and Task-Related Factors. In: Cox, P., Plimmer, B., Rodgers, P. (eds.) Diagrams 2012. LNCS (LNAI), vol. 7352, pp. 35–50. Springer, Heidelberg (2012)

15. Manalo, E., Ueaska, Y., Kriz, S., Kato, M., Fukaya, T.: Science and Engineering Students' Use of Diagrams During Note Taking Versus Explanation. Educational Studies 39, 118–123
16. Van Meter, P., Yokoi, L., Pressley, M.: College Students' Theory of Note-Taking Derived From Their Perceptions of Note-Taking. Journal of Educational Psychology 86, 323–338 (1994)
17. Piolat, A., Olive, T., Kellogg, R.: Cognitive Effort During Note Taking. Applied Cognitive Psychology 19, 291–312 (2005)
18. Hartley, J., Davies, I.K.: Note-taking: A Critical Review. Innovations in Education & Training International 15, 207–224 (1978)
19. Boch, F., Piolat, A.: Note Taking and Learning: A Summary of Research. The WAC Journal 16, 101–113
20. Buzan, T.: Use Your Head. BBC Publications, London (1974)
21. Smith, P.L., Tompkins, G.E.: Structured Notetaking: A New Strategy for Content Area Readers. Journal of Reading 32, 46–53 (1988)
22. How CDs and DVDs Work, `http://www.explainthatstuff.com/cdplayers.html`
23. Chi, M.T.H., Siler, S.A., Jeong, H., Yamauchi, T., Hausmann, R.G.: Learning from Human Tutoring. Cognitive Science 25, 471–533 (2001)
24. Uesaka, Y., Manalo, E.: Task-Related Factors that Influence the Spontaneous Use of Diagrams in Math Word Problems. Applied Cognitive Psychology 26, 251–260

Appendix 1

Specific Components and Idea Units Contained in the CD Passage

Component 1: Information or details about how music is stored on a CD as numbers

- Music is stored on CD as numbers
- Numbers are stored in binary format OR as 0s and 1s
- Music is burned on a CD using a laser beam
- What "bumps" or "pits" are: they are the burned parts; OR they are indentations or etches on the surface of the CD; AND/OR that they represent the number 0
- What "lands" are: where there are no bumps; OR they are the unburned parts without indentations or etching; AND/OR they represent the number 1

Component 2: Information or details about the component parts of a CD player

- Laser beam
- Photocell
- Electronic circuit
- Binary number decoder
- Speaker

Component 3: Information or details about what happens in a CD player that enables music to be produced from a CD

- Laser beam flashes on CD and light is reflected back (or not) depending on pits and lands on surface of CD
- Reflected light is detected by a photocell (which sends current to electronic circuit)
- Electronic circuit generates 0s and 1s, (and sends this pattern to binary number decoder)
- Binary number decoder converts the pattern to pattern of electronic currents (and sends this to speaker)
- Speaker transforms the pattern of electronic currents to music

Appendix 2

Specific Components and Idea Units Contained in the Circulation Passage

Component 1: Information or details about the general purpose and structure of the human circulatory system

- The circulatory system enables the distribution of oxygen, hormones, and nutrients to cells, and the removal of carbon dioxide and wastes
- There are two primary subsystems of circulation (pulmonary circulation and systemic circulation)
- Pulmonary circulation involves the movement of blood from heart to lungs, and back to the heart
- Systemic circulation involves the movement of blood to all parts of the body (except the lungs), and back to the heart

Component 2: Information or details about the structure of the heart and the blood flow through these structures

- The four sections of the heart: left and right sides (divided by the septum), and upper (atrium) and lower (ventricle) chambers of each side
- The direction of blood flow through these chambers (blood enters right atrium; pumped to right ventricle; goes to lungs; comes back to left atrium; pumped to left ventricle; goes to other parts of the body)
- One-way valves separate the chambers of the heart so blood will not flow the wrong way
- The heart sends blood to the lungs for oxygenation
- The heart sends oxygenated blood to cells in other parts of the body
- Blood from other parts of the body (carrying carbon dioxide and wastes) returns to the heart

Component 3: Information or details about the roles and characteristics of the arteries and veins

- Arteries are large blood vessels that carry blood away from the heart
- Arteries are strong and elastic; OR arteries carry blood under high pressure
- A pulse can be felt when the arteries are stretched as blood flows through them
- Veins carry blood back to the heart
- Veins carry blood under less pressure

How Communicative Learning Situations Influence Students' Use of Diagrams: Focusing on Spontaneous Diagram Construction and Protocols during Explanation

Yuri Uesaka[1] and Emmanuel Manalo[2]

[1] Graduate School of Education, The University of Tokyo, Tokyo, Japan
[2] Faculty of Science and Engineering, Waseda University, Tokyo, Japan
y_uesaka@p.u-tokyo.ac.jp, emmanuel.manalo@gmail.com

Abstract. Although diagrams have been shown to be effective tools for promoting understanding and successful problem solving, students' poor diagram use has been identified as a serious issue in educational practice-related reports. To enhance students' diagram construction skills and to address problems in diagram use, creating learning situations that make it inevitable for students to use diagrams would likely be helpful. To realize this, communicative learning situations can be considered a viable option, as students would feel a greater necessity to use diagrams as a consequence of feedback they receive while explaining. Thus, this study examined the hypothesis that an interactive peer instructional learning situation would better promote students' spontaneous diagram use compared to a non-interactive situation. Eighty-eight university students were randomly assigned to one of two conditions: interactive and non-interactive. After reading a passage relating to the science and engineering area, participants in the interactive condition were requested to explain the content of the passage to another participant next to them. In contrast, participants in the non-interactive condition were asked to record an explanation using an IC recorder by imagining that they were explaining to another person. A sheet of paper was provided to participants during the explanation, and diagram use on the paper was analyzed. The results revealed that students' diagram use in the interactive condition was higher than in the non-interactive condition. This indicates that teachers' provision of interactive communication situations can effectively promote students' likelihood of using diagrams spontaneously.

Keywords: communicative learning situation, spontaneous diagram construction, diagram use in explanations, peer interaction, learning strategies.

1 Introduction

1.1 The Efficacy of Diagrams and the Reality of Students' Diagram Use

Diagrams are very powerful tools for learning and problem solving in various domains [e.g., 1-5]. After the seminal paper by Larkin and Simon [2], many studies in psychology have demonstrated empirically that learners benefit from the effective use

T. Dwyer et al. (Eds.): Diagrams 2014, LNAI 8578, pp. 93–107, 2014.

of diagrams. For example, diagrams facilitate the construction of rich mental models when they are incorporated in scientific reading materials [e.g., 4, 5]. Based on the findings from various empirical studies, many educational materials and the instruction that teachers provide in classrooms now include a variety of diagrammatic representations. Students' manifestation of spontaneous use of diagrams (as well as other learning strategies) is generally considered as one important indication of the development of autonomous learning.

While some studies have revealed that constructing diagrams is not always effective in promoting successful problem solving particularly if learners lack necessary skills in diagram use or sufficient domain-related knowledge [6], many other studies have revealed that diagram construction and use brings about huge benefits in problem solving. For example, Hembree [7] demonstrated through a meta-analysis that constructing appropriate diagrams produced the largest effect size compared to other strategies in math word problem solving.

Despite the expectations associated with such findings, the reality is that students often do not use diagrams spontaneously or effectively. Students' poor diagram use has been identified as a problem in various reports based on educational practices: problems like failure in reasoning with diagrams, poor choice in types of diagrams to use, and lack of spontaneity in using diagrams have been pointed out [8, 9]. The existence of these problems suggests that, although the main concern of most diagrams research to date has been to demonstrate the effects of diagrams and elucidate the mechanisms by which diagrams contribute to the enhancement of the quality of learning [1-5], it would be equally important to address through research these problems that have been identified in student diagram use.

Among the various student problems in diagram use, one of the most crucial is the lack of the spontaneity in use. Diagrams are generally considered as effective personal tools for problem solving and learning, and many teachers – especially in science- and math-related subject – use lots of diagrams when teaching in class. When students do not use diagrams spontaneously in learning situations, it has been pointed out that many of them end up failing in their attempts at problem solving and learning. The existence of this problem has been reported in psychological case studies in personal tutoring, as well as in survey-based studies that gauge student learning behaviors [e.g., 10]. This problem strongly suggests a need to bridge the gap between teachers' awareness of the value of using diagrams, and students' lack of spontaneity in using diagrams despite classroom modeling from their teachers.

1.2 The Necessity of Creating Learning Situations in Which Diagram Use Becomes Inevitable

An important question is why student do not use diagrams of their own volition in learning situations when it would be advantageous for them to do so. In a study conducted by Uesaka, Manalo, and Ichikawa [10], one possible explanation was identified: the results they obtained suggest that students have a tendency to perceive diagrams as tools for teachers' instruction, but not as tools for their own (the student's) use. Thus, one possible explanation for the lack of spontaneity in use is that students do not perceive diagrams as clearly beneficial or efficacious for their own use.

However, only enhancing students' perception of the usefulness of diagrams is unlikely to be sufficient in promoting their spontaneity in using diagrams as a strategy: if they cannot construct the necessary diagrams confidently, they would probably be reluctant to use diagrams even if they are aware of their potential usefulness in various learning situations. In other words, students would probably need to possess sufficient drawing skills before they would actually use diagrams. Findings from the same study by Uesaka et al. [10], for instance, showed that students who perceived a higher cost for using diagrams – which probably stemmed from poor skills in constructing diagrams – tended not to use diagrams spontaneously. In a subsequent study, Uesaka, Manalo, and Ichikawa [11] were in fact able to experimentally demonstrate the positive effects of incorporating training in diagram drawing skills in an intervention aimed at improving students' spontaneity in diagram use when solving math word problems. Thus, in order to use diagrams more spontaneously, students may need to also develop the necessary drawing skills in addition to becoming better aware about the benefits of using diagrams.

To promote students' diagram drawing skills, creating learning situations in which students have to use diagrams could be effective. To achieve this goal, one typical teaching strategy is to provide opportunities for training in class. For example, the teacher could provide worksheets for students to use in practicing construction of diagrams for specified purposes, as well as provide feedback about the appropriateness of the diagrams that students produce. The effectiveness of using this kind of instructional strategy has been confirmed [11]. Numerous other studies have also confirmed the effectiveness of utilizing teacher-provided direct instruction in enhancing student skills in using other learning strategies [e.g., 12, 13].

However, the direct instruction approach may not always be effective. Some students believe that getting the correct solution in problem solving is sufficient and they neglect the importance of understanding the processes involved. For such students, teacher encouragement and instruction to use diagrams may not be enough. Uesaka et al. [11] already identified this possibility as they found evidence for a mediation effect of students' outcome orientation on their strategy use. Other studies relating to student beliefs about learning also suggest that students' strategy use is influenced by such beliefs [e.g., 14, 15]. Thus an important challenge is how to construct learning situations in which diagram use would become inevitable – even for such students.

1.3 Diagrams as Communication Tools

Diagrams are not only personal tools for problem solving but also effective interpersonal tools for communication (e.g., [16]; see also the review by Sacchett [17]). For example, Lyon [16] demonstrated that diagrams support communication in adults who have difficulties in communicating because of aphasia. In addition, as pointed out by Dufour-Janvier et al. [9], math teachers use a lot of diagrams in class for explaining how to solve problems and deepening students' understanding. Apart from the research areas of aphasia and math education, psychological researchers have also emphasized the crucial role that diagrams serve in communication, even though many of these claims have been theoretical in approach and not always accompanied by empirical evidence [1]. Diagrams are therefore equally important as tools for communication as they are as tool for problem solving.

When considering ways to create learning situations in which diagram use could become more or less inevitable, focusing on the communicative aspects of diagrams could be useful. If students experience a learning situation in which they need to provide explanations to other students, they are likely to encounter some difficulties when explaining only with the use of verbal expressions. In such situations, they are also likely to receive verbal and non-verbal feedback from their interlocutor, especially concerning aspects of their communication that might seem unclear – and this in turn could 'make them' use diagrams to supplement what they are saying. In fact, Uesaka et al. [10] had previously suggested thc possibility of peer instructional learning situations additionally providing opportunities for improving students' diagram construction skills via the interactions that occur. The effectiveness of using communicative learning situations instead of teacher-direct instruction has also been suggested in the framework of 'reciprocal teaching', proposed by Palincsar and Brown [18] for promoting students' reading strategy use. In this method, students are provided opportunities to use a reading strategy in communication as a kind of training to improve their skills in its use, but the method has been found to be effective also in promoting students' use of the strategy when reading on their own.

There is therefore the possibility that creating communicative learning situations could make diagram use inevitable for students, but this possibility had not been empirically examined, as much of diagrams research on communication has not focused on the production aspect. This study therefore conducted such an examination.

1.4 Overview of the Present Study

The purpose of this study was to examine two hypotheses. The first hypothesis was that a learning situation that requires students to explain the content of what they have learned to other students would promote the generation of diagrams more than a situation without such opportunities to communicate with other students. The second hypothesis was that actual interaction with other students is important, and students who are placed in a learning situation with actual interaction would spontaneously construct more diagrams in explanation than students who are placed in a situation without such interaction. Confirmation of these two hypotheses would in effect propose a practical teaching strategy – easily implementable in real classroom settings – for effectively promoting spontaneity in students' use of diagrams as a learning strategy.

To test these two hypotheses, two conditions were prepared. The first condition was an "interactive condition", in which students were provided an opportunity to explain the passage that they had read to another student. This condition also provided opportunities for receiving questions from the other student during the interaction that ensued. The second condition was a "non-interactive condition", in which students had a chance to explain the passage that they read, but the explanation was recorded on an IC recorder (i.e., "integrated circuit" or digital voice recorder) and provided to the paired student later, so it did not provide opportunities for direct interaction or asking of questions during explanation.

The diagrams that students in these two conditions spontaneously produced were analyzed: if the percentage of diagram use proved to be higher in the interactive condition, the first hypothesis would be confirmed. Such a result would also suggest that, in a communicative learning situation, actual interaction is crucial (i.e., not just the provision of explanation) if the objective is to promote spontaneity in diagram use.

Spontaneous diagram production in the present study pertained to instances during the explanation session when the students constructed and used a diagram – as no specific instruction was provided about how they should explain. In other words, the students could have simply used spoken or written words – and no diagrams – in their explanation. Even in the interactive condition, when the students were responding to their interlocutor, diagram use would have still been optional – and hence considered as spontaneous when it occurred.

This study also analyzed the students' verbal protocol data during the explanation sessions. It was hypothesized that, if feedback from the interlocutor during interaction is a key factor in promoting the use of diagrams, quantity of words uttered and length of speaking time would be greater in the interactive condition than in the non-interactive condition. Also, if active feedback from the interlocutor promotes the desired student behavior during explanation, then it was predicted that pairs of participants who engaged more actively in the communication process would produce more diagrams compared to pairs who were less actively engaged. Moreover, the types of questions used during the communication situations were coded to determine the types that might be most closely associated with the promotion of spontaneous diagram use.

2 Method

2.1 Participants and Experimental Design

Eighty-eight students in a university in Japan participated voluntarily in the study. Participants were randomly assigned to one of two conditions: an interactive communication condition, and a non-interactive (monologue) communication condition. Only one pair of participants in the interactive condition was excluded, as the IC recorder did not work well during their explanation session. As a result, the number of pairs in the interactive condition was 21 (for a total of 42 participants), and the number of pairs in the non-interactive condition was 22 (for a total of 44 participants).

2.2 Materials

The materials used were two types of passages, written in Japanese, which were the same as those used in a study by Manalo and Uesaka [19]. One passage was about the human blood circulation, and the other was about how music is played from a compact disk (CD). The passages contained only words; no diagrams or illustrations were included. Every effort was made to make the two passages as equivalent as possible in length and amount of content. Four booklets were prepared: two containing the CD passage and two containing the circulation passage, with one in each case for use in the interactive condition and the other in the non-interactive condition.

2.3 Procedure

This study was conducted in groups. Students were randomly paired, and students in a pair sat next to each other during the entire experiment. Each pair was randomly assigned to one of the two conditions. The manipulation of the experiment was conducted by providing them the different types of booklet. (However, the students did not know which condition they were assigned to until just prior to the beginning of the explanation phase.) Firstly, during the learning phase, all participants were requested to read the passage provided in their booklet and were informed that they would subsequently be asked to explain its content to another student who did not know about it. One participant in each pair read the circulation passage, and the other read the CD passage (randomly assigned and determined via the distribution of the booklets).

Secondly, during the explanation phase, participants were informed via their booklet which condition they were assigned to. Participants assigned to the interactive communication condition were requested to explain the content of the passage they had read to the participant they were paired with. A sheet of A4 paper was provided and they were permitted to use it freely when explaining the passage. Their verbal interaction was recorded using an IC recorder. The order in which the participants explained was left up to each pair to decide. When one participant had finished, they swapped roles. Each participant had up to a maximum of 5 minutes to explain.

Participants in the non-interactive (monologue) condition were requested to explain to an IC recorder by imagining that they were explaining to another person. In this condition, as in the interactive condition, they were provided a sheet of A4 paper, and permitted to write anything on it during explanation. After recording their explanation, they swapped their IC recorder with the participant they were paired with, and listened to the recording as well as looked at whatever details their partner had put on the A4 sheet. These were undertaken concurrently by each member of a pair. The maximum time given for explaining was again 5 minutes for each participant, and another 5 minutes were provided for listening with the use of a headset to the IC recorder in which his or her pair had recorded the explanation. The interactive and non-interactive conditions were therefore identical in what participants had to do – except for whether direct interaction with the student they were paired with was possible or not during the time they were providing their explanation.

Following completion of the explanation phase, participants were asked to complete a quick comprehension assessment about the passage they had read, comprising of four questions that required short answers (e.g., “Where does the right ventricle pump blood to?” for the circulation passage). This assessment was carried out to check that the participants had sufficiently understood and could answer basic questions about the passage. However, no further consideration will be given to this comprehension assessment in this report as the participants in both conditions obtained high scores in it (suggesting they understood the passage they had read), and no relationship was found between those scores and the students’ diagram use.

3 Results

3.1 Analysis of Diagram Use During Explanation

The number of participants in each condition that used diagrams during the explanation session was examined. What participants put on the A4 sheet was scored according to whether diagrams were used or not. A diagram was defined as any representation, other than representations in the form of words, sentences, or numbers on their own. If a participant was judged as "using diagrams", the variable about diagram use was scored as "1"; otherwise it was scored as "0". (Examples of typical diagrams are shown in Figure 1.) Two raters coded the participants' diagram use independently, and the inter-rater agreement between them was calculated. The kappa coefficient value obtained was .95, which was almost perfectly concordant.

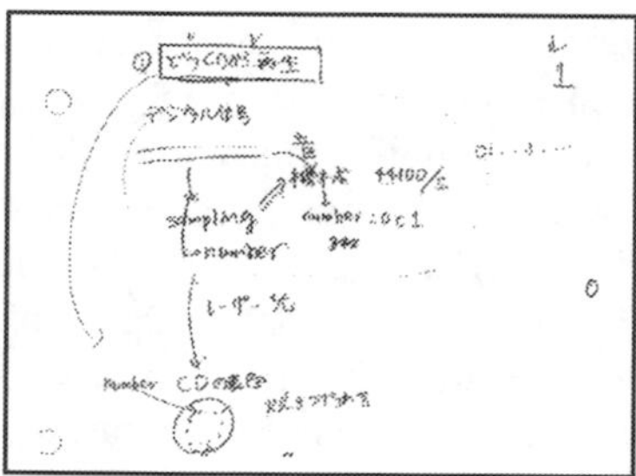

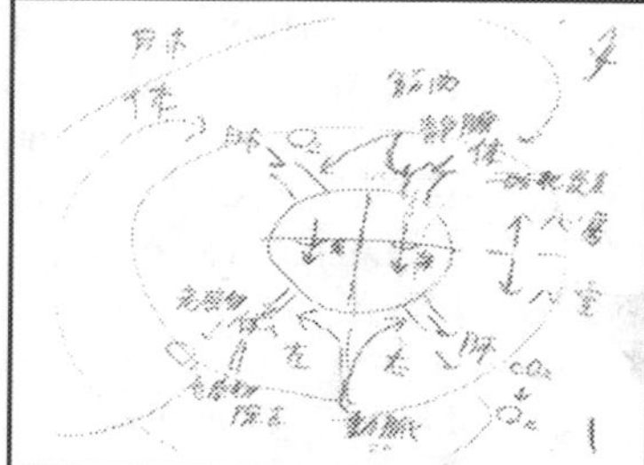

Fig. 1. Examples of Diagrams Participants Constructed During Explanation (left of a flowchart for the CD passage, right of an illustration for the Circulation passage)

Table 1. Percentages of Pairs Using Diagrams in Each Condition During Explanation

Condition	Number of Pairs	% Diagram Use
Interactive Communication	21	72.7% (16/21)
Non-Interactive Communication	22	45.5% (10/22)

In this study, because the participants' responses in the interactive condition were not independent (i.e., the participants influenced each other in what they did), the data were analyzed using the pair as a unit (i.e., in the pairs the participants were randomly assigned to). Thus, if at least one member of a pair used at least one diagram, the pair was coded as "using diagrams"; otherwise the pair was coded as "not using diagrams". The percentages of diagram use in the two conditions, shown in Table 1, were compared using the Chi-square test. This analysis revealed that diagram use in the interactive condition was significantly higher compared to the non-interactive condition, $\chi^2_{(1)} = 4.25$, $p = .04$). (Note that a similar analysis, in which participants' individual responses were assumed to be independent, also revealed that diagram use was significantly higher for participants in the interactive condition compared to those in the non-interactive condition, $\chi^2_{(1)} = 5.57$, $p = .02$.) This result suggests that, when students are placed in a communicative learning situation that requires them to explain to another student, they are more likely to spontaneously use diagrams. The result further emphasizes the necessity for the interactive component to be present for this effect to manifest in such communicative learning situations. Thus, the finding lends support to the first two hypotheses of this study.

3.2 Analysis of Participants' Protocols during Explanation

Analysis of the Total Amount of Words and Time in the Two Conditions. To examine the possible mechanisms that would explain the promotion of diagram use among students in the interactive condition, students' protocol data were analyzed. Before starting analysis, all students' spoken data in the explanation session were transcribed to text data. The students' starting and finishing times for their explanations were also recorded. The total amounts of time and numbers of words spoken were calculated and compared between the two conditions. The means and standard deviations of these are shown in Table 2. The results of *t*-tests revealed that the total amount of time ($t_{(41)} = 4.54$, $p < .001$) and the number of words ($t_{(41)} = 5.14$, $p < .001$) were both significantly greater in the interactive condition compared to the non-interactive condition. These results provide support for the assumption of this study that participants in the interactive condition likely received feedback from their interlocutor, increasing both the amount of time and the number of words used in explanations. The provision of feedback, in turn, probably contributed to awareness of difficulties in explaining only with words, leading to the need to use diagrams.

Table 2. Means (and SDs) of Time and Number of Words During Explanation

Condition	Total Time (sec.)	Total Number of Words
Interactive Communication	528 (82)	2574 (693)
Non-Interactive Communication	407 (78)	1707 (369)

Analysis of the Effect of Questioning in the Interactive Condition. In addition, if the assumption described above is correct, pairs in the interactive condition that communicated more actively (i.e., in which the interlocutor provided greater feedback to the explainer) might be expected to have produced more diagrams. To examine this, the transcripts of spoken communication in the interactive condition were analyzed. Firstly, the existence of questions was coded. If the interlocutor asked at least one question, then the "existence of question" was coded as "1"; otherwise, when no question was asked, it was coded as "0". Two raters independently coded the data, and their coding was perfectly concordant: thus the kappa coefficient was 1. The percentages of diagram use in pairs with questions and without any question were compared using the Chi-square test (see Table 3). The result revealed significantly higher diagram use in pairs in which at least one question was asked (88.2%) than in pairs without any question (25.0%) ($\chi^2_{(1)} = 7.14$, $p = .01$). This result provides support for the mechanism proposed in the previous subsection.

Table 3. Percentages of Diagram Use in the Interactive Condition According to Presence/Absence of Questions and Reference to Diagrams

Category	Present	Absent	Chi-Square Value
All Types of Questions	88.2% (15/17)	25.0% (1/ 4)	7.14**
Clarification Questions	87.5% (14/16)	40.0% (2/ 5)	4.74*
Extension Questions	100.0% (7/7)	64.3% (9/14)	3.24+
Reference to Diagram	100.0% (5/5)	68.8% (11/16)	2.05

+$p < .10$. *$p < .05$. **$p < .01$.

Analysis of the Effect of the Quality of Questions in the Interactive Condition. In a further effort to better understand what was happening in the interactive condition, the quality of the questions that were asked was analyzed. To achieve this goal, the kinds of questions asked were coded. Firstly, the existence of "clarification questions" was coded, in which the interlocutor asked questions to confirm his or her understanding (e.g., "Do you mean that in a CD the information is recorded only as a number?)" or to make clear the information provided by the explainer ("Could you explain again how the 'valve' works?"). If at least one such question was asked, a coding of "1" was given to the pair; otherwise, the pair was given a "0". Two raters independently coded the data, and the coding was again perfectly concordant: thus the kappa coefficient was 1. The percentages of diagram use in groups with and without clarification questions were compared using the Chi-square test. The result was significant ($\chi^2_{(1)} = 4.74$, $p = .03$). As shown in Table 3, this means that spontaneous use of diagrams in explanation was more frequent in the pairs in which at least one clarification question was asked (87.5%), compared to pairs in which no such question was asked (40.0%).

Secondly, the existence of "extension questions" was coded. For the purposes of this study, extension questions were those in which the kind of information sought went beyond clarification of specific content from the passage being explained, and more toward extending the questioner's understanding (e.g., "How is sound recorded in a CD when there are multiple sources of sound coming in?"). If the spoken communication in a pair included at least one extension question, the pair was coded as "1"; otherwise a coding of "0" was given. Two raters independently coded the data, and their coding differed only in one case. The kappa coefficient was 0.89, which was almost perfectly concordant. The percentages of diagram use between groups with and without extension questions were compared using the Chi-square test (see Table 3). Although the result only approached significance ($\chi^2_{(1)} = 3.28$, $p = .07$), it suggests that diagram use in explanation was more frequent in the pairs with at least one extension question (100%) than in the pairs without any such question (64.3%).

Analysis of the Effect of Direct Reference to Diagrams in the Interactive Condition. One of the interesting features of the spoken data in the interactive condition was that sometimes the interlocutor made direct reference to diagrams. In one particular case, for example, the interlocutor directly asked the explainer, "Could you draw a diagram?" Although the number of such references was limited (there were only five such instances in total), the possible effect on diagram use of this type of communication was analyzed. If the spoken communication in a pair in the interactive condition included at least one direct reference to diagrams from the interlocutor, it was coded as "1", otherwise a coding of "0" was given. Two raters independently coded the data, and their coding differed only in one case. The kappa coefficient was 0.88, which was almost perfectly concordant. The percentages of diagram use in groups with and without direct reference to diagrams were compared using the Chi-square test (see Table 3). Although the result was not significant ($\chi^2_{(1)} = 2.05$, $p = .15$) as the number of cases in which direct reference to diagrams was made was limited, all the pairs in which such reference was made manifested diagram use – suggesting the effectiveness of such direct reference to the target strategy. Note however that, in such cases, the interlocutor's reference to diagrams would have provided a direct or indirect hint for the explainer to use a diagram. Strictly speaking, therefore, diagram use in the five such cases would not have been spontaneous.

3.3 Examples of Participants' Protocol Data During Explanation

To better understand the role of feedback in promoting the use of diagrams in explanation, this section introduces some examples of spoken communication in pairs assigned to the interactive condition.

Example of a pair without any question in communication. The first example shown below is from a pair coded as "without any question".

A (explainer): [Explaining the passage] … do you have any question?
B (interlocutor): No. [At this point, the communication finished.]

In such a pair, as shown in Table 1 and Table 3, the percentage of using diagrams was lower (25.0%) than in the non-interactive condition (45.5%). This result suggests that even if an interactive learning situation is set up, diagram use among students would not be enhanced if questioning or feedback in the communication within the pair do not occur.

Example of a pair with clarification questions in communication. The second example shown below is from a pair coded as "with clarification question".

A (explainer): The upper and lower parts of the heart are divided into the atrium and the ventricle.
B (interlocutor): Which is the atrium?
A: The upper part is the atrium and the lower part is the ventricle.

This example shows the interlocutor actively asking questions of the explainer in order to better understand what the explainer described. The finding of higher diagram use in such pairs may be due to the explainer experiencing a need to explain better and in more detail – thus making him or her consider other ways to clarify what he or she just explained. In turn, this may make the explainer consider and use diagrams in such efforts.

Example of a pair with extension questions as well as clarification questions in communication. The following example is one that includes extension questions as well as clarification questions.

A (explainer): … do you have any question?
B (interlocutor): Yes, I've got two. Firstly, I could not really understand what "sampling" means [this is the clarification question] and, secondly, I cannot imagine how closely the pits are etched. The total length of the pits is 6 km, so I wonder how minutely they have to etch in the small area of the disk [this is the extension question].

As shown in this case, most of the pairs in which at least one extension question was asked also included asking of clarification questions (this occurred in 6 out of 7 pairs with extension questions). This type of question might be produced as a result of seeking deeper understanding from the interlocutor. Although the effect found from use of the extension question only approached significance (this may have happened because the number of pairs with this type of question was limited), the percentage of diagram use in pairs with extension questions was 100%. This percentage is higher than the percentage of diagram use in the pairs with clarification questions (with or without other questions) (88%), and in the pairs with only clarification questions (80%). These results suggest the possibility that questions to explore deeper understanding, rather than just seeking clarification of the explanation, may provide more effective feedback and thus better promote the explainer's use of diagrams in explanation.

Example of Pair with Reference to Diagrams in Communication. The final example shown below is from a pair in which a direct reference to diagrams was made.

A (explainer): ... The right side sends the blood to the lungs and the left side sends to the body.
B (interlocutor): The right side goes to the lungs?
A: Yes, the lungs.
…
B: You are not using any diagram and you appear to be speaking just off the top of your head …[here B hinted at experiencing some difficulty in understanding without the inclusion of diagrams].
A: Ok, I will draw. This side goes to the lungs and … [A started to explain by drawing a diagram].

As clearly shown in this example, the interlocutor suggested the use of an external representation to the explainer. He hinted at experiencing some difficulty in understanding without seeing what it looked like, and it prompted the explainer to start drawing a diagram. Although the number of such pairs was limited, in most of the pairs where a reference was made to diagrams, a direct or indirect suggestion was made by the interlocutor to the explainer to use a diagram in explaining.

4 Discussion

4.1 The Effects of Communicative Learning Situations on Diagram Use

The key finding in the present study was that students in the interactive condition used more diagrams than students in the non-interactive condition, even though students in both conditions were required to explain. This suggests that interactivity in communicative learning situations has a unique effect on students' diagram use, an effect that cannot be attributed simply to explaining. This result supports the two hypotheses proposed in the current study, and they reveal positive, beneficial effects of communicative learning situations on diagram use, which had not previously been identified in diagrams research.

In the analysis of the students' protocol data during the explanation sessions, one clear finding was that pairs in the interactive condition in which questions were asked produced more diagrams than the pairs in which no questions were asked. In addition, all pairs in the interactive condition where extension questions were asked used diagrams during explanation, and pairs in which clarification questions were asked produced more diagrams than pairs in which no such questions were asked. These results suggest that the interlocutor's provision of active feedback in order to more deeply understand the information being provided, makes the explainer feel a need to utilize more devices – other than spoken words – in the explanation being provided, and this may make them resort to the use of diagrams.

The process proposed above may appear viable only if the explainer is aware of the benefits that diagram use can bring to explanations. However, analysis of the protocol data obtained in this study also revealed instances where the explainer appeared to have acquired awareness of the value of using diagrams *during the actual process of explaining*. In particular, this appeared to be the case in interactive condition pairs where the interlocutor encouraged the explainer to use a diagram to help him or her to understand more clearly. In such pairs, where a reference to diagram use was made, all explainers ended up using a diagram in their explanation. Thus in the interactive condition, diagram use appeared to have been promoted both by the explainer's own realization of a need to explain more clearly, as well as the interlocutor's encouragement or hint for the explainer to use diagrams in order to clarify the explanation being provided. The effect observed here further highlights a complementarity benefit (i.e., the explainer and the interlocutor complement what each other knows – in this case about the usefulness of constructing a diagram during explanation) that is only possible if the communicative learning situation is interactive.

4.2 Implications for Diagrams and Other Research Areas

The present study revealed that communicative learning situations can promote student motivation to use diagrams when they are explaining. Moreover, it examined the processes that might explain why such situations promote diagram use in explanation by analyzing student protocol data. By doing so, it has put forward a new perspective or approach to examining diagram use in communicative learning situations.

The present study's examination of the construction of diagrams in an actual learning context is another contribution it makes to diagrams research. As previously noted, the majority of studies relating to diagram use have targeted the effect of diagrams provided by the experimenter. However, when considering real learning situations, diagrams are not always provided and the usual expectation is that learners would construct appropriate diagrams by themselves. This aspect of students' spontaneous construction of diagrams – *by themselves* – tend to have been overlooked in this research area, despite its apparent importance in real educational contexts.

The findings of the present study also contribute to other research areas, like that of collaborative learning. Demonstrating the effects of collaborative learning is one important topic in psychology (e.g., [20]). For example, Miyake [21] demonstrated that collaborative learning promotes students' deep understanding of how a machine might work. The Miyake study also revealed the importance of questioning in promoting deep understanding, which is something it has in common with the findings of the present study. However, the Miyake study and other previous studies had not looked into the effect of collaborative learning situations in promoting students' motivation to use external representations, or diagrams, in explanation. Thus, in examining this particular effect, the present study has identified another specific effect or educational benefit of collaborative learning situations. In addition, the findings of this study suggest that diagrammatic notations tend to be automatically shared in pairs (or at least, assumed by members of the pair to be shared), as no questions relating to diagrammatic notations appeared in participants' protocol. This aspect concerning how people acquire and share understanding of diagrammatic notations is an interesting point that warrants examination in future research.

Finally, the present study also proposes a concrete, practical approach to promoting spontaneous strategy use among students. More specifically, it proposes the use of communicative learning situations designed so that students would need to use a target strategy in order to satisfactorily complete the assigned task. This proposed approach is different from instructional approaches in which the teacher simply provides training to students, and the results obtained in this study indicate its efficacy. It can be argued that this approach is more ecologically valid, in that the motivation that is created for spontaneously using the target strategy is instigated by a need that frequently arises *naturally* in real learning contexts.

4.3 Implications for Educational Practices

As previously explained, students tend not to use diagrams of their own volition [e.g., 10], especially students who have an orientation toward simply "obtaining the right answer" (i.e., ignoring the thinking processes involved in obtaining the answer) [11]. For such students, instruction and encouragement to use diagrams are usually insufficient to make them spontaneously use diagrams. To address this problem, this study proposed the use of interactive peer communication situations, and demonstrated that it can produce the desired result of motivating students to use diagrams.

Developing interventions for improving students' ability to use diagrams is important, particularly interventions that can easily be used in real educational settings to address the manifested problems. Among the many student problems that have been identified, the lack of spontaneity in using diagrams has been pointed out as the most serious because teachers themselves do not understand this problem sufficiently even though they may have observed or confirmed its existence among many students. In fact, there is research evidence indicating that teachers are particularly poor at predicting students' spontaneity (or lack of) in using diagrams [22].

Although the present study has not developed an intervention that can improve the spontaneity of diagram use in students' own, independent learning situations, its findings suggest that the spontaneity that students had demonstrated in diagram use in the interactive learning situations could generalize beyond those specific situations and thus contribute more widely to addressing the problem.

4.4 Directions for Future Research

One important topic that ought to be examined in future research is the effect of nonverbal feedback, such as gestures, facial expressions, and possibly even vocal intonations. Although this study proposed the importance of verbal feedback as well as set up communicative learning situations in which verbal feedback was a central component, the likely importance of nonverbal feedback – even in such situations – cannot be ignored or neglected. The kinds of data that had been gathered in the present study did not allow for the investigation of the nonverbal component. However, nonverbal feedback during the student interactions could have also contributed to motivating students to use diagrams in providing their explanations. Hence, examining that potential contribution to the promotion of diagram – and possibly other strategy – use would be important in future research.

In the present study, the actual content of the explanations that participants provided (verbal and written) was analyzed only to measure duration (of speech), and to identify the presence of diagrams and expression of questions. One useful direction to take in future research would be to more carefully examine and analyze the content of what the students represent verbally (i.e., in text and/or speech) and visually (i.e., in diagrams) – particularly the extent to which such representations might complement or duplicate each other [cf. 5]. Such examination could provide useful insights into the functions that text/words and diagrams might be seen to serve in communicating information to others.

Another point which should be examined in future research is the effect of communicative learning situations in promoting students' future spontaneous diagram use, especially in students' own learning situations, when they are working by themselves. Although previous studies have demonstrated that communicative learning situations in math problem solving contexts promote future spontaneous use of diagrams [23], these previous studies included other interventions such as teacher-provided instruction about the efficacy of diagram use. Thus, examining the effect of communicative learning situations on their own would be important to better understand the mechanisms involved in effecting desired change in student learning behaviors.

Acknowledgments. This research was supported by a grant-in-aid (23330207) from the Japan Society for the Promotion of Science. The authors would like to thank Masako Tanaka and Taiki Ishikawa for their help in data collection and/or processing.

References

1. Ainsworth, S., Prain, V., Tytler, R.: Drawing to Learn in Science. Science 333, 1096–1097 (2011)
2. Larkin, J.H., Simon, H.A.: Why a Diagram is (Sometimes) Worth Ten Thousand Words. Cognitive Science 11, 65–99 (1987)
3. Stern, E., Aprea, C., Ebner, H.G.: Improving Cross-content Transfer in Text Processing by Means of Active Graphical Representation. Leaning and Instruction 13, 191–203 (2003)
4. Butcher, K.R.: Learning from Text with Diagrams: Promoting Mental Model Development and Inference Generation. Journal of Educational Psychology 98, 182–197 (2006)
5. Mayer, R.E.: Multimedia Learning. Cambridge Univ. Press, New York (2001)
6. van Essen, G., Hamaker, C.: Using Self-Generated Drawing to Solve Arithmetic Word-Problems. Journal of Educational Research 83, 301–312 (1990)
7. Hembree, R.: Experiments and Relational Studies in Problem-solving: A Meta-analysis. Journal for Research in Mathematics Education 23, 242–273 (1992)

8. Uesaka, Y.: Suugakuteki mondaikaiketu ni okeru zuhyou katuyou no shien (Supporting Students' Use of Diagrams in Mathematics Problem Solving: Development of the "REAL Approach" which Connects Theories and Practices). Kazamashobo, Tokyo (2014)
9. Dufour-Janvier, B., Bednarz, N., Belanger, M.: Pedagogical Considerations Concerning the Problem of Representation. In: Janvier, C. (ed.) Problems of Representation in the Teaching and Learning of Mathematics, pp. 110–120. Erlbaum, Hillsdale (1987)
10. Uesaka, Y., Manalo, E., Ichikawa, S.: What Kinds of Perceptions and Daily Learning Behaviors Promote Students' Use of Diagrams in Mathematics Problem Solving? Learning and Instruction 17, 322–335 (2007)
11. Uesaka, Y., Manalo, E., Ichikawa, S.: The Effects of Perception of Efficacy and Diagram Construction Skills on Students' Spontaneous Use of Diagrams When Solving Math Word Problems. In: Goel, A.K., Jamnik, M., Narayanan, N.H. (eds.) Diagrams 2010. LNCS, vol. 6170, pp. 197–211. Springer, Heidelberg (2010)
12. Chan, L.K.S.: Combined Strategy and Attributional Training for Seventh Grade Average and Poor Readers. Journal of Research in Reading 19, 111–127 (1996)
13. Brown, A.L., Bransford, J.D., Ferrara, R., Campione, J.A.: Tetrahedral Framework for Exploring Problems of Learning. In: Flavell, J.H., Markman, E.M. (eds.) Handbook of Child Psychology, 4th edn. Cognitive Development, vol. 3, pp. 85–106. Wiley, New York (1983)
14. Ueki, R.: Structure of High-School Students' Beliefs About Learning. The Japanese Journal of Educational Psychology 50, 301–310 (2002)
15. Shinogaya, K.: Effects of Preparation on Learning: Interaction With Beliefs About Learning. The Japanese Journal of Educational Psychology 56, 256–267 (2008)
16. Lyon, J.: Drawing: Its Value as a Communication Aid for Adults with Aphasia. Aphasiology 9, 33–50 (1995)
17. Sacchett, C.: Drawing in Aphasia: Moving Towards the Interactive. Human-Computer Studies 57, 263–277 (2002)
18. Palincsar, A.S., Brown, A.L.: Reciprocal Teaching of Comprehension-fostering and Comprehension Monitoring Activities. Cognition and Instruction 1, 117–175 (1984)
19. Manalo, E., Uesaka, Y.: Elucidating the Mechanism of Spontaneous Diagram Use in Explanations: How Cognitive Processing of Text and Diagrammatic Representations Are Influenced by Individual and Task-Related Factors. In: Cox, P., Plimmer, B., Rodgers, P. (eds.) Diagrams 2012. LNCS, vol. 7352, pp. 35–50. Springer, Heidelberg (2012)
20. Snell, M.E., Janney, R. (eds.): Social Relationships and Peer Support. Brookes Publishing, Baltimore (2000)
21. Miyake, N.: Constructive Interaction and the Iterative Process of Understanding. Cognitive Science 10, 151–177 (1986)
22. Uesaka, Y., Manalo, E., Nakagawa, M.: Are Teachers Aware of Students' Lack of Spontaneity in Diagram Use? Suggestions from a Mathematical Model-Based Analysis of Teachers' Predictions. In: Cox, P., Plimmer, B., Rodgers, P. (eds.) Diagrams 2012. LNCS (LNAI), vol. 7352, pp. 312–314. Springer, Heidelberg (2012)
23. Uesaka, Y., Manalo, E.: Peer Instruction as a Way of Promoting Spontaneous Use of Diagrams When Solving Math Word Problems. In: McNamara, D.S., Trafton, J.G. (eds.) Proceedings of the 29th Annual Cognitive Science Society, pp. 677–682. Cognitive Science Society, Austin (2007)

Evaluating the Impact of Clutter in Euler Diagrams

Mohanad Alqadah, Gem Stapleton, John Howse, and Peter Chapman

University of Brighton, UK
{m.alqadah1,g.e.stapleton,john.howse,p.b.chapman}@brighton.ac.uk

Abstract. Euler diagrams, used to visualize data and as a basis for visual languages, are an effective representation of information, but they can become cluttered. Previous research established a measure of Euler diagram clutter, empirically shown to correspond with how people perceive clutter. However, the impact of clutter on user understanding is unknown. An empirical study was executed with three levels of diagram clutter. We found a significant effect: increased diagram clutter leads to significantly worse performance, measured by time and error rates. In addition, we found a significant effect of zone (a region in the diagram) clutter on time and error rates. Surprisingly, the zones with a middle level of clutter had the highest error rate compared to the zones with lowest and the highest clutter. Also, the questions whose answers were placed in zones with medium clutter had the highest mean time taken to answer questions. In conclusion, both diagram clutter and zone clutter impact the interpretation of Euler diagrams. Consequently, future work will establish whether a single, but cluttered, Euler diagram is a more effective representation of information than multiple, but less cluttered, Euler diagrams.

Keywords: Euler diagrams, clutter, diagram comprehension.

1 Introduction

Clutter has long been recognized as a potential barrier to effective communication when visualizing information. In the current climate, where we have access to ever increasing quantities of information, understanding the impact of clutter, and how to reduce it where necessary, is important. Ellis and Dix argue that it is "important to have a clear understanding of clutter reduction techniques in order to design visualizations that can effectively uncover patterns and trends..." [1]. Bertini et al. argue along similar lines, and they devise an approach to analyze and reduce clutter in an information visualization context [2].

This paper focuses on Euler diagrams, which are used for both information visualization and logical reasoning. These diagrams represent sets using closed curves, with their spatial relationships conveying meaning. For instance, two non-overlapping curves represent disjoint sets and so forth. Euler diagrams have been used to visualize gene expression data [3] in the biomedical domain and

T. Dwyer et al. (Eds.): Diagrams 2014, LNAI 8578, pp. 108–122, 2014.

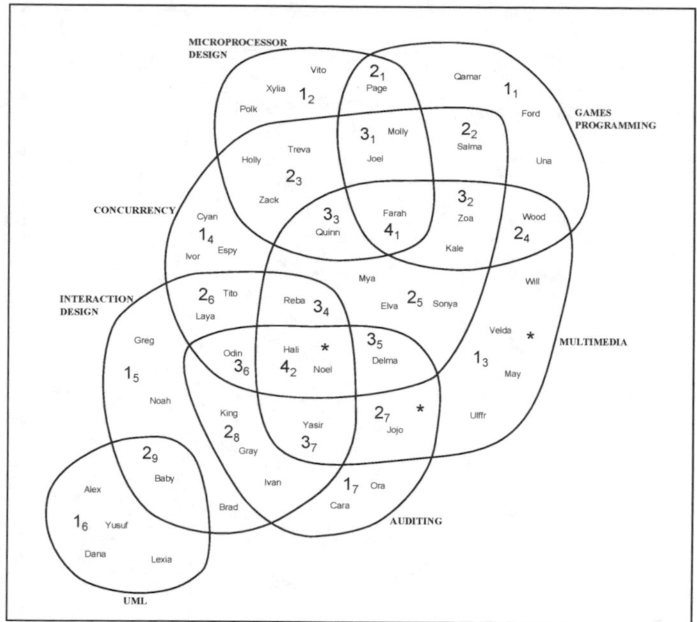

Fig. 1. An Euler diagram annotated with the zone clutter scores

have been used many times in Nature and affiliated journals [4]. With regard to logical systems, contributions include [5–7], which propose methods of reasoning with Euler diagrams. In such systems, items are often visualized within sets to convey information beyond disjointness and subset relationships.

Research has suggested that to use Euler Diagrams successfully we need to understand when they can be effectively interpreted [8–12]. These research contributions tell us that users can understand Euler diagrams but it does not tell us how cluttered an Euler diagram can become before user understanding is significantly reduced. John et al. [13], established a measure of Euler diagram clutter that could be computed from the *zones* in the diagram. Each zone contributes n to the diagram's clutter score, where n is the number of curves that the zone is inside (in [13], this clutter measure is called the *contour score*). For example, the diagram in Fig. 1 has 26 zones and a clutter score of 58. In this diagram, the zones are annotated with their contribution to the clutter score, with the subscripts used to distinguish zones with the same contribution. The diagrams in Figs. 1, 2 and 3 exhibit three different clutter scores, with Fig. 1 being considered least cluttered and Fig. 3 being most cluttered. John et al. empirically established that this way of measuring Euler diagram clutter met with user expectations: diagrams with a higher clutter score were generally considered more cluttered by the study participants. However, no attempt has been made to establish the impact of Euler diagram clutter on user comprehension.

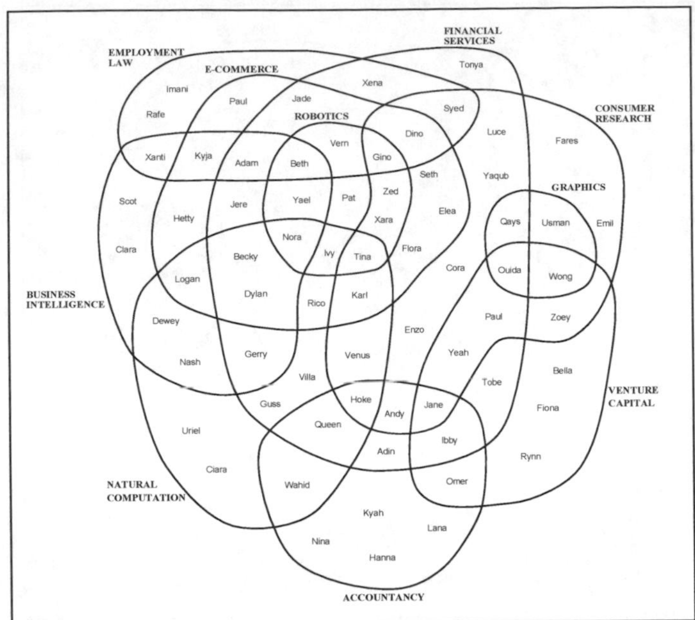

Fig. 2. An Euler diagram with medium clutter

This paper focuses on two distinct features with respect to clutter in the context of interpreting Euler diagrams. First, we see whether Euler diagrams are significantly harder to interpret as the clutter score increases. As well as the clutter score for the diagram, each zone can be thought of as having a contribution to the clutter score. By asking questions of the information contained in the diagram, users have to focus their attention on the zones. The placement of the information needed to answer the question is contained within zones and could impact on the speed and accuracy with which the information is extracted. We therefore set out to determine whether the clutter score attributable to the zone in which the desired information for answering a question is located significantly impacts user performance. In section 2 we present the experiment design. Section 3 describes how we executed the experiment and section 4 presents the experiment results. In section 5 we offer conclusions and discussion on future work. The diagrams used in the study, along with the raw data collected can be found at `https://sites.google.com/site/msapro/evaluating-the-impact-of-clutter-in-euler-diagrams`.

2 Experiment Design

This paper now proceeds to present an empirical study undertaken to answer the questions: "does Euler diagram clutter affect user comprehension?" and "does zone clutter affect user comprehension?" The study follows a within-group design, where each participant is shown a set of diagrams with varying amounts

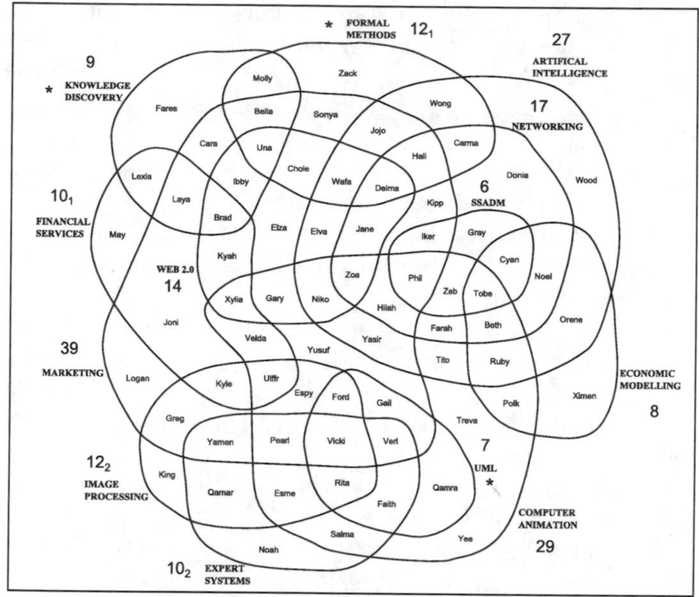

Fig. 3. An Euler diagram annotated with the curve clutter scores

of clutter and is asked questions about the information in these diagrams. Consistent with other research contributions towards understanding user comprehension [10, 14–16], two primary variables were recorded: the time taken to answer the question and whether the question had been answered correctly. Moreover, we also set a limit of 2 minutes to answer questions and all timeouts were recorded. If clutter level affects comprehension then we would expect to see a significant difference between the means of the time taken to answer the questions by the participants or the error rates over the different levels of clutter.

2.1 Study Details and Question Types

The participants of the study were students from the School of Computing, Engineering and Mathematics in University of Brighton. Consistent with Rodgers et al. [12], the diagrams were drawn with pseudo-real data that represent students taking university modules. This ensured that the diagrams were relevant to the participants, by representing information about hypothetical Computer Science modules (the curve labels) taken by students (represented by names inside the curves). This choice will reduce any risk that could arise if the participants needed to learn about the information represented by the diagrams and, since the context is hypothetical, ensure that prior knowledge cannot be used. The modules names were based on modules commonly found in British undergraduate computing courses. We avoided using module names that look similar or share the first syllable within the same diagram, because that could confuse the participants, as was found in the pilot study. Furthermore, the module names were placed outside the relevant curve avoiding places that could lead to confusion as much as possible (such as identifying which label was associated with which curve).

The student names were decided to be first name only, a randomised mixture of both male and female names across a variety of ethnicities to reduce any bias that could occur if we chose the names manually. The length of the student names were 3 to 5 letters, to reduce the risk of having a name which could be identified without the need for full reading of the diagram: names of particularly different lengths would be obvious.

Three types of questions were used in the study: "Who", "How many", and "Which", following Rodgers et al. [12] and Blake et al. [10] who conducted user performance studies on Euler diagrams. All the questions were multiple choice and had five choices of answers. Examples of the questions are:

1. **Who** of the following is taking MULTIMEDIA but not AUDITING, CONCURRENCY or GAMES PROGRAMMING?
2. **How many** students are taking ACCOUNTANCY and NATURAL COMPUTATION but not FINANCIAL SERVICES?
3. **Which** one of the following modules is being taken by 7 students?

The questions 1, 2, and 3 were asked of the diagrams in Figs. 1, 2, and 3 respectively. When participants were asked a "How many" question, they were required to type in the names of the students they had counted. For the response to be classed as correct, the set of names has to be correct.

2.2 Levels of Diagram Clutter

Previous research gave us a method to measure the complexity of Euler diagrams, called the *contour score* (CS) [13]. The CS for a diagram is the sum of all *zone scores* (ZSs) of the diagram: for each zone, the ZS is the number of curves (often called *contours*) that contain the zone. Each zone is uniquely described by the curves that it is inside. For example, the diagram in Fig. 4 has a zone inside the two curves A and C (but outside B). This zone can be described by AC, however to differentiate zones from curves, we write this as ac and blur the distinction between the zone itself and this description. This diagram, thus, has zones $\{a, b, c, ab, ac, bc, abc\}$ and the score for each zone is: $1, 1, 1, 2, 2, 2, 3$. Therefore, the CS for this diagram is 12. Another method of calculating the CS of a diagram is by counting how many zones are inside each curve within the diagram, we call this measure the *curve contour scores* (CCS), and summing these across all curves in the diagram. For example, each of the 3 curves in Fig. 4 has 4 zones. Both methods produce the same score.

In order to establish whether *diagram clutter* impacts performance when interpreting diagrams, we used three different levels of diagram clutter in the study. It was felt important that the diagrams were sufficiently complex to demand notable cognitive effort by the study's participants. Given this, the diagrams

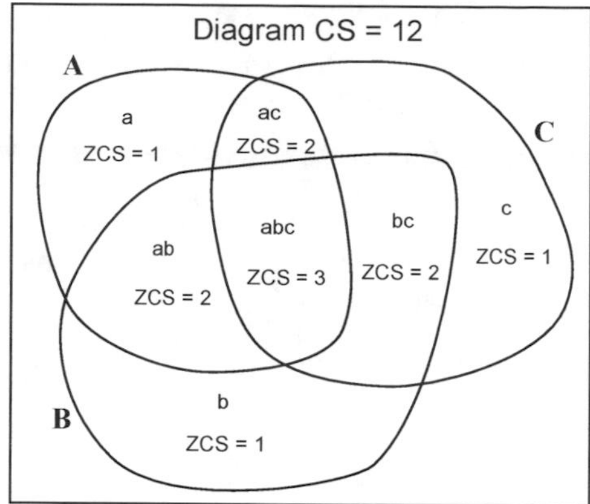

Fig. 4. How to calculate the CS

were chosen so that their CSs were: $50 \pm 20\%$ (low CS), $150 \pm 20\%$ (medium CS), and $250 \pm 20\%$ (high CS). The three diagrams in Figs. 1, 2, and 3 have CSs of 54, 145, and 200 respectively. For the study, nine diagrams were drawn which represent the three levels of diagram clutter. These diagrams were rotated to create 27 diagrams in total, explained further below.

2.3 Levels of Zone Clutter

With regard to whether the *zone clutter* impacts performance when interpreting diagrams, questions were devised so that the zones that required attention to extract the answer also exhibited three clutter levels. For example, in Fig. 4 we see the zones $\{a, b, c\}$ have a ZS of 1, and the zones $\{ab, ac, bc\}$ have a ZS of 2 and the zone $\{abc\}$ has a ZS of 3. So, there are three levels of zone clutter within the diagram. The diagrams used in the study were divided into three sets reflecting three different levels of ZS. This method was used for two types of question "How many" and "Who", since those question types require users to focus on particular zones.

The diagrams used for investigating the impact of ZS were generated using six diagrams, which are designed for questions type "How many" and "Who", of the nine diagrams created by considering the CS. This was achieved by randomly rotating the original six diagrams, noting the result Blake *et al.* [10] which empirically established that diagram orientation does not significantly affect user comprehension of Euler diagrams. In order to avoid learning effect, the curves in the diagrams were relabelled after rotation. Each diagram thus yielded two more diagrams, giving six sets of three diagrams (18 diagrams in total for "How many" and "Who" questions). One of these three sets of diagrams would be subject to a question whose answer placement was a zone with low ZS, one with a medium ZS and, lastly, with a high ZS. In Fig. 1, there are zones with ZS of $1, 2, 3$ and 4. We partitioned these scores into three categories (called *low*, *medium* and *high*), in this example $\{1\}, \{2, 3\}$ and $\{4\}$. In the example in Fig. 1, there are 7 zones categorised as having low ZS, 16 zones categorised as having medium ZS (9 with a ZS of 2, and 7 with a ZS of 3), and 2 zones categorised as having high ZS.

The actual zone for which the question would be asked was then chosen randomly from within the respective zone clutter group. The zones selected for the shown diagram and its rotated versions were: zone number 1_3 (low), zone number 2_7 (medium), and zone number 4_2 (high). The stars ($\star$) in Fig. 1 identify these zones. If zone clutter has an impact on comprehension then we expect to see significant differences between performance data when grouping the 18 diagrams by ZS.

2.4 Levels of Curve Clutter

The third type of question "Which", requires participants to count the number of items in a curve. So, we need to focus on the whole curve instead of the zone to get the answers of the questions. For example, in Fig. 3 we see the curve UML has a CCS of 7, and the curve FINANCIAL SERVICES has a CCS of 10 and so on. Thus, there are different levels of curve clutter within these diagrams. For "Which" questions we designed another three diagrams reflecting three different levels of CCS (9 diagrams in total after rotation). We decided to only ask "Which" questions of the curves that have a maximum of 15 zones, to prevent the task becoming too difficult. In Fig. 3, there are 13 curves and the curves that have 15 zones or fewer have the CCSs (6, 7, 8, 9, 10, 10, 12, 12, and 14). We partition the curves in the manner as we partitioned zones in section 2.3, labelling a curve as having either low, medium or high curve clutter. The actual curve from the respective curve clutter group was chosen randomly. The curves selected for the shown diagram and its rotated versions were: curve number 7 (low), curve number 9 (medium), and curve number 12_1 (high). The stars ($\star$) in Fig. 3 represent the chosen curves.

2.5 Euler Diagram Layout and Characteristics

As can be seen from table 1, the diagrams were divided into nine characteristic types; the number of sets represented is the number of curves and the number of student names is the number of items. For each characteristic type there were three diagrams: an original diagram and its two rotated versions. The three diagrams of each characteristic type were allocated the same type of question but with three different ZSs for the answer placement.

In all the diagrams, data items were roughly evenly distributed within the zones, and at least one data item was placed within each zone. Requiring that each zone contained at least one item placed a lower bound on the total number of items to be used. The number of data items was chosen to be large enough so that, again, answering the question required cognitive effort. Furthermore, the number of data items in the diagrams designed for "Which" questions was 70 with maximum of 15 in each curve for all the three levels of diagram clutter: we did not want the number of items to affect the time taken to answer the

Table 1. The Characteristics of first set of diagrams

Questions types	Who	How many	Which
Clutter Score \ Number of Curves	7 Curves	10 Curves	13 Curves
$50 \pm 20\%$	50 items	50 items	70 items
$150 \pm 20\%$	70 items	70 items	70 items
$250 \pm 20\%$	80 items	80 items	70 items

questions. This decision was made because when too many items were in each curve, answering the question could take more time and result in counting errors.

When drawing the diagrams for the study, we were careful to control their layout features, to minimize (so far as practically possible) unintended variation. First, the diagrams are designed to be well-matched [17] and well-formed [18] which is known to produce more effective diagrams [12]. Additionally, to strengthen the study and to make sure all the diagrams have consistent layout features that correlate positively with usability, we adopted the following six drawing guidelines, following [10] and using results on comprehension from [8]:

1. All diagrams were drawn with black curves and were drawn in an area of 765×765 pixels.
2. The stroke width for the curves was set to 2 pixels.
3. The curve labels were written using upper case letters in Time New Roman, 14 point size, font in bold.
4. Data items were written using lowercase letters, except that the first letter was capitalized, and with Ariel 12 point size font.
5. Data items were evenly distributed within the zones.
6. Each curve label was placed outside the curve and placed close to the curve.

Guidelines 1-6 are from [10], but they are not sufficient to control variability across diagrams in our study. We further stipulate the following layout guidelines:

7. All curves drawn using irregular shapes but with smooth and diverging lines.
8. The curve labels do not have the same first word.
9. There is at least one data item within each zone.

We used smooth curves as jagged lines can negatively affect user comprehension [8], irregular curves so that we could make zone areas roughly equal, and diverging lines to make it easier to see the path taken by each curve.

2.6 Research Software

To collect the data from the participants, we extended the software developed by Blake et al. [10]. The software is used to display the diagrams and the questions, gather the answers given to the questions by the participants and measure the time taken to answer each question. The time taken to answer each question was counted starting from the instant a question was displayed until the instant a participant had selected an answer to the question, measured in units of $1/60^{th}$ of a second. The software allowed the participant to rest between the questions and they could continue with the next question whenever they were ready. There was a time limit of two minutes for each question. The participants had to answer the question in this time and if they failed to answer, the software registered a timeout. To present the second type of question, "How many", the time taken to answer the question also stopped when the answer was selected from the multiple choice list and did not record the time taken to enter the names. The software was used for two phases of the experiment, a training phase and the main study phase. The training phase was to make sure the participants knew how to use the software, and to make them familiar with the Euler diagrams and the questions. In the training phase, the software presented three Euler diagrams with their questions in a random order. The diagrams used in the training phase were chosen from the rotated version of original diagrams but the rotation was different to the diagrams used in the main study. The participants were exposed to one example of each level of Euler diagram CS and question type (Who, How many, and How). In the main study phase, the research software presented all of the 27 diagrams to the participants one at a time in random order.

3 Experiment Execution

Initially five participants (3 M, 2 F, ages 19-44) took part in a pilot study. A problem was revealed with curves names that look similar or share the first syllable: this could confuse the participants and cause high error rate. However, during the pilot, all the questions were answered in the two minutes time allowed. Before we executed the main experiment the curves labels were changed to make them easier to distinguish. The main experiment was carried out with 30 participants (21 M, 9 F, ages 17-65). All the participants were students from the School of Computing, Engineering and Mathematics at University of Brighton. The experiment was performed on the university campus on a weekday between 9am and 5pm. The participants performed the experiment within a usability laboratory that provided a quiet environment free from interruption and noise. Each participant was alone during the experiment, except that the experimental facilitator was present, to avoid any interruption. The same computer and monitor were used by all participants. All the participants were treated in the same way during the experiment. Each participant was informed not to discuss the detail of the study with the other participants who are yet to perform the experiment. Each participant was paid £6, in the form of a canteen voucher, to take part in the study.

There were three phases for the experiment. First is the introduction and the paper based training phase, the experimental facilitator used the introduction script to introduce the participant to Euler diagrams and the type of questions they will see during the experiment. The participants then started the second, computer based training, phase where the participants had the chance to use

Table 2. ANOVA for time log (Diagram Clutter)

Source	DF	SS	MS	F	P
Participant	29	8.65	0.30	12.50	0.000
Diagram Clutter	2	6.95	3.48	139.78	0.000
Error	232	5.27	0.02		
Total	809	50.54			

the experiment software to answer three questions in the two minutes given for each question without interruption from the experiment facilitator. If a training question was answered incorrectly, the experimental facilitator went through the question with the participant. If the participant wished to continue then the third, main study, phase where we collected the quantitative data for our analysis. The participant was not interrupted or given any help during this phase.

4 Experiment Results

Given the collected data (time and error), there are two types of analysis we can perform. For diagram clutter, we can determine whether significant differences exist between each CS group (comprising nine diagrams) and both time taken and error rate. Likewise, the nine diagrams for each ZS group can be analyzed with respect to time taken and error rate. The results presented in this section are based on the data collected from the 30 participants in the main study phase, each answering 27 questions, yielding $27 \times 30 = 810$ observations.

To analyze the statistical distribution of time data, in both cases, we used an ANOVA test. In order to carry out this test, the time data needs to be normal. We normalized the data by applying the log function to the time taken. The statistical tests of significance presented below are derived from this normalized data.

4.1 Analysis of Time Data

To analyse the time data we used the ANOVA calculations presented in table 2.

Diagram Clutter: The observed means are 29.20 seconds for low diagram clutter, 38.50 seconds for medium diagram clutter, and 49.77 seconds for high diagram clutter. The ANOVA test results in table 2, in the row for diagram clutter, we can see the p-value is 0.000. We therefore reject the hypothesis that "the mean time taken to interpret Euler diagrams does not alter significantly as diagram clutter increases". Pairwise comparisons between means for the three diagram clutter levels yield p-values of 0.000, so all three means of the time taken to answer the questions are significantly different from each other. In conclusion, diagram clutter does impact the time taken to interpret Euler diagrams. Unsurprisingly, as clutter increases, so does time taken.

Table 3. ANOVA for time log (Diagram Clutter and Zone Clutter)

Source	DF	SS	MS	F	P
Participant	29	5.75	0.20	9.23	0.000
Diagram Clutter	2	8.54	4.27	171.85	0.000
Zone Clutter	2	0.38	0.19	8.92	0.000
Diagram Clutter * Zone Clutter	4	3.14	0.78	46.25	0.000
Error	116	2.14	0.02		
Total	539	33.29			

Zone Clutter: The observed means are 40.09 seconds for low zone clutter, 47.06 seconds for medium zone clutter, and 43.98 seconds for high zone clutter. The ANOVA test results in table 3 in the row of zone clutter, the p-value is 0.000. We therefore reject the hypothesis "the mean time taken to interpret diagrams does not alter significantly as the clutter of the zone increases in which the answer is placed". Pairwise comparisons between the three zone clutter levels shows that the mean time taken with low zone clutter is significantly different from the mean times taken with medium and high zone clutter. However, although the results show a difference between the mean times taken with medium zone clutter and high zone clutter, the p-value is 0.052 (which is greater than 0.05), so this result is (just) not statistically significant. In conclusion, zone clutter does impact the time taken to interpret Euler diagrams. Surprisingly, medium levels of clutter increase time taken.

Interaction between Diagram Clutter and Zone Clutter: We showed that each of diagram clutter and zone clutter have their own effect on user comprehension of Euler diagrams. However, it could be that we have to consider both of diagram clutter and zone clutter to improve the comprehension of Euler diagrams. Fig. 5 shows the interaction between diagram clutter and zone clutter.

The statistical results of the pairwise comparisons of the interaction between zone clutter and diagram clutter shows that if the zone clutter is low there is significant difference between all three diagram clutter levels (the observed means are 30.17 seconds for low diagram clutter, 35.81 seconds for medium diagram clutter, and 54.28 seconds for high diagram clutter). Moreover, when the zone clutter is medium there is significant difference between all the three diagram clutter levels (the observed means are 20.61 seconds for low diagram clutter, 53.42 seconds for medium diagram clutter, and 67.14 seconds for high diagram clutter). Additionally, if zone clutter is high there is a significant difference between the time taken to answer the questions of all diagrams with high diagram clutter (52.33 seconds) and the time taken to answer the questions of all diagrams with low (37.35 seconds) or medium diagram clutter (42.26 seconds), because the p-value in both cases is 0.000. Nevertheless, there is no significant difference between the time taken to answer the questions of all diagrams with low diagram clutter and the time taken to answer the questions of all diagrams with medium diagram clutter when zone clutter is high, because the p-value is

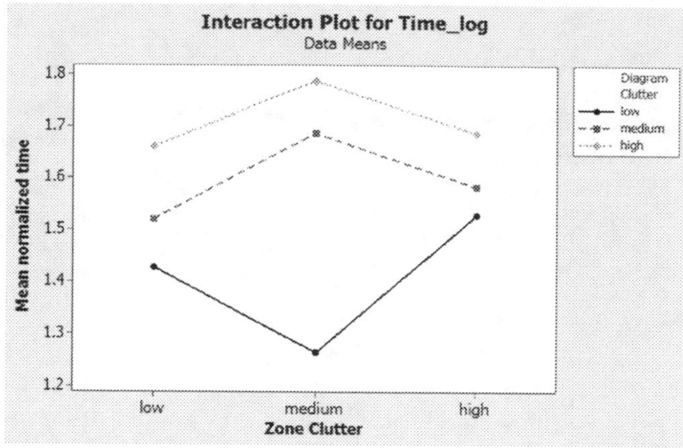

Fig. 5. The interaction between diagram clutter and zone clutter

0.073 (which is higher than 0.05). The results suggest that the combination of diagram clutter and zone clutter impacts on the time taken to interpret Euler diagrams.

4.2 Analysis of Errors and Timeouts

Diagram Clutter: Of the 810 observations (30 participants × 27 questions) there were a total of 65 errors and 14 timeouts which gives an error rate of 8.03% and a timeout rate of 1.73%. Table 4 shows the expected and the observed errors for each diagram clutter type. The errors column shows the total of the errors when the participants chose wrong answers (including errors when the participants typed the wrong set of names when the right answer was chosen with the "How many" question type). We eliminated the timeout data from this test because neither a correct nor an incorrect answer was provided. The expected error rate was between 22.05 and 21.07. However, the observed errors increase when diagram clutter gets higher. The observed errors are 9 for low diagram clutter, 22 for medium diagram clutter and 34 for high diagram clutter. We have performed a χ^2 test on the values shown in table 4 to test the effect of diagram clutter on user comprehension of Euler diagrams. The p-value is 0.000, so we conclude that diagram clutter affects the comprehension of Euler diagrams.

Zone Clutter: Only two of the question types, "How many" and "Who", asked in the study focused on the zone clutter to place the answers of the questions. Therefore, we run the error analysis this time without "Which" question type data which brings the number of observations down to 540 (30 participants × 18 questions). Of the 540 observations there were a total of 56 errors and 12 timeouts which gives an error rate of 10.37% and a timeout rate of 2.22%. Table 5 shows that the highest observed error rate was when the zone clutter is medium. The

Table 4. The expected and observed errors for each diagram clutter type

Diagram Clutter Type		Correct	Errors	Total
Low Clutter	Expected	247.95	22.05	270
	Observed	261	9	
Medium Clutter	Expected	246.12	21.88	268
	Observed	246	22	
High Clutter	Expected	236.93	21.07	258
	Observed	224	34	
Total		731	65	796

next high observed error rate was when zone clutter is high, then the low zone clutter. This means it was more difficult to interpret a diagram when a question asked of medium zone clutter than the other zone clutter types. We have done a χ^2 test on the values showed in table 5 to test the effect of zone clutter on user comprehension of Euler diagrams. Again,we have eliminated the timeout data. The p-value is 0.000, so we conclude that zone clutter affect the comprehension of Euler diagrams.

Table 5. The expected and observed errors for each zone clutter type

Zone Clutter Type		Correct	Errors	Total
Low Clutter	Expected	156.44	18.56	175
	Observed	170	5	
Medium Clutter	Expected	155.54	18.45	174
	Observed	145	29	
High Clutter	Expected	160.01	18.98	179
	Observed	157	22	
Total		472	56	528

4.3 Discussion

We now seek to explain the findings of the study. We have demonstrated that extracting information from a highly cluttered diagram takes longer, and is more error prone, than extracting information from a relatively uncluttered diagram. That more cluttered diagrams are, by our measures, more difficult to interpret than less cluttered diagrams fits existing models of cognition, and existing results (albeit in the context of general images [19]).

The zone clutter results, it could be argued, are less intuitive, because the worst time performance was with the medium clutter level. Using the definition of ZS as counting the number of curves a zone is inside would suggest that the most cognitive effort is needed to identify a high-cluttered zone, since a participant could need to find all these curves. However, the situation of a high cluttered zone could also be seen as follows: a participant only needs to identify the curves the zone is *outside* to find the zone. Given there are a fixed number of curves in a diagram, a high-cluttered zone will be outside few curves. With this observation, the result about medium-cluttered zones makes sense. To find a low- or high-cluttered zone requires identifying only a small number of curves. In the former case, the searched-for zone will be inside these curves (resp. outside these curves for high-cluttered zones). In the case of medium-cluttered zones, however, the searched-for zone may be inside 4 or 5 curves, for instance, and outside 4 or 5 curves. To identify the zone, a larger number of curves, relatively speaking, must be processed whichever approach (searching inside or outside) is taken. Visually, low-cluttered zones are around the edges of the diagram (generally speaking) and are so easy to find. High-cluttered zones are (usually) in the centre of the diagram, and are so easy to find (but not as easy as those around the edges). The medium-cluttered zones are neither central, nor peripheral, to the diagram, and so it takes longer to find them.

It remains to explain the result of low diagram clutter interacting with medium zone clutter. The mean time taken to answer such questions was the lowest across all types of diagram-zone clutter combination. We cannot explain this, and the study was not designed to give insight into seemingly anomalous result. A further study, with the sole intention of either confirming our observations, or providing evidence that they are as a result of some unintended aspect of the diagrams, is needed. Finally, the results for contour clutter using the "Which" question type used only 9 diagrams and therefore the results concerning contour clutter were not statistically significant. Of course, these 9 diagrams contributed to the diagram clutter results.

5 Conclusion

In this paper, we wanted to determine how clutter affects the understandability of Euler diagrams. The main result of the study was the perhaps unsurprising finding that the mean time taken to interpret Euler diagrams increases significantly as diagram clutter increases. Additionally, the study showed that the mean time taken to interpret Euler diagrams alters significantly as the clutter of the zone in which the answer is placed increases. The surprising result being that diagrams in which the answer was placed in zones with medium clutter took longest to interpret. This result may indicate a further study is required to determine whether the zone clutter score corresponds to the human perception of a cluttered zone.

These results are complementary to [13] and inform the direction of further research into the effect of the complexity of Euler diagrams on comprehension. For instance, it would be interesting to establish whether these results still hold if more naive participants take part in the study. The next step for this research is to study Euler diagram components to determine how to balance the number of Euler diagrams used to convey information versus the amount of clutter present in the diagrams, using the results in this paper. We will then build on these studies to consider the effect of clutter in concept diagrams which are based on Euler diagrams.

Acknowledgements. The authors wish to express their gratitude to Andrew Blake for his assistance with this study and to Liz Cheek for statistical advice. We also thank the study participants.

References

1. Ellis, G., Dix, A.: A taxonomy of clutter reduction for information visualisation. IEEE Trans. Visualization and Computer Graphics 13(6), 1216–1223 (2007)
2. Bertini, E., Dell'Aquila, L., Santucci, G.: Reducing infovis cluttering through non uniform sampling, displacement, and user perception. In: SPIE Proceedings. Visualization and Data Analysis, vol. 6060 (2006)
3. Kestler, H., Muller, A., Gress, T., Buchholz, M.: Generalized Venn diagrams: A new method for visualizing complex genetic set relations. Bioinformatics 21(8), 1592–1595 (2005)
4. Wilkinson, L.: Exact and approximate area-proportional circular Venn and Euler diagrams. IEEE Trans Visualization and Computer Graphics 18(2), 321–331 (2012)
5. Howse, J., Stapleton, G., Taylor, J.: Spider diagrams. LMS Journal of Computation and Mathematics 8, 145–194 (2005)
6. Mineshima, K., Okada, M., Takemura, R.: A diagrammatic inference system with Euler circles. Journal of Logic, Language and Information 21(3), 365–391 (2012)
7. Swoboda, N., Allwein, G.: Heterogeneous reasoning with Euler/Venn diagrams containing named constants and FOL. In: Proc. Euler Diagrams 2004. ENTCS, vol. 134 (2005)
8. Benoy, F., Rodgers, P.: Evaluating the comprehension of Euler diagrams. In: IEEE Information Visualization, pp. 771–778 (2007)
9. Blake, A., Stapleton, G., Rodgers, P., Cheek, L., Howse, J.: The Impact of Shape on the Perception of Euler Diagrams. In: Dwyer, T., Purchase, H.C., Delaney, A. (eds.) Diagrams 2014. LNCS (LNAI), vol. 8578, pp. 124–138. Springer, Heidelberg (2014)
10. Blake, A., Stapleton, G., Rodgers, P., Cheek, L., Howse, J.: Does the orientation of an Euler diagram affect user comprehension? In: Proc. Visual Languages and Computing, DMS, pp. 185–190 (2012)
11. Fish, A., Khazaei, B., Roast, C.: User-comprehension of Euler diagrams. Journal of Visual Languages and Computing 22(5), 340–354 (2011)
12. Rodgers, P., Zhang, L., Purchase, H.: Wellformedness properties in Euler diagrams: Which should be used? IEEE Trans Visualization and Computer Graphics 18(7), 1089–1100 (2012)
13. John, C., Fish, A., Howse, J., Taylor, J.: Exploring the notion of 'Clutter' in Euler diagrams. In: Barker-Plummer, D., Cox, R., Swoboda, N. (eds.) Diagrams 2006. LNCS (LNAI), vol. 4045, pp. 267–282. Springer, Heidelberg (2006)
14. Isenberg, P., Bezerianos, A., Dragicevic, P., Fekete, J.: A study on dual-scale data charts. IEEE Trans. Visualization and Computer Graphics 17(12), 2469–2478 (2011)
15. Puchase, H.: Which aesthetic has the greatest effect on human understanding? In: DiBattista, G. (ed.) GD 1997. LNCS, vol. 1353, pp. 248–261. Springer, Heidelberg (1997)
16. Riche, N., Dwyer, T.: Untangling Euler diagrams. IEEE Trans. Visualization and Computer Graphics 16(6), 1090–1099 (2010)
17. G.C.: Effective diagrammatic communication: Syntactic, semantic and pragmatic issues. Journal of Visual Languages and Computing 10(4), 317–342 (1999)
18. Stapleton, G., Rodgers, P., Howse, J., Taylor, J.: Properties of Euler diagrams. In: Proc. Layout of Software Engineering Diagrams, pp. 2–16. EASST (2007)
19. Rosenholtz, R., Li, Y., Nakano, L.: Measuring Visual Clutter. J. Vision 7(2) (2007)

The Impact of Shape on the Perception of Euler Diagrams

Andrew Blake[1], Gem Stapleton[1],
Peter Rodgers[2], Liz Cheek[1], and John Howse[1]

[1] University of Brighton, UK
{a.l.blake,g.e.stapleton,liz.cheek,john.howse}@brighton.ac.uk
[2] University of Kent, UK
p.j.rodgers@kent.ac.uk

Abstract. Euler diagrams are often used for visualizing data collected into sets. However, there is a significant lack of guidance regarding graphical choices for Euler diagram layout. To address this deficiency, this paper asks the question 'does the shape of a closed curve affect a user's comprehension of an Euler diagram?' By empirical study, we establish that curve shape does indeed impact on understandability. Our analysis of performance data indicates that circles perform best, followed by squares, with ellipses and rectangles jointly performing worst. We conclude that, where possible, circles should be used to draw effective Euler diagrams. Further, the ability to discriminate curves from zones and the symmetry of the curve shapes is argued to be important. We utilize perceptual theory to explain these results. As a consequence of this research, improved diagram layout decisions can be made for Euler diagrams whether they are manually or automatically drawn.

Keywords: euler diagram, shape, visualization, perceptual organisation.

1 Introduction

An Euler diagram represents sets using graphical elements called closed curves. The interior of each curve represents elements that are in the set [18,20]. Fig. 1 shows a staff hierarchy within an academic institution. The curve labelled 'Managers' intersects with 'Academics' meaning that there are some managers who are academics. 'Researchers' is a subset of 'Academics' meaning that all researchers are academics. 'Researchers' is disjoint from 'Managers' so there are no researchers who are managers. Euler diagrams are often regarded a natural [20] and effective [19] way to depict sets. Their numerous application areas include the natural sciences [12], art and architecture [1], education [10], criminology [7], computer file organisation [5] and classification systems [24]. This provides clear and strong motivation for the need to better understand how the choices made when laying out (drawing) Euler diagrams impact on user comprehension. Providing such an understanding will improve the effectiveness of these

T. Dwyer et al. (Eds.): Diagrams 2014, LNAI 8578, pp. 123–137, 2014.

diagrams as a mode of information visualization with, potentially, wide ranging benefits.

Reflecting their widespread use, recent times have seen a variety of methods derived to automatically produce Euler diagrams. These methods make varying choices of the topological and graphical properties to be possessed by the diagrams that they produce. For example, Wilkinson's method uses circles, motivated by the fact that 72 Euler diagrams used in articles appearing in Science, Nature and online affiliated journals, 90% use circles [26]. Wilkinson is not alone in choosing to use circles, with other automated drawing methods doing the same, such as [22]. However, Micallef and Rodgers [14] prefer the use of ellipses while Riche and Dwyer [17] prefer the use of rectangles, albeit in a stylized form. Other methods make no preference towards any particular geometric shapes at all [21] but some of them, including [23], aim to minimize or avoid certain topological properties, such as the use of non-simple curves. There is a need to derive more informed automated layout methods that take proper account of user comprehension when drawing Euler diagrams. At present, many of the graphical choices are based on assumptions about what yields an effective diagram.

As there are a range of automated Euler diagram drawing methods, using a variety of shapes, it seems important to establish the relative performance of users with different shapes. Providing such an understanding will allow those devising drawing methods to prioritize efforts towards the use more effective shapes. The contribution of this paper is guidance on shape choice by conducting an empirical study. The next section provides a discussion on related work on Euler diagram drawing choices, covering the state-of-the-art in terms of layout guides. Section 3 presents the experiment design, followed by section 4 where experiment execution is described. The data are analyzed in section 5. We interpret the significant results with reference to perceptual theory in the section 6. Threats to the validity of the experiment are described in section 7, after which we conclude in section 8. All of the diagrams used in our study, and the data collected, are available from `http://www.cem.brighton.ac.uk/staff/alb14/experimental_resources/shape/shape.html`

2 Euler Diagram Background

Given a data set to be visualized, there are numerous choices of Euler diagram that can be drawn. We categorize these choices into three types: descriptional

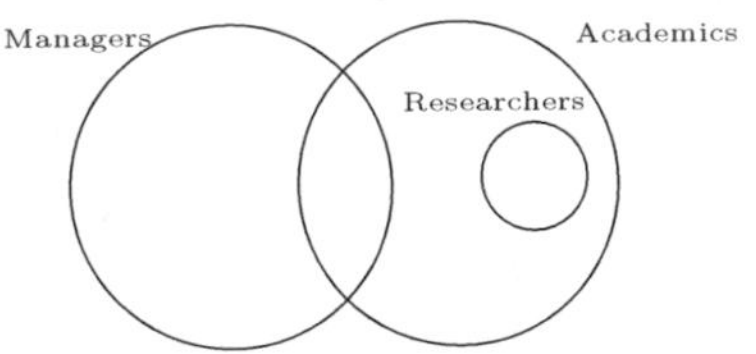

Fig. 1. Staff roles

(the abstract syntax level), topological and graphical (both at the concrete syntax level). To illustrate, Figs 2 and 3 show four Euler diagrams that each represent the same information yet vary the choices made. When starting with the information that we wish to visualize using an Euler diagram, the first choice that must be made is descriptional, at the so-called abstract syntax level. Descriptional choices determine the zones that must be present in the Euler diagram [22]. To illustrate, suppose we wish to represent the following syllogism:

1. All Students are People
2. All People are Mammals
3. (therefore) All Students are Mammals.

The Euler diagram on the left of Fig. 2 represents the above syllogism. The curve labelled 'S' represents students, 'P' represents people and 'M' represents mammals. "All Students are People" is expressed by drawing 'S' inside 'P'. Similarly, "All People are Mammals" is expressed by drawing 'P' inside 'M'. From this diagram we can 'read off' "All Students are Mammals". This diagram can be described by an abstract syntax, which is a list of zone descriptions: $\emptyset$, M, MP, and MPS. For instance, $\emptyset$ describes the zone that is outside all three curves, and MP describes the zone that is inside 'P' and 'M' but outside 'S'. Other Euler diagrams can represent the same syllogism. They have zones for $\emptyset$, M, MP, and MPS as well as additional zones which are shaded to assert no elements are in the set they represent, an example of which is illustrated on the right of Fig. 2. The left diagram is considered, by Gurr [8], to be the most effective, precisely encapsulating the semantics of the syllogism. This diagram is said to be *well-matched* to its meaning as it has no additional zones. Gurr explains:

> "The transitive, irreflexive and asymmetric relation of set inclusion is expressed via the similarly transitive, irreflexive and asymmetric visual of proper spatial inclusion in the plane [8]."

To summarise, we define guide 1 for Euler diagram drawing:

Guide 1 (Well-matched). *Draw well-matched Euler diagrams (i.e. no extra zones).*

Being able to draw a well-matched diagram only solves part of the problem of ascertaining an effective layout. Fig. 3 illustrates two further well-matched

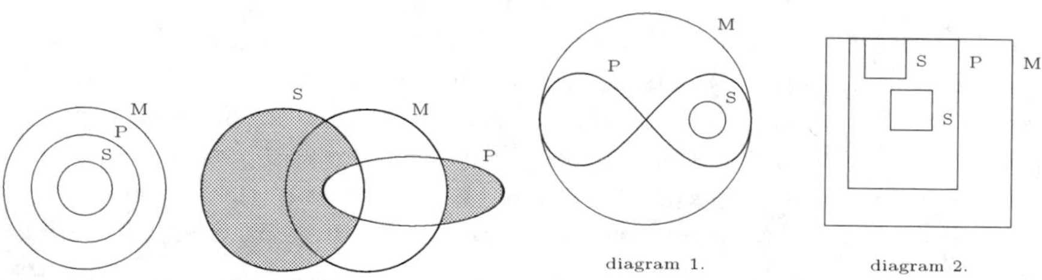

Fig. 2. Choices of abstract syntax

Fig. 3. Choices of topology

diagrams. However, both diagrams exhibit topological properties, our second category of choice type, known to inhibit effectiveness. Diagram 1 shows the following topological properties. Curves 'P' and 'M' illustrate brushing points, defined as two or more curves which meet but do not cross at a point. Curves 'P' and 'M' give rise to a disconnected zone, defined as a zone that consists of one or more minimal regions. Curve 'P' is a non-simple (self-intersecting) curve. In diagram 2, curve 'P' and one curve labelled 'S' have concurrency, defined as two or more curve segments sharing the same line. Curves 'M', 'P' and one labelled 'S' illustrate a triple point, defined as three curves meeting at a point. The two curves labelled 'S' illustrate duplicated curve labels, where two or more curves represent the same set. Diagrams exhibiting none of these properties are *well-formed*. Rodgers et al.'s study [18] considers the impact of well-formed properties, summarised here as guide 2:

Guide 2 (Well-formed). *Draw well-formed Euler diagrams.*

Irrespective of laying out well-matched and well-formed Euler diagrams, there still exist numerous graphical choices to be made, our third category of choice type. Benoy and Rodgers [2], with their work on aesthetics, acknowledged the importance of making the correct graphical choices when drawing Euler diagrams. They conducted a study that focused on the jaggedness of curves, zone area equality and the closeness of one closed curve to another. To summarise their results, we define three further guides:

Guide 3 (Smooth curves). *Draw Euler diagrams with smooth curves.*

Guide 4 (Zone area equality). *Draw Euler diagrams with zone area equality.*

Guide 5 (Diverging lines). *Draw Euler diagrams with diverging lines.*

Benoy's and Rodgers' work, while valuable, remains limited. There are many other graphical choices that might be considered. Bertin [3] identifies both retinal and planar variables, which constitute a variety of graphical choices, to which we are known to be perceptually sensitive. With respect to planar variables, we established that the effect of an Euler diagram's orientation does not impact on users' comprehension [4], leading to:

Guide 6 (Orientation). *Draw Euler diagrams without regard to orientation.*

Retinal variables include shape and colour. With resect to shape, Figs 4, 5, 6 and 7 illustrate four equivalent diagrams each adhering to guides 1 to 6 but drawn using different shapes. None of the aforementioned automated layout methods have been shown to adhere to all six guides. The remainder of this paper describes our study undertaken to establish whether the shape of a closed curve – in particular squares, circles, ellipses and rectangles – affects users' comprehension.

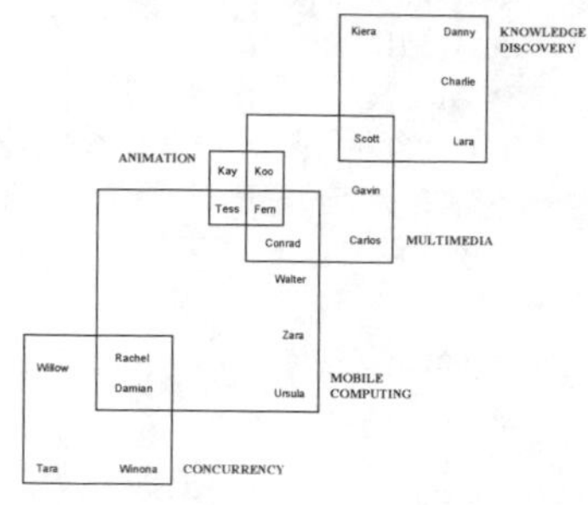

Fig. 4. Type 1 diagram: squares

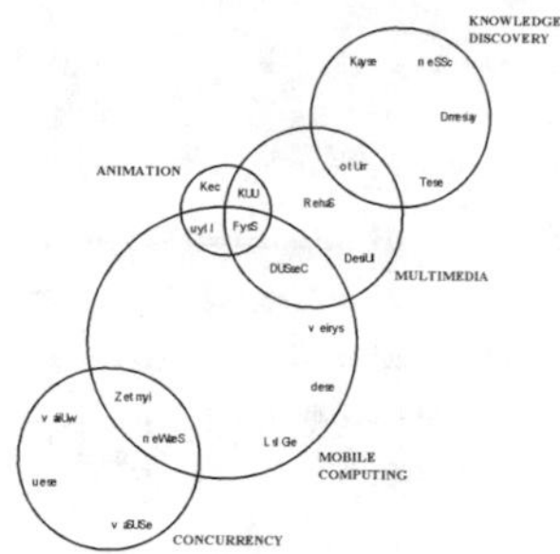

Fig. 5. Type 1 diagram: circles

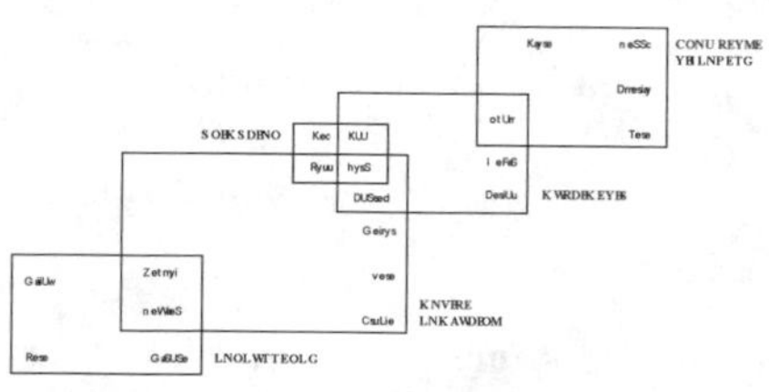

Fig. 6. Type 1 diagram: rectangles

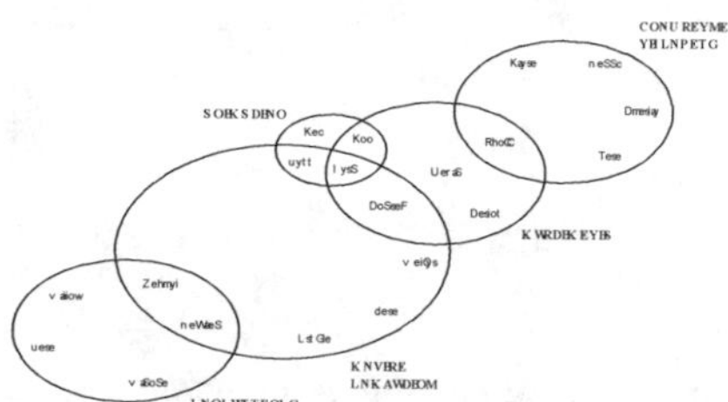

Fig. 7. Type 1 diagram: ellipses

3 Experimental Design

For the purposes of this study, congruent with previous studies [4,11,15,16,17], we view comprehension in terms of task performance: one diagram is more comprehensible than another diagram if users can interpret it, on average, more quickly and with fewer errors. We compare squares, circles, ellipses and rectangles. These four shapes were identified because they are pervasive in existing Euler diagram layout work and are widely used. We adopted a between group design consisting of four participant groups. All participants were asked questions of 18 Euler diagrams. If shape impacts on comprehension then we would expect to see, for some diagram, significant differences between time taken or errors accrued. From this point forward, all Euler diagrams in the paper and are scaled versions of those used in the study. Next, we discuss a series of factors considered during the design. All of the diagrams in the study were drawn sensitive to the six layout guides and the following drawing conventions, to ensure that different diagrams had consistent layout features:

1. all diagrams were drawn using 3 sized curves: small, medium and large,
2. the medium and large curves were scaled 200% and 300%, respectively, relative to the small curve,
3. rectangles and squares were drawn with their sides parallel to the x and y axes and, similarly, so were the major and minor axes of the ellipses,
4. each rectangle and ellipse was drawn adhering to the golden ratio,

5. all closed curves were drawn with a 2 pixel stroke width,
6. all diagrams were monochrome, drawn in an area of 810 × 765 pixels,
7. the curve labels were written using upper case letters in Times New Roman, 14 point size, font in bold,
8. data items were written using lowercase letters, except that the first letter was capitalised, and with Ariel 12 point size font,
9. each curve label was positioned closest to its corresponding curve, and
10. data items were evenly distributed within each zone.

We also required a variety of diagrams to be drawn. To this end, diagrams were drawn pertaining to the following three characteristic types:

1. Type 1: 5 curves (1 large, 3 medium and 1 small), 11 zones and 20 data items (Figs 4 to 7),
2. Type 2: 7 curves (2 large, 3 medium and 2 small), 15 zones and 30 data items (Fig. 8) and
3. Type 3: 9 curves (3 large, 3 medium and 3 small), 19 zones and 40 data items (Fig. 9).

In our study, and congruent with [18], a real-world scenario was employed as it was regarded pertinent to the reader. Further, abstract representations were considered a barrier to the participants understanding. Therefore, information was visualized about fictional university modules and the students. It was anticipated that participants used in the study would be university students and therefore would have a reasonable understanding of this information. The module names were based on those commonly found in British undergraduate computing courses. Student names were first names only, a mixture of both male and female names, and reflected a variety of ethnicities. Following [18], three styles of question were specified, 'Who', 'Which' and 'How', that allowed us to elicit the following type of information:

1. Who is taking ANIMATION, MULTIMEDIA and MOBILE COMPUTING?
2. Which module is being taken by 5 students?

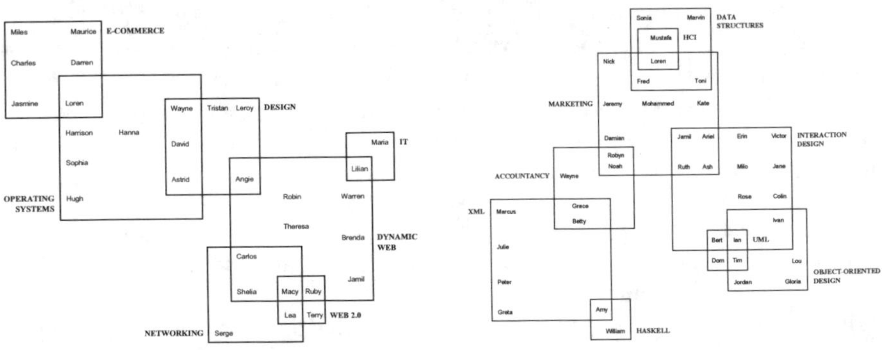

Fig. 8. Type 2 diagram: squares

Fig. 9. Type 3 diagram: squares

3. How many students are taking DATA STRUCTURES and HCI but not MARKETING?

The above questions were asked of Figs. 4, 8 and 9 respectively. All questions had a choice of five answers, a unique one of which was correct. The diagrams were divided equally into the three characteristic types, each allocated two of each question style.

Initially, 18 diagrams were drawn using squares and data items were added to each diagram. The diagrams drawn for the other three shapes were equivalent in terms of their underlying description and topology as those drawn using squares. Given a curve in a 'squares' diagram, the corresponding curves in the other three diagrams had the same area. Careful attention was given to how the syntax was positioned: the placement of data items and curves labels was preserved, as much as possible, across equivalent diagrams.

For the collection of performance data, we used a software tool (called the research vehicle) to display the diagrams and questions, to gather answers and the time taken. The time taken to answer a question was determined from the instant a question was presented until the instant a participant had selected an answer to the question. Each time the participant answered a question, the research vehicle would ask them to indicate when they were ready to proceed to the next question, thus allowing a pause between questions. There was a maximum time limit of two minutes for each question to ensure that the experiment did not continue indefinitely. The 18 diagrams were presented in a random order.

To collect preferential data, participants were asked to rank four diagrams based on their shape. A scale of 1 to 4 was employed with 1 being the most preferred shape and 4 being the least preferred shape. If desired, shapes could be ranked equally. Participants were also required to explain the reasoning for their ranking. Initially, participants were asked to rank each diagram based on their aesthetic preference. They were then asked to rank each diagram based on how they perceived the ease of answering a 'Who', 'Which' and 'How' style question respectively.

4 Experiment Execution

Having conducted two pilot studies and implemented a number of minor adjustments, 80 participants were recruited for the main study. All 80 participants were randomly allocated to groups each with 20 participants. They were all students from the University of Brighton's School of Computing, Engineering and Mathematics and they spanned both undergraduate and postgraduate levels. The experiment was performed within a usability laboratory which affords a quiet environment free from noise and interruption. The same computer and monitor was used by each participant. All participants were alone during the experiment, in order to avoid distractions, with the exception of an experimental facilitator who was present throughout. All participants that took part in the study successfully completed the experiment. The experiment took approximately 1 hour per participant and they were given £6 for their contribution to the research. Next we discuss the four phases of the experiment.

The first phase of the experiment was initial training. All participants were asked whether they were familiar with Euler diagrams to which the response was no but some stated they had seen Venn diagrams. Consequently, all participants were treated as having no previous experience of Euler diagrams and given the same training. Training began by introducing participants to the notion of Euler diagrams and the types of questions to be asked. This was achieved using hard copy printouts of the diagrams, one for each style of question. Participants were given a few minutes to study the diagrams and questions, after which the experimental facilitator explained how to answer the questions. The second phase of the experiment provided participants with further training on the notion of Euler diagrams as well as how to use the research vehicle. Participants were presented with six questions, one at a time. If a question was answered incorrectly the facilitator went through the question with the participant. The third phase of the experiment is where we collected performance (time and error) data. Lastly, if desired, participants took a short break before entering the final, fourth phase of the study where we collected preferential data.

5 Statistical Analysis

The following analysis of the performance data is based on 80 participants each attempting to answer 18 questions: $80 \times 18 = 1440$ observations. The grand mean was 20.35 seconds and the mean times taken to answer questions by shape are: squares 19.48; circles 15.96; ellipses 21.96; and rectangles 24.01.

In order to establish if there is significant variation across shapes, we conducted an ANOVA. The results of the ANOVA are summarised in table 1, which uses log (base 10) of time to achieve normality. If $p \leq 0.05$ then we regarded the result as significant. The row for shape, with a p-value of 0.008, tells us that there are significant differences between the mean times of at least one pair of shapes. Hence, we can see that shape impacts user comprehension. The row for question, with a p-value less than 0.001, tells us that there are significant differences between the mean times of at least one pair of questions. This shows robustness, in that the questions are sufficiently varied, requiring different amounts of user effort to answer. The row for shape*question, with a p-value of 0.042, tells us that there are significant differences between the mean times for at least one question and an associated pair of shapes. This allows us to deduce that there is a question for which the mean times taken for the associated four diagrams are significantly different. Again, this reinforces the observation from row 1 that shape impacts user comprehension. Further, when removing timeouts from the data (i.e. when participants failed to provide an answer within the two minutes allowed), of which there were only two, the above results are unchanged.

Next, we performed a Tukey simultaneous test which compares pairs of shapes, to establish whether one mean is significantly greater than another in order to rank the shapes. P-values of less than 0.01 were regarded to be significant, given multiple comparisons being made on the same data. All comparison returned a p-value of less than 0.001 with the exception of the comparison between ellipses and rectangles ($p = 0.066$). Table 2 presents the rankings of shape. We see that circles allow participants to perform significantly faster than with squares which, in turn, are significantly faster than both ellipses and rectangles. However, there was no significant difference in performance between ellipses and rectangles with respect to time taken.

Where there were significant differences between pairs of shapes, the magnitude of these differences is reflected in the following effect sizes, given in table 3. For example, the largest effect size tells us that 73% to 76% of participants were faster interpreting diagrams drawn with circles than the average person using rectangles. By contrast the smallest (significant) effect size is between squares and ellipses at 54% to 58%, which is still a substantial difference. Thus, we conclude that circles are the most effective shape, with respect to time taken, followed by squares and then, jointly, by ellipses and rectangles.

Regarding errors, of the 1440 observations there were a total of 62 errors (error rate: 4%). The errors were distributed across the shapes as follows: squares 6; circles 14; ellipses 16; and rectangles 26. We performed a chi-square goodness-of-fit test to establish whether shape had a significant impact on the distribution of errors. The test yielded a p-value of 0.003. However, we noted an outlier with respect to rectangles. A single participant yielded approximately one third of all the errors accrued under rectangles, thus potentially biasing the results. Moreover, one question accounted for 16 of the errors. Consequently, we treat this p-value with caution. Re-running the test with the outliers removed yields a p-value of 0.094, leading us not to reject the null hypothesis: shape does not impact on error rate. Taking both error rate and time into account we have, therefore, evidence that circles are significantly more effective than squares, ellipses and rectangles.

The following analysis of the preferential data concerning shape is based on 80 participants answering 4 questions. Recall, participants were asked to rank the shapes by aesthetic preference, and the perceived ease of answering the three styles of question. To analyze the preferential data, we performed four Friedman tests. The p-values are: aesthetic 0.013; who 0.006; how 0.049; and which 0.000. Hence there are significant differences between the shapes in all four cases. Consequently, four Wilcoxon signed ranked tests were performed to identify the significant differences between pairs of shapes. As with the pairwise

Table 1. ANOVA for the log of time

Source	DF	MS	F	P
shape	3	1.41310	4.22	0.008
question	17	1.08369	48.88	0.000
shape*question	51	0.03057	1.38	0.042
participant(shape)	76	0.33513	15.12	0.000
Error	1292	0.02217		
Total	1439			

Table 2. Pairwise comparisons

Shape	Number	Mean (log)	Ranking
Circle	360	1.161	A
Square	360	1.224	B
Ellipse	360	1.276	C
Rectangle	360	1.304	C

Table 3. Effect sizes

Shape	Circle	Square	Ellipse	Rectangle
Circle	–	58%-62%	66%-69%	73%-76%
Square	–	–	54%-58%	62%-66%

Table 4. Wilcoxon tests

Shape	Aesthetic	Who	How	Which
Circle	A	A	A	B
Square	B	B	A B	A
Ellipse	B	A B	B	C
Rectangle	A B	A B	A B	A

comparisons above, p-values of less than 0.01 are regarded to be significant because multiple comparisons being made on the same data. The rankings are presented in table 4. To summarize, circles were ranked top in all cases except for the 'Which' case. For 'Which' style questions, participants perceived square and rectangles to be preferable to circles and ellipses. Analysing the qualitative data gathered in phase 4, participants' reason for this preference was that they found it easier to count data items that were "listed" or in a "tabular" form, as it was presented for squares and rectangles, as opposed to data items that were considered to be "randomly" distributed within circles or ellipses; participants' reasons were not primarily based on the shapes of the curves.

To investigate this further, we performed another pairwise comparison restricted to just the time data for 'Which' questions. This revealed that circles were still significantly more effective than all other shapes, whereas rectangles were the least effective shape. There was no significant difference between ellipses and squares. Given that the study was not designed for this analysis, only having six 'Which' style questions asked of the diagrams, it would not be robust to accept this as a significant difference. Thus, a new study may be required to establish whether differences in performance exist when data items are "listed" or in "tabular" form in all shapes, not just for squares and rectangles.

6 Interpretation of Results

Our results suggest that, overall, circles are significantly more effective than the other shapes. They were significantly faster for accessing information and in doing so accrued an insignificant number of errors, comparatively. For all curve shapes, information is accessed from within a closed curve or from zones. With this in mind, we next refer to perceptual theory to provide insight into the manifestation of our results. We place emphasis upon Gestalt principles [13,25] regarded to be central to shape discrimination. We also consider similarity theory [6,9] as it identifies important constraints upon the speed of a visual search.

First, we specifically refer to the principle of good continuation and the time taken, during a visual search, due to any similarity between targets and distracters. The principle of good continuation states that shapes consisting of smooth curves are easier for the eye to follow than those shapes made up of 'hard' or jagged contours. With respect to curve smoothness, we have already seen guide 3 (smooth curves). Circles and ellipses both adhere to this guide. However, squares and rectangles do not because they have corners, meaning that the principle of good continuation is contravened.

Largely because they are smooth, both circles and ellipses give rise to zones that take rather different shapes to the curves; the boundaries of the zones are typically not smooth. In particular, the point at which one circle/ellipse intersects another manifests a sudden large change in good continuation: the two intersecting smooth curves are abruptly discontinued at the point at which they intersect. Large changes in good continuation are said to promote shape discriminability and, during the process of figure-ground segregation, we posit that intersections between circles/ellipses stand out to become salient. By contrast, the intersections in diagrams drawn with squares and rectangles are not salient: they look similar to the corners of the curves, as illustrated in Fig. 4, and are not therefore easily discriminable from the curves. These observations have implications for discriminating the zones from the curves. We posit that, due to the similarity between squares/rectangles and their (rectilinear) zones, the zones and curves are not discriminable. By contrast, the zones in diagrams drawn with circles and ellipses are discriminable.

Now, similarity theory [6] states that search time increases based on two criteria. The first criteria pertains to the degree of similarity between targets and distractors and the second being the degree of similarity between distractors themselves. With respect to our study, targets can be regarded as either closed curves or zones. If a target is a zone, as many were in the study, the rectilinear shape of a target zone, as illustrated in Fig. 4, is very similar in shape to the majority of distractor zones, as well as being similar in shape to square closed curves themselves. Therefore, it is not unreasonable to liken the task of identifying a zone in an Euler diagram drawn using squares or rectangles to that of searching for one square/rectangle among many similar looking shapes. These observations regarding shape disciminability are summarised in the 'Shape Discrimination' column of table 5, where the shapes are listed in order of mean performance time.

As squares performed better than ellipses, this suggests that a further graphical property is more influential than shape discrimination. In addition, circles performed better than all other shapes, leading us to seek a graphical property possessed only by diagrams drawn with circles. The principle of contour completion states that shapes made up of smooth curves and exhibit a constant rate of change are easier to interpolate into shapes than those that do not. As we are dealing with 2-dimensional geometric shapes, only circles can support contour completion as a constant rate of change must be exhibited. Thus, this explains why circles outperform all other shapes and ellipses in particular, captured by the 'Contour Completion' column in table 5.

Table 5. Perceptual properties of shapes

	Shape Discrimination	Contour Completion	Highly symmetric
Circle	Y	Y	High
Square	N	N	Medium
Ellipse	Y	N	Low
Rectangle	N	N	Low

We can explain why squares perform better than ellipses and rectangles by appealing to the principle of proximity. A square has four equal sides at four equal angles and the proximity of each side, relative to one and other, is constant. The principle of proximity states that the strength of grouping between elements and their properties increases as these elemental properties are brought nearer to each other. Unlike squares (and circles), the shape of an ellipse contravenes the principle of proximity. Following the smooth curvature of an ellipse, starting at an antipodal point of a minor axis, the rate of change of its curvature tends the eye away from its centre until we arrive at the antipodal point of a major axis. Similarly, the shorter sides a rectangle tend away from each other as the eye follows the longer sides. These observations, based on the principle of proximity, are embodied by the fact that ellipses and rectangles exhibit only two lines of symmetry while a square exhibits four and a circle exhibits an infinite number. Whilst there are degrees of symmetry, the last column of table 5 crudely identifies circles and squares as highly symmetric as compared to rectangles and ellipses. The higher degree of symmetry possessed by a circle over a square further supports our statistical results.

We have observed our perceptual sensitivity to the graphical properties of shapes in Euler diagrams. In summary, the smoothness of circles permits effective shape discrimination, they support contour completion, and they are highly symmetric, captured in table 5, distinguishing them from the other three shapes. Further, squares only exhibit high symmetry, distinguishing them from ellipses and rectangles. Table 5, supported by perceptual theory, leads us to posit a further three guides, in an order of priority:

Guide 7 (Shape). *Draw Euler diagrams with circles.*

Guide 8 (Symmetry). *Draw Euler diagrams with highly symmetrical curves.*

Guide 9 (Shape Discrimination). *Draw Euler diagrams so that the zones are discernable from the curves via their shape, but not at the expense of symmetry.*

The first of these guides is strongly supported by the statistical results. The remaining guides require further validation, particularly that for shape discrimination which is only weakly supported by our data. In this case, shape discrimination is the only identified perceptual difference between ellipses and rectangles; we could not conclude that a significant difference exists between these shapes.

7 Threats to Validity

Threats to validity are categorized as internal, construct and external [16]. Internal validity considers whether confounding factors, such as carry-over effect, affects the results and, if so, to what extent. Construct validity examines whether the independent and dependent variables yield an accurate measure to test our hypotheses. External validity considers the extent to which we can generalise the results. The following discusses the primary threats to validity that were considered and addressed to ensure the study is robust and fit for purpose. With regard to internal validity, following two factors were among a number that were considered in an attempt to manage potential disadvantages of our study design:

Carry-over effect: in a repeated measure experiment this threat occurs when the measure of a treatment is effected by the previous treatment. To manage this effect a between group design was employed. Each participant group i.e. square, rectangle, circle and ellipse, was exposed to four different treatments.

Learning effect: the learning effect was considered a threat if participants were not given appropriate training prior to the data collection phase. To reduce this effect, training was given to the participants and they attempted questions used before answering the 18 questions used in the statistical analysis.

Next we consider construct validity by focusing on our dependent variables (error rate, false negatives, and time) and independent variables (diagram and shape), respectively, and examine their rigour for measuring comprehension:
Error rate: all diagrams were drawn to adhere to the six orignal layout guides as well as the layout characteristics detailed above. This drawing approach minimised the possibility of confounding variables creeping into each diagram.

False negatives: to minimise false negatives i.e. a participant selecting the wrong answer while reading it to be the correct answer, the similarity of module and student names was minimised during all phases of the experiment.

Time: to ensure the rigour of time measurements, consideration was paid to the precise duration elapsed interpreting a diagram as well as the units employed to measure time. Further, participants used the same PC located in the same laboratory with no applications running in the background.

Diagram: it was considered a threat if participants did not spend time reading and understanding the diagrams. To manage this threat *diversity* was introduced in the diagrams so that participants had to read and understand each diagram before being able to answer the posed question. It was also considered a threat if the diagrams were regarded as trivial; having only a few curves, zones, or data items was deemed insufficient to yield noticeable differences in response times, should they exist. To manage this, diagrams were designed to exhibit an appropriate level of complexity in order to demand cognitive effort.
Shape: it was essential that the process of drawing equivalent diagrams was carefully planned and executed in order to minimise the threat of unwanted variances between pairs of diagrams.

The following factors consider the limitations of the results and the extent to which the they can be generalised, thus examining their external validity:

Curve shape: Shapes were limited to squares, rectangles, circles and ellipses.
Set theoretic concepts: Euler diagrams conveyed set disjointness, subset and intersecting relationships.
Question styles: three styles of questions were asked: 'Who', 'Which' and 'How'.

Participant: participants were representative of a wider student population.

Thus, the results should be taken to be valid within these constraints.

8 Conclusion

The six guides identified in section 2 render effective Euler diagrams. However, irrespective of this guidance, we have observed that the shape of a closed curve significantly affects user comprehension. Consequently, to further improve the effectiveness of an Euler diagram we posited three new guides, focusing on shape, symmetry and shape discrimination. These guides support the common use of circles in both manually drawn Euler diagrams, where they are commonly used [26], and in automated drawing methods such as [22]. Similarly, it may indicate that drawing methods using other shapes, such as rectangles, are less effective. Further, not all data sets can be visualized with Euler diagrams that are both well-formed and drawn using circles. Thus, the two new guides on symmetry and shape discrimination can be employed for drawing diagrams when circles cannot be used.

There is still scope, however, to improve existing drawing methods by gaining further insight into which layout choices lead to more effective diagrams. In particular, there remain numerous graphical properties to which were are known to be perceptually sensitive that have not been given serious consideration when drawing Euler diagrams. As we have demonstrated with shape, the choice of graphical property can have a significant impact on the effectiveness of an Euler diagram. Colour, and its effect on a user's comprehension, is of particular interest when drawing and laying out Euler diagrams. Colour is widely used when automating Euler diagram layout as well as for visualising data. Bertin [3] realised that colour can be used to promote the interpretation of both quantitative and qualitative information. Colour value or lightness is typically prescribed for presenting quantitative information while colour hue is commonly used to represent qualitative information. Consequently, the immediate future direction of our work will be to investigate the best use of colour when visualising data using Euler diagrams.

References

1. Architectural Association, London (April 2013), `http://www.aadip9.net/shenfei/`
2. Benoy, F., Rodgers, P.: Evaluating the comprehension of Euler diagrams. In: 11th International Conference on Information Visualization, pp. 771–778. IEEE (2007)
3. Bertin, J.: Semiology of Graphics: Diagrams, Networks, Maps. University of Wisconsin Press (1983)
4. Blake, A., Stapleton, G., Rodgers, P., Cheek, L., Howse, J.: Does the orientation of an Euler diagram affect user comprehension? In: 18th International Conference on Distributed Multimedia Systems, International Workshop on Visual Languages and Computing, pp. 185–190. Knowledge Systems Institute (2012)
5. DeChiara, R., Erra, U., Scarano, V.: A system for virtual directories using Euler diagrams. Proceedings of Euler Diagrams. Electronic Notes in Theoretical Computer Science 134, 33–53 (2005)

6. Duncan, J., Humphreys, G.: Visual search and stimulus similarity. Psychological Review 96, 433–458 (1989)
7. Farrell, G., Sousa, W.: Repeat victimization and hot spots: The overlap and its implication for crime control and problem-oriented policing. Crime Prevention Studies 12, 221–240 (2001)
8. Gurr, C.: Effective diagrammatic communication: Syntactic, semantic and pragmatic issues. Journal of Visual Languages and Computing 10(4), 317–342 (1999)
9. Healey, C., Enns, J.: Attention and visual memory in visualization and computer graphics. IEEE Transactions on Visualisation and Computer Graphics 18, 1170–1188 (2011)
10. Ip, E.: Visualizing multiple regression. Journal of Statistics Education 9(1) (2001)
11. Isenberg, P., Bezerianos, A., Dragicevic, P., Fekete, J.: A study on dual-scale data charts. IEEE Transactions on Visualization and Computer Graphics, 2469–2478 (2011)
12. Kestler, H.: Vennmaster: Area-proportional Euler diagrams for functional GO analysis of microarrays. BMC Bioinformatics 9(1)(67) (2008)
13. Koffka, K.: Principles of Gestalt Pschology. Lund Humphries (1935)
14. Micallef, L., Rodgers, P.: Drawing Area-Proportional Venn-3 Diagrams Using Ellipses. Tech. Rep. TR-3-11, School of Computing, University of Kent, Canterbury, UK (2011), `http://www.cs.kent.ac.uk/pubs/2011/3118`
15. Purchase, H.: Which aesthetic has the greatest effect on human understanding? In: DiBattista, G. (ed.) GD 1997. LNCS, vol. 1353, pp. 248–261. Springer, Heidelberg (1997)
16. Purchase, H.: Experimental Human Computer Interaction: A Practical Guide with Visual Examples. Cambridge University Press (2012)
17. Riche, N., Dwyer, T.: Untangling Euler diagrams. IEEE Transactions on Visualisation and Computer Graphics 16, 1090–1097 (2010)
18. Rodgers, P., Zhang, L., Purchase, H.: Wellformedness properties in Euler diagrams: Which should be used? IEEE Transactions on Visualization and Computer Graphics 18(7), 1089–1100 (2012)
19. Rodgers, P., Stapleton, G., Howse, J., Zhang, L.: Euler graph transformations for Euler diagram layout. In: Visual Languages and Human-Centric Computing (VL/HCC), pp. 111–118. IEEE (2010)
20. Simonetto, P., Auber, D.: Visualise undrawable Euler diagrams. In: 12th International Conference on Information Visualization, pp. 594–599. IEEE (2008)
21. Simonetto, P., Auber, D., Archambault, D.: Fully automatic visualisation of overlapping sets. Computer Graphics Forum 28(3) (2009)
22. Stapleton, G., et al.: Automatically drawing Euler diagrams with circles. Journal of Visual Languages and Computing 12, 163–193 (2012)
23. Stapleton, G., et al.: Inductively generating Euler diagrams. IEEE Transactions of Visualization and Computer Graphics 17(1), 88–100 (2009)
24. Thièvre, J., Viaud, M., Verroust-Blondet, A.: Using Euler diagrams in traditional library environments. In: Euler Diagrams. ENTCS, vol. 134, pp. 189–202. ENTCS (2005)
25. Wagemans, J., Elder, J., Kubovy, M., Palmer, S., Peterson, M., Singh, M.: A century of gestalt psychology in visual perception: I. perceptual grouping and figure-ground organisation. Computer Vision, Graphics, and Image Processing 31, 156–177 (1985)
26. Wilkinson, L.: Exact and approximate area-proportional circular Venn and Euler diagrams. IEEE Transactions on Visualisation and Computer Graphics 18, 321–330 (2012)

Alternative Strategies in Processing 3D Objects Diagrams: Static, Animated and Interactive Presentation of a Mental Rotation Test in an Eye Movements Cued Retrospective Study

Jean-Michel Boucheix and Madeline Chevret

University of Burgundy, LEAD-CNRS, Dijon, France
Jean-Michel.Boucheix@u-bourgogne.fr,
Madeline.Chevret@hotmail.fr

Abstract. Spatial abilities involved in reasoning with diagrams have been assessed using tests supposed to require mental rotation (cube figures of the Vandenberg & Kruse). However, Hegarty (2010) described alternative strategies: Mental rotation is not always used; analytical strategies can be used instead. In this study, we compared three groups of participants in three external formats of presentation of the referent figure in the Vandenberg & Kruse test: static, animated, interactive. During the test, participants were eye tracked. After the test, they were interrogated on their strategies for each item during the viewing of the replay of their own eye movement in a cued retrospective verbal protocol session. Results showed participants used varied strategies, part of them similar to those shown by Hegarty. Presentation format influenced significantly the strategy. Participants with high performance at the test used more mental rotation. Participants with lower performance tended to use more analytical strategies than mental rotation.

Keywords: Mental rotation, Strategy, Presentation formats, Eye movements.

1 Introduction

Designers, learners, are increasingly working with complex 3D diagrams: for example, students in the medical area often use on screen presentations of virtual organs, sometimes in a user controllable modality. Previous research showed that processing successfully complex diagrams is positively correlated with spatial abilities [1, 2]. In previous research on diagrams processing and comprehension, spatial ability is often assessed with mental rotation test. One of the tests commonly used is the Vandenberg and Kruse test [3] inspired by the principle of the cube figures of Schepard & Metzler, fig. 1. People are shown a standard figure on the left and 4 items on the right. Their task is to decide which of the 4 objects on the right has the same shape as the object on the left. For participants, this task is supposed to involve mental rotation processes. In order to compare the configuration of two figures, the subject needs to create an internal representation of the targeted figure, and is supposed to internally simulate the rotation of the figure to bring it to an angle which allows a comparison with the reference object [1, 2].

T. Dwyer et al. (Eds.): Diagrams 2014, LNAI 8578, pp. 138–145, 2014.

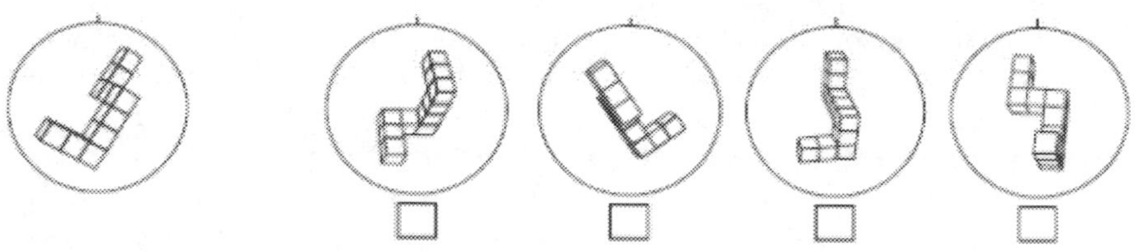

Fig. 1. Example of an item from the Vandenberg and Kruse Mental Rotation test

However, previous research has shown that participants are not always making a mental rotation, but alternative strategies can be used instead [1, 4, 5]. In their study, Hegarty, De Leeuw & Bonura [4], see also Stieff, Hegarty & Dixon [5], used the Vandenberg and Kruse test. Students took a first session of each item of the test. Then gave a think-aloud protocol while solving the items in a second session. Hegarty and colleagues [4, 5] identified several strategies that students used to solve the items. Four main strategies were found. (i) *Mental Rotation* consisted in manipulating or simulating the rotation of the figure. (ii) *Perspective taking* consisted in *"imagining the objects in the problem as stationary, while they moved around the objects to view them from different perspectives"* [5]. Mental rotation and perspective taking are *"imagistic strategies"* [5]. Two others more "analytic" strategies were discovered. (iii) In the "*comparing arm strategy*" participants compared the relative directions of the arms in the standard reference figure to that in each of the four answer choice. (iv) In the "*counting cubes strategy*" participants tried to count the number of cubes in each segment of the reference object and compare that to the other objects in the problem. Participants reported also they used several mixed strategies.

1.1 Strategies and Presentation Format of the 3D Objects

In the present study we followed the line of research initiated by Hegarty and colleagues [1, 2, 4, 5, 6]. The goal was to examine the possible effect of different external presentation formats of the standard reference figure of the Vandenberg test on the strategies used for solving the problem items. Three different presentation formats of each item of the test were designed and proposed to three different groups of participants. In the first format, the left figure was the standard static presentation, fig. 1. In the animated format, the 3D left figure rotated dynamically (and slowly) on itself showing all the different sides of the figure. In the interactive animated format, the participant could rotate the figure on itself, to view all the sides of the object in a user controllable modality. In all formats, the four figures on the right side remained static. The size of the figures was the same in all formats. One main expectation (H1) was that different external formats could elicit varied strategies. Some of them may be different from those found in previous research [4, 5]. Three complementary hypotheses were proposed. (i) (Hc2), the effect of the formats on strategies could be different depending of the participant's performance level at the Vandenberg and Kruse test. (ii) (Hc3) whatever the condition, strategies used at the Vandenberg test could vary according to performance level at this test (high *vs.* low), but also according to the performance level at another different spatial ability test which measures the spatial orientation abilities, a competence found dissociated from mental rotation [6]. Kozhevnikov & Hegarty [6] showed participants with higher spatial ability level used often the *"comparing arm strategy"* than participants with lower ability level. (iii) (Hc4), animated and interactive figures might "facilitate" mental imagery and improve the performance at the Vandenberg test. However, recent results from Keehner, Hegarty, Cohen, Khooshabeh & Montello [7] showed clearly that *"what's matter is what you see, not whether you interact"* with the diagram.

2 Method

2.1 Participants, Experimental Design and Materials

Fifty eight students participated in the study (*M*age = 20.6 years, 46 females). They were randomly assigned to three groups (4 males in each group) in each format: Static presentation *vs.* Animated, *vs.* user-controlled animated presentation of the left reference figure of the Vandenberg & Kruse test. A computer screen version of the 19 items of the test (plus training items) was designed from the French adapted version. Participant's task is to decide which 2 of the 4 objects on the right have the same shape as the object on the left. Eye movements of participants were recorded during the test (Tobii 120 hz). Responses and reaction times to each item were recorded. A think-aloud verbal protocol second session was undertaken after the end of the test. A microphone connected to the computer recorded the verbal explanation given by each participant about its strategy for each item. Finally the spatial orientation test (Kozhevnikov & Hegarty [6]) was used. This test measures the ability to imagine different perspectives or orientations in space (by drawing arrows). Response accuracy is measured with the size of the angular deviation from the right perspective.

2.2 Procedure

After having completed the training items of the Vandenberg test, each participant eyes were calibrated for eye tracking recording. Then each participant individually undertook the test itself (one item at once). For control matter of the time across the formats, the presentation time of each item was fixed; here 30 sec. per item was given. Participants were told to take their decision as fast as possible. After the test, each participant was shown the replay (slow speed) of his (her) own eye movements. Participants were told to rely on the replay of eye movements, for each item, to think-aloud and explain in detail their own procedure used during the test to find the right figures (25-35 min./subject). This *cued retrospective verbal protocol procedure* based on the traces of dynamic eye movements could strengthen the reliability of the participant's explanation. Finally, each participant completed the spatial orientation test.

2.3 Coding and Analyses Criteria for Dependant Variables

The following variables were analyzed (i) the verbal explanations about strategies (two independent raters), (ii) the performances (means, *SD,* medians) at the Vandenberg test (each correct response was awarded 1 point), (iii) and reaction time (per item). Eye movements were also analyzed as well as the performances at the spatial orientation test. This paper will be focused on the presentation of the results about (i) the strategies at the mental rotation test; (ii) their relation with presentation formats and (iii) with the performance levels at the Vandenberg test and at the spatial orientation test.

3 Results

Varied strategies were found, and the number of times they occur counted. Some strategies were similar to those described by Stieff & al., [5], others were specifically related to the different presentation formats of the standard reference figure.

3.1 Strategies and Presentation Formats

Three "types" of strategies have been distinguished. The First type called "global" strategy was composed of five strategies some of them based on the use of mental rotation of the object: (i) *Mental rotation.* Subjects manipulate and imagine the rotation of the figure as exemplified by the following eye movements' cued retrospective protocol: *"I turned it* (the figure)*, I looked if it corresponded when I turned it in my head"*. (ii) *Looking at the middle* of the figure and turning around it, which seems similar to the *"perspective taking"* strategy. The following protocol is an example: "(here) *I look at all the figure, I look around at the overall shape and I compare (to the other figures)"*. (iii) *Mental Completion of the rotation* from the animation. This strategy is taken only by participants' in the animated format. The subjects used the initiation of the rotation of the figure as "a priming" for completing mentally themselves the rotation of the figure. This is exemplified by the following protocols: "*When it* (the figure*) turns, I look if it falls similarly and then I incline it in my head*"; *"I wait until it is running and I anticipate the rotation"* (iv). *Matching*. In the animated condition, participants looked at the external rotation of the left figure and compared each of the position (shape) of the object with the figures on the right. Two examples of participant's protocols using this strategy are as follow: *"I waited and when I saw that the figure arrived at the same position (shape) as another, I selected it"; "I waited until it turns in the same figure"*. (v). *User-control of the rotation.* In the interactive animated format, participants initiated and controlled the rotation of the figure step by step. This strategy was used for aligning the shape of the standard figure on the shape (position) of the figures on the right. The following protocol excerpts are examples: "(here) *I try to put it (the figure) in the same position*"; *"I try to place it to see all*". Secondly, two more "analytical" strategies were found. (i) "*Comparing arm axes*". This strategy was similar to Hegarty & al. [1, 5]. Participants compared the direction of the cubes arms of the left figure to the directions of the arms of the figure on the right. Here is an example of a protocol: *"I looked more at the orientation ends, I looked at the ends and I tried to see where it was going; here, I focused on the bottom part to see and compare to the top part whether it went over to the right or left most"*. (ii). *"Counting cubes"*. This strategy is also similar to the strategy found by Hegarty [1, 5]. Participants counted the number of cubes of the standard figure and compared it with the cubes of the other figures as exemplified in the following protocol: *"I counted the cubes, I looked at the layouts and I counted"*. Finally, a third strategy type is the use of *Mixed strategies*. In the mixed strategy type, participants used at least an analytical and a global strategy. The number of time each strategy was reported to be used by each subject was counted for each of the 19 items of the test. The data are presented in Table 1.

Table 1. Mean number of times (and SD) each type of strategies was used across conditions

Strategy/ Format	Mental Rotation	Looking Middle	Complet. Rotation	Match	Control rotation	Comp. Arms	Count Cubes	*Mixed*
Static	14.65 (5.16)	3.65 (4.40)	0 (0)	0 (0)	0 (0)	9.60 (6.83)	4.20 (5.97)	*9.70 (6.83)*
Animated	0.79 (1.93)	1.53 (2.96)	0.47 (1.42)	8.90 (5.54)	0 (0)	9.31 (5.42)	7.00 (6.93)	*5.68 (4.70)*
Interactive	6.15 (5.56)	1.68 (3.05)	0 (0)	0 (0)	13.05 (5.69)	10.00 (6.29)	2.05 (2.77)	*3.37 (4.24)*

One way ANOVAs comparing the single individual strategies by condition were performed. For static format, the analysis showed significant differences between strategies in favor of mental rotation, $F\ (3, 57) = 16.1$, $p < .001$, $\eta^2 = .45$. For the animated format, significant differences were found between strategies in favor of matching and comparing arms, $F\ (5, 90) = 14.80$, $p < .001$, $\eta^2 = .45$. For the interactive format also differences were observed in favor of the control of rotation and comparing arms ($F\ (12, 330) = 23.80$, $p < .0001$, $\eta^2 = .46$). This result supports H1. A mean of 3.50 ($SD = 0.97$) strategies per subject was used. A complementary ANOVA on the number of mixed strategies showed the differences between formats was significant ($F\ (2, 55) = 6.86$; $p = .002$ $\eta^2 = .20$).

3.2 Performances at the Test and Presentation Formats

Results for scores and reaction times at the Vandenberg test are presented in Table 2.

Table 2. Mean scores (and SD) and time at the French version of Vandenberg & Kruse test

Format	Static	Animated	Interactive
Scores (out of 38)	18.45 (8.90)	17.21 (8.69)	16.26 (6.03)
Time (in second per item)	19.97 (5.07)	17.91 (3.88)	24.42 (2.39)

The analysis of performances and times results (ANCOVA with time as covariate factor, groups as between subject factor and performance as the dependant factor) indicated that there was no effect of the presentation formats on performances (F (2, 54) = 1.89, p = .16, η^2 = .06). A significant effect of reaction times on the scores was found (F (1, 54) = 5.72, p = .02, η^2 = .10). For results on times, a one factor ANOVA (formats as the between subject factor and reaction times as the dependant variable) revealed a significant effect of formats (F (2, 55) = 13.44, p< .0001, η^2 = .33). Regarding the scores at the spatial orientation test, a one factor ANOVA (with formats of the Vandenberg test as the between factor and the mean size of the angular deviation from the right perspective at orientation test as the dependant variable) showed there was no effect of presentation formats (F (2, 55) = 0.75, *ns*.; M*Static*, 32° (SD = 18.10), M*Animated*, 40.3° (SD = 23.3°); M*Intercative*, 35.23° (SD = 22.01).

3.3 Strategies and Spatial Ability Performances

To explore the relations between the strategies and performances at the Vandenberg test, the distribution of scores at this test was split into high and low performances groups (median for each presentation format) Table 3. Further, in order to examine the relations between strategies used in the Vandenberg "mental rotation" test and performances at spatial orientation test the same technique was undertaken, Table 4.

Table 3. Mean number of times (SD) each type of strategy was used for high and low spatial groups at the Vandenberg test

Strat/Gr		Rot	Look M	Compl	Match	Contr	Arms	Count	Mixed
Stat.	H	15.6 (4.87)	3.08 (3.05)	0 (0)	0 (0)	0 (0)	9.33 (6.64)	2.33 (3.34)	9.25 (6.82)
	L	13.25 (5.6)	4.50 (6.05)	0 (0)	0 (0)	0 (0)	10.0 (7.56)	7.0 (8.01)	10.37 (7.27)
Anim.	H	1.00 (1.91)	0.28 (0.75)	0.57 (1.5)	9.86 (5.58)	0 (0)	9.14 (4.81)	7.14 (9.2)	5.0 (4.86)
	L	0.66 (2.01)	2.25 (3.54)	0.42 (1.44)	8.33 (5.69)	0 (0)	9.42 (5.69)	6.92 (5.71)	6.08 (4.78)
Intera	*H*	*8.37 (4.8)*	*1.50 (3.46)*	*0 (0)*	*0 (0)*	*13.00 (6.3)*	*5.37 (4.17)*	*0.75 (1.38)*	*3.12 (3.39)*
	L	*4.54 (5.73)*	*1.81 (2.89)*	*0 (0)*	*0 (0)*	*13.09 (5.52)*	*13.36 (5.44)*	*3.00 (3.19)*	*3.54 (4.92)*

For the Vandenberg test, a repeated measures ANOVA was conducted on the number of times strategies were used, with spatial group's levels at the test as the between subjects factor and the strategy types (the seven singles) as the within factor. As expected (H3c), the interaction between group's levels and strategies was significant F (6, 336) = 2.45, p = .025, η^2 = .05. High level performers used more Mental rotations (F (1,56) = 5.60, p = .021); and less analytical strategies (comparing arms axes, counting cubes) (F (1,56) = 7. 04, p = .01) than lower level performers. A similar analysis was conducted with spatial group's level at the orientation test, table 4. The interaction between group's levels and strategy types was not significant (F (6,336) = 0.43, *ns*). Consistent with previous research by Hegarty & al., [1, 2, 4, 5, 6, 7], this result tends to confirm that the orientation test measures a specific competence which is different from mental rotation. The correlation between the two tests, was significant (r (df 58) =-.53, p < .05), this relation was not very high.

Table 4. Mean number of times (SD) each type of strategy was used at the Vandenberg test for high and low performance groups at the Hegarty's spatial orientation test

Strat/Gr		Rot	Look M	Comp 1	Match	Contr	Arms	Count	Mixed
Stat.	H	13.8 (6.52)	2.7 (2.98)	0 (0)	0 (0)	0 (0)	11.9 (6.8)	2.2 (3.76)	10.2 (7.71)
	L	15.5 (3.47)	4.6 (5.48)	0 (0)	0 (0)	0 (0)	7.3 (6.36)	6.2 (7.22)	9.2 (6.21)
Anim.	H	0.70 (1.63)	1.00 (1.94)	0.40 (1.26)	9.70 (6.25)	0 (0)	10.40 (5.5)	7.30 (8.42)	6.10 (5.91)
	L	0.89 (2.31)	2.11 (3.85)	0.55 (1.66)	8.0 (4.84)	0 (0)	8.11 (5.39)	6.67 (5.13)	5.22 (3.15)
Interact	*H*	*5.40 (5.6)*	*2.90 (3.87)*	*0 (0)*	*0 (0)*	*12.50 (6.83)*	*8.50 (5.19*	*1.60 (1.83)*	*2.40 (2.54)*
	L	*7.00 (5.72)*	*0.33 (0.5)*	*0 (0)*	*0 (0)*	*13.67 (4.41)*	*11.66 (7.28)*	*2.55 (3.6)*	*4.44 (5.54)*

4 Conclusion and Further Analyses

The goal of this study was to compare the effect, on the use of strategies, of the conditions of presentation of the standard figure in the Vandenberg test: static, animated and user controllable formats. Participants were interrogated on their strategies during the viewing of a replay of their eye movements. Results showed participants used varied strategies, part of them similar to those shown by Hegarty and others new. An animation favors matching; an interactive presentation favors the control of rotation. Participants with high performance at the test used more mental rotation and less analytical strategies than participants with lower performance.

References

1. Hegarty, M.: Components of spatial intelligence. In: Ross, B.H. (ed.) Psychology of Learning and Motivations, vol. 52, pp. 265–297. Elsevier, Amsterdam (2010)
2. Hegarty, M., Waller, D.: A dissociation between mental rotation and perspective-taking spatial abilities. Intelligence 32, 175–191 (2004)
3. Vandenberg, S.G., Kruse, A.R.: Mental rotations, a group of three dimensional spatial visualization. Perceptual and Motor Skills 47, 599–604 (1978)
4. Hegarty, M., De Leeuw, K., Bonura, B.: What do spatial ability test really measure. In: Proceedings of the 49th Meeting, Psychonomic Society, Chicago, IL (2008)

5. Stieff, M., Hegarty, M., Dixon, B.: Alternative strategies for spatial reasoning with diagrams. In: Goel, A.K., Jamnik, M., Narayanan, N.H. (eds.) Diagrams 2010. LNCS (LNAI), vol. 6170, pp. 115–127. Springer, Heidelberg (2010)
6. Koshevnikov, M., Hegarty, M.: A dissociation between object manipulation and perspective taking spatial abilities. Memory & Cognition 29, 745–756 (2001)
7. Keehner, M., Hegarty, M., Cohen, C., Khooshabeh, P., Montello, D.: Spatial reasoning with external visualizations: What matters is what you see, nor whether you interact. Cognitive Science 32, 1099–1132 (2008)

Visualizing Sets: An Empirical Comparison of Diagram Types

Peter Chapman[1], Gem Stapleton[1], Peter Rodgers[2], Luana Micallef[2], and Andrew Blake[1]

[1] University of Brighton, UK
{p.b.chapman,g.e.stapleton,a.l.blake}@brighton.ac.uk
[2] University of Kent, UK
{p.j.rodgers,l.micallef}@kent.ac.uk

Abstract. There are a range of diagram types that can be used to visualize sets. However, there is a significant lack of insight into which is the most effective visualization. To address this knowledge gap, this paper empirically evaluates four diagram types: Venn diagrams, Euler diagrams with shading, Euler diagrams without shading, and the less well-known linear diagrams. By collecting performance data (time to complete tasks and error rate), through crowdsourcing, we establish that linear diagrams outperform the other three diagram types in terms of both task completion time and number of errors. Venn diagrams perform worst from both perspectives. Thus, we provide evidence that linear diagrams are the most effective of these four diagram types for representing sets.

Keywords: set visualization, linear diagrams, Venn diagrams, Euler diagrams.

1 Introduction

Sets can be represented in both sentential (textual) and visual forms and the latter is often seen as cognitively beneficial but only if the visual form is effective [1]. To-date, various different visualizations of sets have been proposed, but there is little understanding of their relative effectiveness. This paper addresses this knowledge gap by empirically comparing four visualizations: Venn diagrams, Euler diagrams with shading, Euler diagrams without shading, and linear diagrams. We do not consider the relative effectiveness of these diagrams with traditional sentential notations (such as $(A \cap B) - C = \emptyset$) as it was felt the latter would be too hard for many people to understand in a short time frame.

The Venn and the Euler variants will be familiar to most readers. All three use curves to represent sets: the area inside a curve with label A represents the set A. *Venn diagrams* (upper left, Fig. 1) require that every possible intersection between curves is present. In order to assert that sets are empty, the appropriate regions (often called zones) are shaded. *Euler diagrams with shading* (upper right, Fig. 1), by contrast, can either not include or shade zones which represent the empty set. *Euler diagrams without shading* (lower left, Fig. 1) provide a

T. Dwyer et al. (Eds.): Diagrams 2014, LNAI 8578, pp. 146–160, 2014.

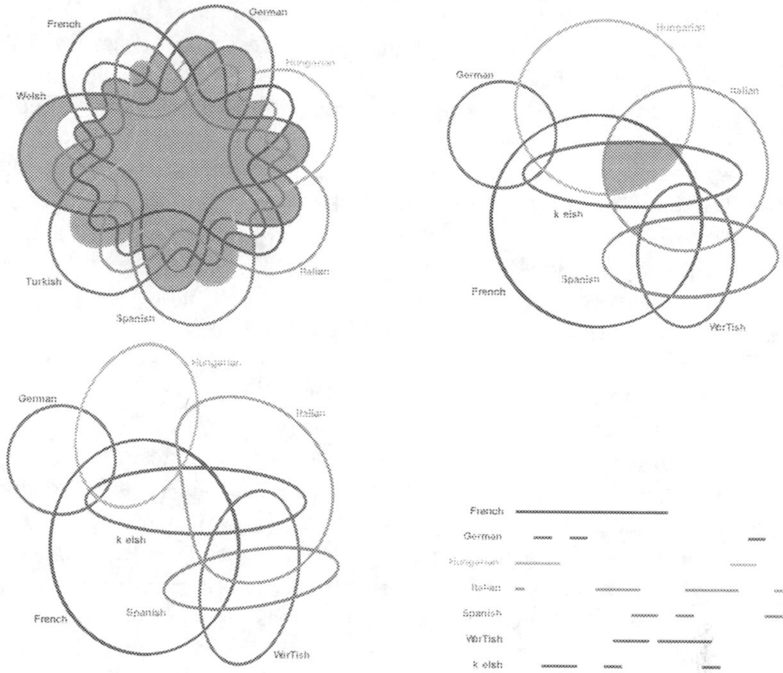

Fig. 1. The four diagram types considered in this paper

minimal representation of the underlying sets: all, and only, zones that represent non-empty sets are included. Minimality of representation can necessitate the presence of diagrammatic features considered sub-optimal for cognition, such as three curves meeting at a point [2]; in this example, Hungarian, Italian and Welsh form such a 'triple' point. Across these three diagram types, any pair of diagrams expressing the same information will have the same number of unshaded zones; only the number of shaded zones can differ.

Linear diagrams were introduced by Leibniz [3], with parallel bargrams [4] and double decker plots [5] being similar. Each set is represented as one or more horizontal line segments, with all sets drawn in parallel. Where lines overlap, the corresponding intersection of sets contains an element that is not in any of the remaining sets. Moreover, between them all of the overlaps represent all of the non-empty set intersections. As an example, consider the linear diagram in the lower right panel of Fig. 1. Since there is a region of the diagram where the lines French, Italian, Turkish and Welsh (and only those lines) overlap, the intersection of those four sets, less the union of Spanish, German and Hungarian, is non-empty. Further, it is not the case that Hungarian is a subset of French, because part of the line representing Hungarian does not overlap with that for French: although one segment of the Hungarian line completely overlaps with the French line, the other segment does not. Also, because there is no overlap involving all seven sets, we can infer that no element is in all of the sets.

Most existing research on diagram effectiveness evaluates notations against some cognitive framework for what *should* constitute a good diagram, such as the Physics of Notations [6] or empirically determines which aspects of a *particular* notation are most effective (e.g. [2],[7],[8]). There are some exceptions, such as [9] which shows that the most effective diagram type is task dependent. Following [9], we perform an empirical comparison of different notations. Similar studies do exist, and, by necessity, are task specific. In [10] Euler diagrams, Venn diagrams and linear diagrams were compared in the context of syllogistic reasoning (i.e. the interactions between three sets). In a more general reasoning context, a study between Euler and Venn diagrams was undertaken in [11]. In both studies, Venn diagrams were least effective, and in [10], linear diagrams were as effective as Euler diagrams. Our study is the first to assess the effectiveness of the four diagram types for visualizing sets and the first of its kind to be conducted on a large and diverse group of participants through crowdsourcing.

The structure of the paper is as follows. In section 2 we describe the experimental design, including drawing criteria for the diagrams and the crowdsourcing data collection methodology. Further details on maintaining quality of data are given in section 3, and the results are analyzed in section 4. In section 5 and section 6, we provide a discussion of the results and their validity, respectively. Finally, we conclude in section 7. All of the diagrams used in our study, and the data collected, are available from `www.eulerdiagrams.com/set`.

2 Experimental Design

We are aiming to establish the relative impact on user comprehension of four different diagrams types that visualize sets. For the purpose of this study, as with previous studies, e.g. [12–14], we measure comprehension in terms of task performance using time and error data. We adopted a between group design with one participant group for each diagram type to reducing learning effect. A further advantage of a between groups design was that participants only had to be trained in one notation. We recorded two dependent variables: the time taken to answer questions and whether the answer was correct. Each participant group was shown a set of diagrams about which they were asked a set of questions. If diagram type impacts on comprehension then we would expect to see significant differences between time taken to answer questions or error rates.

2.1 Sets to Be Visualized

Each diagram represented a collection of sets with varying relationships between them. Each such collection involved either three, five, or seven sets, and we had six collections of each number of sets (thus, 18 in total). This was to ensure that the questions exhibited a range of difficulties, thus requiring varying levels of cognitive effort to answer the questions. The study included 18 questions – one for each collection of sets – and, therefore, 18 diagrams of each of the four types. Further, we wanted to ensure that, for each question, the Venn diagram, Euler

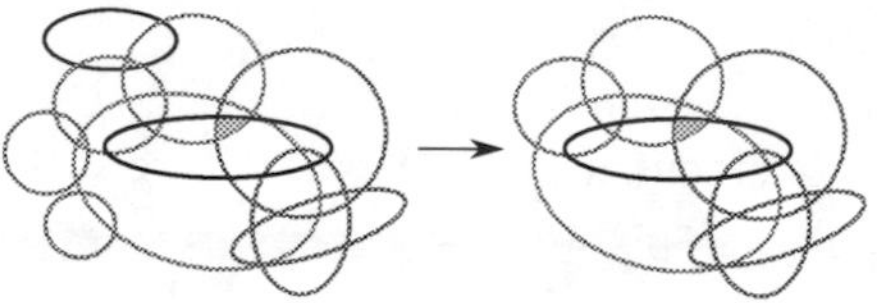

Fig. 2. Generating diagrams from real examples

diagram with shading, and the Euler diagram without shading were different from each other. For diagrams representing three sets, we chose six combinations of three sets that ensured the diagrams were different.

The choice of diagrams representing five sets and seven sets, respectively, was not significantly limited. Rather than generate random combinations of sets, which might be unlikely to arise in real situations, we turned to Google images to choose diagrams for the five and seven set cases. We searched for examples of all diagram types that people had drawn to visualize data; we could not find any actual examples of linear diagrams and we excluded any diagrams that had been drawn by the authors of this paper. Some of the diagrams returned in this search represented more than five or seven sets. In these cases, some sets were removed, to yield the required number, in such a way as to keep the diagram connected (the curves formed a connected component) whilst ensuring that the number of zones remained as high as possible. This set removal method meant that the diagram was, roughly speaking, close in complexity to the original. An example of a diagram found through Google images can be seen on the left of Fig. 2 (re-drawn here for copyright reasons, approximating the colours used and adding shading to the zones representing empty sets; see `http://govwild.hpi-web.de/images/govwild/overlapLegalEntity.png`). It represents ten sets. Three sets were identified for removal, to yield a 7-set diagram, shown on the right. The reduced 7-set diagram corresponds to the four diagrams shown in Fig. 1 that were used in the study.

Since displaying real data can lead to bias (through the potential for prior knowledge), the names of the sets were changed to a pseudo-real context, focused on three domains: film collections, subjects studied, and languages spoken. It was anticipated that participants would have a reasonable preconception of this kind of information, but no prior knowledge of the (fictional) information visualized.

2.2 Study Questions

As this study aims to establish which of the four notations is most effective for accessing information about sets, statements made about the diagrams were chosen to adhere to the following *templates*:

1. **Simple question templates:**
   (a) **Intersection:** Some $\langle$elements in X are$\rangle$ also $\langle$in $Y\rangle$.
   (b) **Subset:** Every $\langle$element in X is$\rangle$ also $\langle$in $Y\rangle$.
   (c) **Disjointness:** No $\langle$element in X is$\rangle$ also $\langle$in $Y\rangle$.

2. **Complex question templates:**
   (a) **Intersection:** Some ⟨elements in X and Y are⟩ also ⟨in Z⟩.
   (b) **Subset:** Every ⟨element in both X and Y are⟩ also ⟨in Z⟩.
   (c) **Disjointness:** No ⟨elements in both X and Y are⟩ also ⟨in Y⟩.

Every statement was prefixed with "This diagram shows ⟨some contextual text goes here⟩. Is the following statement true?" and participants were asked to choose the answer 'yes' or 'no'. The statement templates were populated with context-specific text by randomly choosing the sets for X, Y and, where necessary, Z. The actual phrasing of the individual statements was far less mathematical in style than the templates just given. One of each type of complex question is given here, one from each domain. The four diagrams associated with question 3 are in Fig. 1:

1. **Intersection:** This diagram shows the subjects studied by Mrs Robinson's students. Is the following statement true? Some of those studying both Geology and History are also studying Music.
2. **Subset:** This diagram shows the classifications of films owned by Grace. Is the following statement true? Every film classified as both Action and Thriller is also classified as Period.
3. **Disjointness:** This diagram shows the languages spoken by employees at Interpro Translators. Is the following statement true? No one who speaks both Welsh and Italian also speaks Turkish.

2.3 Diagram Specification and Layout Characteristics

All of the diagrams were drawn sensitive to various layout guides, used to minimize variability across types. These guides also helped ensure that each diagram type was not compromised by bad layouts, but to-date only some of these guides have been verified by empirical testing. The following conventions were adopted:

1. Curves/lines were drawn with a 6 pixel stroke width.
2. Diagrams were drawn in an area of 810 by 765 pixels.
3. Curves/lines representing a particular set were given the same colour. No two sets, appearing in the same diagram, had same colour.
4. Set names had an using upper case first letter in Sans font, 24 point size.
5. Set names were positioned closest to their corresponding curve/line and took the same colour.
6. The set names used in any one diagram started with a different first letter.
7. The same colour (grey) shading was used across diagrams where relevant.

A palette of seven colours was generated using colorbrewer2.org (accessed November 2013), in a similar fashion to [15]. Colour generation using the Brewer colour palette is recognized as a valid approach for empirical studies, such as in the context of maps [16]. So that the colours were distinguishable, but not sequential or suggestive (e.g. increasingly vivid shades of red used to denote heat), they were generated using the 'qualitative' option, based on work by Ihaka [17].

In Fig. 1 the diagrams exhibit all of these layout choices (although they are scaled). Further layout conventions were adopted for each diagram type, using results from the literature that guide us toward effective layouts [2],[7],[8],[18]. The conventions were as follows:

Venn Diagrams:
1. The curves were drawn smoothly.
2. The overlaps were drawn so that the zone areas are similar.
3. The diagrams were drawn well-formed; see [19].
4. The diagrams had rotational symmetry, using layouts given in [20].

Euler Diagrams with Shading:
1. The curves were circles where possible, otherwise ellipses were used.
2. The diagrams were drawn well-formed.
3. The number of shaded zones was kept minimal.

Euler Diagrams without Shading:
1. The curves were drawn smoothly, with recognizable geometric shapes (such as circles, ellipses or semi-circles), or with rectilinear shapes.
2. The diagrams were drawn as well-formed as possible, aiming to minimize (in this order of priority, based on [2]): concurrency between curves, non-simple curves, triple points, and brushing points (points were two curves meet but do not cross).

Linear Diagrams
1. The number of line segments representing each set was kept small.
2. Favour layouts where, when reading from left to right, the number of overlapping line segments changes minimally.

In order reduce the number of line segments, the set with the largest number of intersections with the other sets was drawn using a single line segment.

2.4 Data Collection Methods

For this study, we adopted a crowdsourcing approach and we used Amazon Mechanical Turk (MTurk) [21, 22] to automatically out-source tasks to participants. In MTurk, the tasks are called HITs (Human Intelligence Tasks) which are completed by anonymous participants (called workers) who are paid if they successfully complete the HIT. The use of crowdsourcing platforms for conducting research-oriented studies is becoming more popular, as this method for collecting data has now gained recognition within the scientific community. In particular, there is evidence that it is a valid approach, where [22] compared lab-based experiments with MTurk, showing that no significant differences arise in the results. Moreover, MTurk has been specifically used to collect performance data in other scientific studies in the visualization field, such as [23].

The MTurk HITs were based on the templates provided by Micallef et al.[24], at http://www.aviz.fr/bayes. Every question, in both the training and the main study, was displayed on a separate page of the HIT. Previous pages could not be viewed and subsequent pages were not revealed until the question on the current page was answered. The questions in the main study were randomly ordered to reduce ordering effects.

After every five study questions, participants were asked to answer a question designed to identify inattentive workers (spammers) and those that had difficulties with the language. In MTurk there is little control over who participates in the study and, so, some workers may fail to give questions their full attention [21]. A recognized technique for identifying workers who cannot understand the language used is to include questions that require careful reading, yet are very simple to answer (e.g, [25]). In our study, these questions asked participants to click on a specific area in the diagram, whilst still presenting the participants with (redundant) radio buttons for the 'yes' and 'no' answers seen for the 18 main study questions. Participants were classified as spammers if they clicked a radio button on more than one of the four spammer-catching questions included in the study. All data obtained from spammers were removed before analysis.

3 Experiment Execution

Initially 20 participants took part in the pilot study (1 spammer). The pilot study proved the experimental design to be robust, with a few minor changes made to the wording of the questions (mostly due to typographical errors). A further 440 participants were recruited for the main study. Of note is that we only allowed MTurk workers with a HIT approval rate of at least 95% to participate. All participants were randomly allocated to one of the four diagram types in equal numbers. There were 16 participants identified as spammers, leaving each participant group with the following number of participants: Venn diagrams 107, Euler diagrams with shading 109, Euler diagrams without shading 106, and linear diagrams 102. The ID of all the workers that either completed one of our HITs or started and returned the HIT before completion was recorded. A worker whose ID was previously recorded was not assigned a HIT, so preventing multiple participation. The participants performed the experiment at a time of their choosing, in a setting of their choosing. They were told that the experiment would take approximately 20 minutes, based on participants' performance in the pilot study, and were paid $1 to take part (this was reduced from $1.50 for the pilot study, as all 20 HITs were completed within 30 minutes). For the main study, the data were collected within 24 hours, with HITs made available in sequential batches of 100, and a final batch of 40. We also note that $1 for approximately 20 minutes work is higher than is typical for MTurk workers, for example [22].

At the beginning of the study, each participant was told that they could only participate once in the study and instructed to read the questions carefully. They were further advised that they had to answer 75% of "key questions" correctly in order to be paid (i.e. not classified as a spammer). They were further advised that the first five pages of the HIT were training, which was the first phase of the experiment. During this phase, participants attempted questions and were told whether they had answered correctly, with the answer was explained to them. An example of a training page can be seen in Fig. 3. The super-imposed rectangles highlight the two radio buttons and show the text displayed *after* the participant had clicked 'reveal answer'.

The participants then entered the data collection phase. This began with three questions, in addition to the 18 main questions. The data relating to these first three questions was not included in the analysis, in order to reduce the impact of any learning effect. Consequently, the following results are based on 424 participants each answering 18 questions, giving $18 \times 424 = 7632$ observations.

4 Statistical Analysis and Results

We now proceed to analyze the data collected by considering time taken and error rate. The raw data are available from `www.eulerdiagrams.com/set`, allowing the computation of non-essential basic statistics omitted for space reasons.

4.1 Time Data

The grand mean was 20.452 seconds (standard deviation: 23.620) and the mean times taken to answer questions by diagram type are: Venn diagrams 24.477 (sd: 24.158); Euler diagrams with shading 20.532 (sd: 26.996); Euler diagrams without shading 19.810 (sd: 20.121); and linear diagrams 16.813 (sd: 21.843). In many tables below, we abbreviate the diagrams types' names to Venn, shaded ED, unshaded ED and linear respectively. In order to establish if there is significant variation across diagram types, we conducted an ANOVA. The results of the ANOVA are summarized in table 1, which uses log (base 10) of time. Although the data are not normal, even after taking logs, the skewness of the logged data is 0.39 and the sample size is 7632, making the ANOVA test robust; the raw data have a skewness of 12.08, which is outside the permissable range

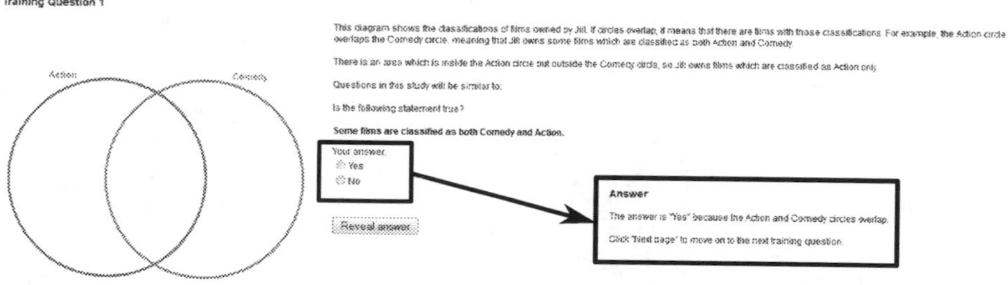

Fig. 3. The first training page for the Venn diagram group

of ±2 for the ANOVA to be valid. We regarded p-values of less than 0.05 as significant, unless stated otherwise.

The row labelled question, with a p-value of 0.000, tells us that there are significant differences between the mean times of at least one pair of questions. This shows robustness, in that the questions are sufficiently varied in that they required different amounts of user effort to answer.

The row for diagram, with a p-value of 0.000, tells us that there are significant differences between the mean times taken to answer questions across the diagram types. That is, diagram type significantly impacts on task performance. Next, we performed a Tukey test to compare pairs of diagram types, thus establishing whether one mean is significantly greater than the other, in order to rank the diagram types. Any p-values of less than 0.01 were regarded to be significant, given multiple comparisons being made on the same data. Table 2 presents the rankings of diagram type by mean time taken. We see that **linear diagrams allow participants to perform significantly faster** on tasks than all other diagram types. There is no significant difference between shaded Euler diagrams and Euler diagrams without shading, whereas **Venn diagrams cause participants to perform significantly slower** on tasks. Thus, we conclude that linear diagrams are the most effective diagram type, with respect to time taken, followed by both unshaded Euler diagrams and shaded Euler diagrams and, lastly, Venn diagrams.

The magnitude of the significant differences is reflected in the effect sizes given in table 3. For example, the largest effect size tells us that 62% to 66% of participants were faster interpreting linear diagrams than the average person using Venn diagrams. The effect sizes all suggest that not only are the differences in mean time taken significant but, taken with the mean times, real differences in task performance will manifest through their use in practice.

Continuing now with our interpretation of the ANOVA table, the row for diagram*question, with a p-value of 0.000, tells us that there is a significant interaction between diagram type and question: the diagram type used impacts user performance for at least one question. We further investigated this manifestation by running another ANOVA, looking for an effect of question type and an interaction between diagram type and question type. This would establish whether there was any obvious systemmatic way of describing the interaction

Table 1. ANOVA for the log of time

Source	DF	MS	F	P
question	17	4.3966	108.32	0.000
diagram	3	7.5847	13.52	0.000
diagram*question	51	0.2155	5.31	0.000
participant(diagram)	420	0.5584	13.76	0.000
Error	7140	0.0406	–	–
Total	7631			

Table 2. Pairwise comparisons

Diagram	Mean	Rank
Venn	24.477	C
Shaded ED	20.532	B
Unshaded ED	19.810	B
Linear	16.813	A

between diagram type and question. The ANOVA showed no significant effect of question type ($p = 0.201$) and no interaction between diagram and question type ($p = 0.171$). This implies that **task performance is not affected by question type**.

4.2 Error Data

Regarding errors, of the 7632 observations there were a total of 1221 errors (error rate: 16%). The errors were distributed across the diagram types as follows: Venn diagrams 391 out of 1926 observations; Euler diagrams with shading 377 out of 1962; Euler diagrams without shading 258 out of 1908; and linear diagrams 195 out of 1836. We performed a χ^2 goodness-of-fit test to establish whether diagram type had a significant impact on the distribution of errors. The test yielded a p-value of 0.000. Thus, the number of errors accrued is significantly affected by diagram type. Investigating further, table 4 summarizes where significant differences exist. We conclude that **linear diagrams accrued significantly fewer errors** than all other diagram types. Moreover, **significantly more errors were accrued using Venn diagrams and shaded Euler diagrams** than the other two diagram types.

It is natural to ask whether question type impacts error rate by diagram type. Table 5 summarizes the raw data for error counts for each diagram type, broken down by question type. Statistically analyzing these data, by question type, reveals significant differences in all cases. In particular, our analysis revealed that, for all question types, linear diagrams lead to significantly fewer errors. For intersection and subset questions, Venn diagrams make a large contribution to the χ^2 statistic, indicating that they account for a significantly large number of errors. Lastly, for disjointness questions, Euler diagrams with shading make a large contribution to the χ^2 statistic, indicating that they account for a significantly large number of errors. In summary, our analysis of the errors suggests that linear diagrams are the most effective diagram type, with Venn diagrams being the worst, except for questions on disjointness where Euler diagrams with shading are worst.

4.3 Summary of Results

Linear diagrams allow users to perform most effectively in terms of both completion time and correctness: the mean time taken was significantly faster than for all other diagram types and the number of errors was significantly lower. By contrast, Venn diagrams were ranked bottom for both time taken and, jointly with

Table 3. Effect sizes

Diagram	Unshaded	Shaded	Venn
Linear	54%-58%	54%-58%	62%-66%
Shaded	–	–	54%-58%
Unshaded	–	–	54%-58%

Table 4. Error differences

Diagram	Unshaded	Shaded	Venn
Linear	Y	Y	Y
Unshaded	–	Y	Y
Shaded	–	–	N

Table 5. Error counts and significance

	Intersection		Subset		Disjointness	
Diagram	Error	Correct	Error	Correct	Error	Correct
Venn	152	490	107	535	132	510
Shaded	122	532	96	558	159	495
Unshaded	100	536	70	566	88	548
Linear	70	542	51	561	74	538
p-values	0.000		0.000		0.000	

shaded Euler diagrams, error rate. Thus, the error analysis allows us to distinguish Euler diagrams with shading from Euler diagrams without shading, which were not significantly different in terms of time taken. We can therefore give a ranking of diagram types using both time and errors, in order of effectiveness:

1. linear diagrams,
2. Euler diagrams without shading,
3. Euler diagrams with shading,
4. Venn diagrams.

5 Subjective Discussion

We now seek to explain our results in the context of theories about diagrams, cognition and perception. One feature of diagrams that is thought to correlate with their effectiveness is *well-matchedness*, introduced by Gurr [18]. A diagram is *well-matched* if its syntax directly reflects its semantics. Euler diagrams without shading are well-matched because the spatial relationships between the curves directly mirror the relationships between the sets they represent. Linear diagrams are also well-matched because the spatial relationships between the lines also directly mirror the relationships between the sets. For example, in Fig. 1, the lines representing German and Italian do not overlap, and the represented sets are disjoint. By contrast, Euler diagrams with shading are not well-matched because of the additional zones used to represent empty sets. Likewise, Venn diagrams that include shading are not well-matched. That is, the spatial relationships between their curves does not directly mirror the relationships between the represented sets. Thus, our results - with linear diagrams and Euler diagrams without shading being more effective than Euler diagrams with shading and Venn diagrams, support Gurr's theory.

An interesting point is that the Euler diagrams with shading and the Venn diagrams used in the study were all *well-formed*, whereas those that did not use shading were all non-well-formed (see e.g. [2]). Rodgers et al., in [2], established that well-formed diagrams are more effective than equivalent diagrams that are not well-formed (the diagrams in [2] did not use shading). Thus, our results indicate that being well-matched is more important than being well-formed.

Another way of examining differences between diagram types is through *visual complexity*. For the Venn and Euler family, one measure of visual complexity

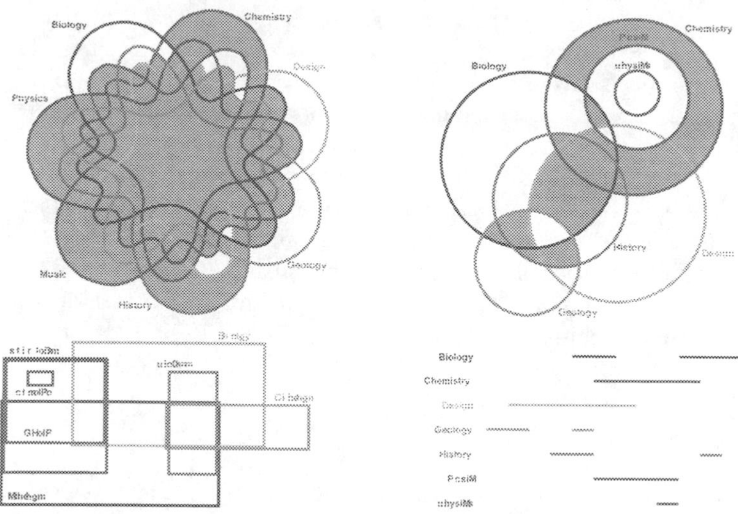

Fig. 4. Diagram complexity

arises through the number of crossings between curves. In our study, Venn diagram exhibited more crossings than Euler diagrams with shading which, in turn, had more than Euler diagrams without shading. By contrast, linear diagrams have no line crossings. For example, in Fig. 4, the Venn diagram has 121 crossings between curves, the Euler diagram with shading has 20, the Euler diagram without shading has 9 (and a further 4 points where curves meet but do not cross), and the linear diagram has none. The results of our study lead us to hypothesize that this measure of visual complexity, at least for the diagrams in this study, correlates with diagram effectiveness and, moreover, helps to explain why linear diagrams are the most effective.

We can further explain why linear diagrams are interpreted more quickly, and with fewer errors than all other diagram types and, in particular, Euler diagrams without shading. With regard to linear diagrams, we quote Wagemans et al. "[the] comparison of features lying on pairs of line segments is significantly faster if the segments are parallel or mirror symmetric, suggesting a fast grouping of the segments based on these cues [26]", who reference Feldman [27] as the source of this insight. As we have seen, linear diagrams use parallel line segments and so are thought to be effective for this reason. However, as Wagemans et al. consider the relative effectiveness of diagrams drawn with lines, this alone does not explain why linear diagrams are more effective than the other diagram types.

To gain more insight into our observations, we consider the work of Bertin who describes graphical features consisting of elements and properties [28]. Closed curves and lines can be regarded as elements. Properties include shape and size. Bertin, recognizing our visual sensitivity to graphical properties, proposes eight visual variables. Two of these, called planar variables, are the x and y coordinates in the plane. Venn and Euler diagrams are not constrained by planar variables: drawing their closed curves is not constrained by x or y coordinates. Therefore, each closed curve's position in the plane is arbitrary other than topological constraints imposed by the sets being visualized. Conversely, linear diagrams are constrained by planar variables. Lines are ordered vertically and run horizontally, in parallel. In the context of this study, a top down hierarchy is imposed, along the y axis, based upon the alphabetical order of set names, and this layout feature is thought to aid reading the diagram. There is no such prescribed order in the Euler and Venn family. Moreover, relationships between combinations of sets can be 'read off' along the x axis. Consequently, the prescriptive planar layout of linear diagrams, as opposed to the 'free' (disordered) positioning of curves in the other diagram types, is thought to aid comprehension.

6 Threats to Validity

Threats to validity are categorized as internal, construct and external [29]. The following discusses threats to validity, focusing primarily on those arising from using a crowdsourcing approach, that were considered and addressed to ensure the study is robust and fit for purpose. With regard to internal validity, the following factor was among a number that were considered in our study design:
Laboratory: ideally, all participants undertake the study in the same environment, ensuring each participant was exposed to the same hardware, free from noise and interruption. By adopting a crowdsourcing approach, we had no control over the environment in which each participant took part. To reduce the effect of this compromise, a large data set was collected, with over 400 participants.

Now we consider construct validity, examining the rigour of our dependent variables and independent variables for measuring comprehension:
Time: to ensure the rigour of time measurements, consideration was paid to the precise duration elapsed interpreting a diagram as well as the units employed to measure time. As we used a crowdsourcing approach, there was little control over any distractions impacting the time taken by each participant on each question. To manage this, a large sample size was used.
Question: it was considered a threat if participants did not spend time reading and understanding the questions and diagrams. To manage this threat *diversity* was introduced in the diagrams so that participants had to read and understand each diagram before being able to answer the posed question. It was also considered a threat if the diagrams were regarded as trivial; having only a few sets was deemed insufficient to yield noticeable differences in response times, should they exist. To manage this, diagrams represented three, five or seven sets in order to demand cognitive effort. Lastly, the study included questions to allow spammers to be identified, catching those who did not read questions carefully.

The following factor considers the limitations of the results and the extent to which the they can be generalized, thus examining their external validity:
Participant: participants were representative of the wider population, being MTurk workers. They were predominately from the USA or India. Thus, the results should be taken to be valid within these constraints.

7 Conclusion

In this paper we have examined four diagram types that are used for visualizing sets. By conducting an empirical study, we have established that task performance is significantly better when using linear diagrams over the Euler diagram family, comprising Venn diagrams, Euler diagrams with shading and Euler diagrams without shading. Furthermore, the Euler diagrams variants that we tested can be ordered: Euler diagrams without shading were most effective, Euler diagrams with shading were next and, finally, Venn diagrams proved to be least effective, having both poor time and accuracy performance. Given the prevalent use of Euler and Venn diagrams for visualizing sets, and the relative lack of use of linear diagrams, these results have implications for the use of diagrams in practice. Our results suggest that linear diagrams should be more widely adopted, at least for use by the general population. It would be interesting to establish whether these results manifest for expert users also.

Looking to the future, we plan to conduct further studies that augment the syntax of these diagrams with data items (i.e. set elements). Many diagrammatic systems, such as spider diagrams and Euler/Venn diagrams, exploit Euler diagrams with graphs to represent sets and their elements. It will be interesting to establish whether linear diagrams should instead be adopted for representing this more complex data. Moreover, the result suggest that the number of shaded zones, in the Euler and Venn diagram family, could impact on performance. However, the current study did not control this variable and further study will be needed to gain insight into their effect.

References

1. Larkin, J., Simon, H.: Why a diagram is (sometimes) worth ten thousand words. J. of Cognitive Science 11, 65–99 (1987)
2. Rodgers, P., Zhang, L., Purchase, H.: Wellformedness properties in Euler diagrams: Which should be used? IEEE Trans. on Visualization and Computer Graphics 18(7), 1089–1100 (2012)
3. Couturat, L.: Opuscules et fragments inédits de Leibniz. Felix Alcan (1903)
4. Wittenburg, K., Lanning, T., Heinrichs, M., Stanton, M.: Parallel Bargrams for Consumer-based Information Exploration and Choice. In: 14th ACM Symposium on User Interface Software and Technology, pp. 51–60 (1985, 2001)
5. Hofmann, H., Siebes, A., Wilhelm, A.: Visualizing Association Rules with Interactive Mosaic Plots. In: ACM SIGKDD Int. Conf. on Knowledge Discovery and Data Mining, pp. 227–235 (2000)
6. Moody, D.: The "Physics" of Notations: Toward a Scientific Basis for Constructing Visual Notations in Software Engineering. IEEE Trans. on Software Engineering 35(6), 756–779 (2009)
7. Benoy, F., Rodgers, P.: Evaluating the comprehension of Euler diagrams. In: 11th Int. Conf. on Information Visualization, pp. 771–778. IEEE (2007)
8. Blake, A., Stapleton, G., Rodgers, P., Cheek, L., Howse, J.: Does the orientation of an Euler diagram affect user comprehension? In: 18th Int. Conf. on Distributed Multimedia Systems, pp. 185–190. Knowledge Systems Institute (2012)

9. Grawemeyer, B.: Evaluation of ERST – an external representation selection tutor. In: Barker-Plummer, D., Cox, R., Swoboda, N. (eds.) Diagrams 2006. LNCS (LNAI), vol. 4045, pp. 154–167. Springer, Heidelberg (2006)
10. Sato, Y., Mineshima, K.: The Efficacy of Diagrams in Syllogistic Reasoning: A Case of Linear Diagrams. In: Cox, P., Plimmer, B., Rodgers, P. (eds.) Diagrams 2012. LNCS, vol. 7352, pp. 352–355. Springer, Heidelberg (2012)
11. Sato, Y., Mineshima, K., Takemura, R.: The Efficacy of Euler and Venn Diagrams in Deductive Reasoning: Empirical Findings. In: Goel, A.K., Jamnik, M., Narayanan, N.H. (eds.) Diagrams 2010. LNCS, vol. 6170, pp. 6–22. Springer, Heidelberg (2010)
12. Isenberg, P., Bezerianos, A., Dragicevic, P., Fekete, J.: A study on dual-scale data charts. IEEE Tran. on Visualization and Computer Graphics, 2469–2478 (2011)
13. Puchase, H.: Which aesthetic has the greatest effect on human understanding? In: DiBattista, G. (ed.) GD 1997. LNCS, vol. 1353, pp. 248–261. Springer, Heidelberg (1997)
14. Riche, N., Dwyer, T.: Untangling Euler diagrams. IEEE Tran. on Visualization and Computer Graphics 16(6), 1090–1099 (2010)
15. Harrower, M., Brewer, C.: ColorBrewer.org: An Online Tool for Selecting Colour Schemes for Maps. The Cartographic Journal 40(1), 27–37 (2003)
16. Silva, S., Madeira, J., Santos, B.S.: There is more to color scales than meets the eye: A review on the use of color in visualization. In: Information Visualization, pp. 943–950. IEEE (2007)
17. Ihaka, R.: Colour for presentation graphics. In: 3rd Int. Workshop on Distributed Statistical Computing (2003)
18. Gurr, C.: Effective diagrammatic communication: Syntactic, semantic and pragmatic issues. J. of Visual Languages and Computing 10(4), 317–342 (1999)
19. Stapleton, G., Rodgers, P., Howse, J., Taylor, J.: Properties of Euler diagrams. In: Proc. of Layout of Software Engineering Diagrams, pp. 2–16. EASST (2007)
20. Ruskey, F.: A survey of Venn diagrams. Electronic J. of Combinatorics (1997), `http://www.combinatorics.org/Surveys/ds5/VennEJC.html`
21. Chen, J., Menezes, N., Bradley, A., North, T.: Opportunities for crowdsourcing research on amazon mechanical turk. Human Factors 5(3) (2011)
22. Paolacci, G., Chandler, J., Ipeirotis, P.G.: Running experiments on amazon mechanical turk. Judgment and Decision Making 5(5), 411–419 (2010)
23. Heer, J., Bostock, M.: Crowdsourcing graphical perception: Using mechanical turk to assess visualization design. In: ACM Conf. on Human Factors in Computing Systems, pp. 203–212 (2010)
24. Micallef, L., Dragicevic, P., Fekete, J.D.: Assessing the Effect of Visualizations on Bayesian Reasoning through Crowdsourcing. IEEE Trans. on Visualization and Computer Graphics 18(12), 2536–2545 (2012)
25. Oppenheimer, D., Meyvis, T., Davidenko, N.: Instructional manipulation checks: Detecting satisficing to increase statistical power. J. of Experimental Social Psychology 45(4), 867–872 (2009)
26. Wagemans, J., Elder, J., Kubovy, M., Palmer, S., Peterson, M., Singh, M.: A century of gestalt psychology in visual perception: I. perceptual grouping and figure-ground organisation. Computer Vision, Graphics, and Image Processing 31, 156–177 (1985)
27. Feldman, J.: Formation of visual "objects" in the early computation of spatial relations. Perception and Psychophysics 69(5), 816–827 (2007)
28. Bertin, J.: Semiology of Graphics. Uni. of Wisconsin Press (1983)
29. Purchase, H.: Experimental Human Computer Interaction: A Practical Guide with Visual Examples. Cambridge University Press (2012)

Recognising, Knowing and Naming: Can Object Picture Processing Models Accommodate Non-Picture Visuals?

Richard Cox

Monash Adaptive Visualisation Lab (MArVL)
Monash University, Caulfield
Victoria 3145, Australia
Richard.Cox@monash.edu

Abstract. This paper provides an overview of information processing accounts of pictures of objects and of non-picture visuals (NPVs) such as graphs and diagrams, including theories of graph comprehension. Compared to the study of objects, there appear to be rather few information processing studies of NPVs. An NPV corpus was developed and items were used as visual stimuli in four cognitive tasks. The tasks assessed perceptual level processing (NPV recognition), semantic knowledge and lexical production (naming). The results are discussed in relation to several questions: How well do models of object *picture* processing accommodate the findings from this study of NPV processing? To what extent can NPVs be considered to be another class of object pictures? Are well-established phenomena in the visual object domain such as frequency and age of acquisition effects observed for NPVs? How do patterns of performance on the perceptual, semantic and naming tasks differ across NPV item sub-classes? The results show that performance patterns across a range of cognitive tasks utilizing NPV stimuli are - to some degree - similar to those seen in object picture processing. Age of acquisition effects were also observed. It is concluded that the use of experimental paradigms from studies of object picture processing are useful for understanding how people understand and use non-pictorial graphical representations such as diagrams.

Keywords: information processing, cognitive processing, diagrams, external representations, graph comprehension, object recognition, picture naming, age of acquisition.

1 Introduction

Diagrams and pictures of objects are both forms of graphical external representation and both have been studied from cognitive information processing perspectives. A large body of research has shown that people can accurately categorise and name a wide variety of pictures of living things (animals, fruits, body parts, marine creatures,...) and non-living things (furniture, tools, vehicles,...). Object recognition researchers have developed comprehensive cognitive

T. Dwyer et al. (Eds.): Diagrams 2014, LNAI 8578, pp. 161–175, 2014.

processing models informed by experimental studies (e.g. [1]). They describe basic discrimination and feature analysis (perceptual level), various types of semantic processing through to the levels of processing required for naming the depicted item (e.g. lexical access).

Information processing accounts of 'non-picture' stimuli such as diagrams, graphs, charts, and maps have also been developed (e.g. [2, 3]). These theories tend to focus on either bottom-up early feature analyses such as graph feature extraction or top-down processes (template or schema invocation). In [2], Shah et al. (2005) state, "...there are two major classes of models of graph interpretation. First, ...models provide detailed descriptions of simple graph..tasks.., such as the ability to retrieve simple facts.....or the ability to discriminate between two proportions....A second class of models ...focuses on more general process analyses...but is less precise in its predictions about specific processes..." (p. 429).

Compared to the study of objects, there appear to be rather few studies about semantic and lexical processing of *non-object picture* visual (NPV)[1] stimuli such as graphs, charts, network, tree, set diagrams, notations, lists, tables and text. Little is known about people's ability to accurately categorise and name NPV stimuli. How is an individual's ability to recognise diagrams related to their ability to name the diagrams or related to their knowledge or understanding of the diagram and its purpose? Are these processes similar or different to their counterparts in object knowledge? The answers to these questions have implications for how we understand NPVs as images and how we comprehend diagrams as a language of spatial relationships. At one level diagrams can be argued to be just another exemplar of pictorial images, yet at another level the rule constraints of diagrams, maps and tables (required for their comprehension) make them more akin to domains such as reading where decoding knowledge has to be brought to the comprehension process.

In the next sections, the literature on object recognition models will be briefly summarised followed by a review of cognitive approaches to processing non-picture visuals (e.g. models of graph comprehension).

1.1 Information Processing Models of Object Pictures

A widely-cited model of object picture recognition [1] includes linked perceptual and cognitive subsystems for processing visual input, controlling attention, allocating processing resources and controlling responses or output. The model is hierarchical and includes visual and verbal processing modalities. Object picture processing begins with sensory feature registration allowing 'figure' to be distinguished from 'ground' and the processing of basic object features such as size, colour, length, orientation and location. Next, the visual features are integrated and grouped on the basis of similarity, facilitating perception.

Objects are recognised even if viewed from unusual angles. This is achieved by mapping perceptual descriptions built up in visuospatial working memory

[1] The term 'non-picture visual' is due to Cripwell [4]. It is a useful term because it includes linguistic forms of external representation (text, tables, notations) in addition to non-picture graphical forms such as diagrams, graphs, charts and maps.

[5] to structural descriptions retrieved from LTM (picturecon). At still higher levels of processing, semantic information is also retrieved (e.g. is the object a tool, an animal?) to allow for comprehension of what is perceived. A naming (production) response is possible when lexical access occurs.

Information processing models of object picture representations are evolved on the basis of empirical evidence provided by experimentation. A common paradigm in this research is to develop a large pool of normed stimuli such as pictures of objects like the Snodgrass & Vanderwart corpus [6][2] to enable experiments to test the independence of different types of knowledge about these objects.

Information processing models allow us to investigate the independence of discrete cognitive processes and make fairly accurate predictions of performance based on our knowledge about how earlier or later processes operate. A process model of object picture analysis - such as the one presented in Figure 1 (right) - provides testable hypotheses concerning whether or not being able to recognise an object is a necessary precursor to comprehending what it is and how it is used. According to that model, one cannot describe how a saucepan is used unless one recognises a saucepan in the first place. Similarly the model predicts that being able to name a saucepan is contingent upon having some knowledge of what the saucepan is for. The model also predicts that we might observe phenomena such as frequency effects or age of acquisition effects. Frequency of experience is correlated with the age at which the object was first encountered, hence effects such as 'age of acquisition' are known to significantly influence naming performance. Objects that have names learned early in life are responded to more quickly (e.g. [8]). The age of acquisition effect is specific to picture naming - object *recognition* does not show this effect (see [9] for a review).

An interesting question concerns whether similar effects are found for NPV naming. The UK National Curriculum was analysed by Garcia Garcia & Cox [10] for all subjects taught in schooling years 1 to 9 (ages 5 to 14 years). The forms of representation that are introduced to students aged between 5 and 9 years old include illustrations, tables, lists, bar charts, maps, flow charts and Venn diagrams. Forms introduced later (9-14 years) include entity-relation diagrams, network diagrams, tree diagrams, pie charts and Cartesian coordinate forms such as line graphs. Responses to these two classes of item will be examined in this study in order to examine whether age of acquisition effects are observed for NPVs.

1.2 Cognitive Models of Non-Picture Visuals Processing

Cognitive models of NPVs such as graphs and diagrams are less completely specified than object picture processing models. In the NPV area, studies tend to focus on cartesian (X-Y axis) charts and various kinds of graph [3, 11–18].

[2] See [7] for a review of sets of object-naming stimuli.

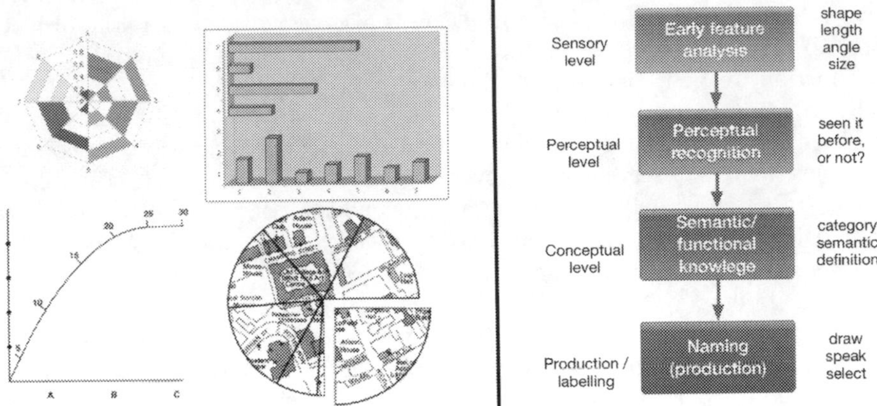

Fig. 1. Four examples of pseudo-diagrams (left) used in the recognition task. Cognitive model of visual object processing (right).

In general, the NPV models propose that bottom-up and top-down processes occur to some extent in parallel and involve three major cognitive subsystems - perceptual, short-term memory (STM) and long-term memory (LTM). First, low level visual features of, say, a bar chart are discriminated and encoded (X-Y axes, number of bars, their relative heights). These have been referred to as 'elementary perceptual tasks' [11] and as 'visual primitives' [15]. Examples include perceiving line lengths, areas, shading, curvature, position relative to a scale on an axis, angles, etc. Performance varies across these perceptual tasks. For example, position along a common scale is judged more accurately than the size of an enclosed area [19]. The relative lengths of bars in a bar chart are generally more accurately discerned than the relative areas of two circles in a set diagram or a comparison of 'slice' sizes in a pie chart. Features are believed to be chunked according to Gestalt perceptual features such as continuity, proximity and form [3]. Some features can be perceived in parallel but some require serial scanning. Cycles of scanning and rescanning build up a 'visual description of the graph' in STM, which has an approximately 7 second capacity for retention [15]. Visuospatial working memory is where visual descriptions in STM invoke schemas from LTM and where rehearsal and attentional resource allocation is managed via a central executive [5].

Trickett & Trafton [20] call for models of graph comprehension to incorporate more spatial processing in addition to visual feature processing. They argue that spatial processing occurs particularly when implicit information in a graph requires inferential processing in order for the required information to be extracted. They also observe gestures during communication of graph-derived information and they argue that this provides further evidence of the importance of spatial cognition.

Schemas are 'standard, learned procedures for locating and decoding information presented in the graph' ([15], p. 357). Schemas are used as the basis for mapping graph features to meaning (e.g. [12, 13]). They result from prior experience and contain domain knowledge, knowledge of representations and knowledge about which representations to use for particular tasks (i.e. 'applicability conditions' [17]; 'decomposition' [21]). Schema and semantic representations in LTM reflect an individual's level of expertise. Cox & Grawemeyer [28] used an NPV card-sorting paradigm and cluster analyses to assess the mental organisation of semantic information. It was shown that people who performed well[3] on diagrammatic tasks tended to sort graphs, charts, diagrams, etc into fewer, more cohesive and accurately-named categories than people who performed less well on diagrammatic reasoning tasks.

Novices lack schema knowledge and are familiar only with a narrow range of graphical and diagrammatic forms. Errors such as the 'graph-as-picture misconception' (GAPm) have been observed in students [22, 23]. For example, a line graph of a quadratic function or relationship between two variables might be interpreted as as a picture of a mountain or a distance-time travel graph is interpreted as a map of an actual journey. Garcia Garcia & Cox [23] found that approximately 25% of children in a sample of 48 elementary school students manifested such misconceptions. Network graphs, pie charts and set diagrams were associated with higher rates of GAPm than bar charts, line graphs or tables. Leinhardt et al. [22] review GAPm phenomena but do not speculate on their origin in cognitive information processing terms. Information processing accounts offer a way of characterising graphical misconceptions such as the GAPm more theoretically. In terms of the model, a GAPm arises from visual errors at the recognition level - it is not a semantic error.

1.3 Ecological Validity of Information Processing Tasks Using NPVs

Previous work has established that NPV tasks such as those used in the present study are useful for investigating perceptual, semantic and naming processes and that they correlate with real-world reasoning tasks, including diagrammatic reasoning. They therefore have a degree of criterion validity. Some of the tasks employed in this study were used in studies that also collected data on reasoning with external representations. Cox et al. [24] showed that computer science students' functional knowledge of, and ability to name representations relevant to the software engineering domain[4] correlated more highly with program debugging performance than years of previous programming experience (the next highest correlate). Garcia Garcia & Cox [23] developed a simplified version of

[3] i.e. who could assign an effective diagram to particular forms of information, or who could draw diagrams that were effective in aiding their problem solving.

[4] These were: lists, node and arrow (network) diagrams, tables, textual/linguistic representations (*e.g.* handwriting, typed text), trees and notations and formulae such as program code

the NPV recognition task for use with young students. It was administered to 45 students in grades 4,5 and 6 using an interactive, touch surface computer. The task identified students who had a 'graph-as-picture' misconception (validated against an independent external assessment). It was further shown that the so-called 'graph as picture' misconception should perhaps be renamed the 'diagram as picture' misconception since students were observed to misclassify as pictures NPVs other than graphs. Grawemeyer & Cox [25] administered the NPV tasks to subjects who also performed a demanding representation-selection task which required them to select data displays that matched both the information-to-be-displayed and the requirements of the response task. The semantic NPV tasks were shown to be good predictors of display selection accuracy and reasoning performance.

1.4 Aims

At a theoretical level, questions about non-picture visuals include 'are they just another category of picture visuals?' and, if not, how are they represented and processed as a class of objects? This paper explores the extent to which NPVs function as a class of object pictures. Is accuracy greater on perceptual level tasks than on tasks involving more cognitive subsystems - i.e. 'higher level' semantic and naming tasks? To what extent do forms of NPV (graphical, linguistic) differ in terms of perceptual, semantic and naming task performance? How does performance differ at perceptual, semantic and naming levels of processing? Does performance on classifying NPVs correlate more highly with naming performance or with functional knowledge? Are 'age of acquisition' effects observed on NPV naming tasks? Answers to these questions will inform a model of NPV processing that would have potential practical use - e.g. as a basis for developing educational instruments for assessing graphical literacy and detecting representational misconceptions. To address these issues, a corpus of NPV images was used in four tasks.

2 Method

2.1 Participants

Fifty-four participants took part in the study. They were recruited from a UK university and consisted of undergraduate and postgraduate students, research assistants and their friends. There were 38 males and 16 females. Nine males and two females were aged between 16-20 years, 9 males and 1 female were aged between 21-25 years, 8 males and 5 females were aged between 26 and 30 years, 4 males and 3 females were aged between 31 and 35 years, 4 males and 2 females were aged 36-40 years and 4 males and 3 females were aged over 40 years. Twenty-eight had undergraduate degrees (computer science, psychology, theology, or arts), 12 had masters degrees and 14 were qualified at PhD level.

2.2 The NPV Corpus

The corpus consisted of 90 graphical representations of a wide variety of forms: **Maps** (street, topographic, weather, strip map (6 items)); **Set diagrams** (Venn, Euler (5 items)); **Text items** (newsprint, handwriting, email message, computer program,poem, telephone directory extract, questionnaire (8 items)); **Lists** (shopping, medical, recipe,telephone directory (5 items)); **Tables** (timetables, bi-directional matrix, truth table, contingency matrix (8 items)); **Graphs & charts** (lattice, X/Y function, scatter plot, pie chart, bar chart, box plot, bubble chart, candlestick. frequency histogram, PERT (16 items)); **Tree diagrams** (organisation chart, decision tree, genealogical, dendrogram (4 items)); **Node & arrow** (probability diagram, flow chart, Gantt, control, electrical circuit, entity-relation, UML, London underground (8 items)); **Plans** (building floor, house floor, dressmaking pattern, blueprint, seating (8 items)); **Notations & symbols** (logic notation, pictographic sequences, maths formulae, geometry, music, choreograpy (5 items)); **Scientific Illustrations** (exploded mechanical diagrams, engineering drawing,laboratory apparatus, physics, mechanics, geometry anatomy (7 items)); **Icons** (sport activity logos, commercial logos, packaging logos, road signs, audiovisual logos, Chinese logograph (10 items)). The categories and item examples within categories were derived from various taxonomic studies of external representations [26–29]) and from information visualisation sourcebooks [30, 31].

Examples from each of the major categories are shown in Figure 2. In addition to the 90 NPVs, there were 22 'pseudo' NPVs. These were designed for use as 'fake' items in the decision task (described below). Four examples are shown on the left side of Figure 1. There were also 'fake' non-graphical NPVs - a graph with tick marks and axis labels on the function curve instead of on the axes, invented math-like notations, an ill-formed syllogism, a 'scrambled' recipe (textual), *etc.* These items were designed to be analogous to pictures of 'unreal' or 'chimeric' objects used in experimental neuropsychology (e.g. [32]).

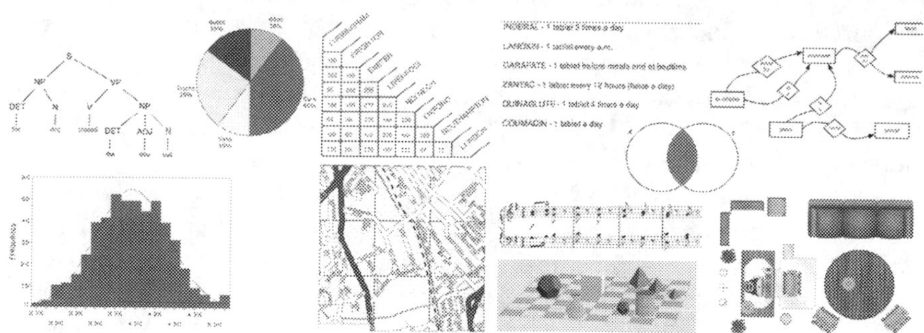

Fig. 2. Examples from the NPV corpus used in recognition, categorisation, functional knowledge & naming tasks

Care was taken to ensure that the fake items were not discriminable on the basis of spurious perceptual factors such as font or color.

2.3 The Perceptual, Semantic and Lexical Tasks

Assessing recognition knowledge via a decision task Decision tasks are used as a paradigm for establishing an individual's recognition knowledge in a domain (e.g. picture decision). The task required respondents to indicate whether or not the non-picture visual stimulus was a valid and authentic example. The response mode was forced-choice, with response options 'real' or 'fake'.

Semantic knowledge tested via categorisation task A single NPV image was presented on each trial (presentation order randomised across tasks and subjects). Subjects were prompted *How would you categorise this representation?* Twelve response options were provided: 1. Graph or Chart; 2. Icon/Logo; 3. List; 4. Map; 5. Node & arc/network; 6. Notation system/formula; 7. Plan; 8. Maths or scientific figure/illustration; 9. Set diagram; 10. Table; 11. Text/linguistic and 12. Tree.

Second Semantic task (knowledge of NPVs' function) This task assessed subjects' functional knowledge of each non-picture visual in the corpus. Each item was displayed and the subject instructed: *"... What is its function? (select one). Note: if you think it may have more than one function, decide which is the most important or primary function."* The 12 response options were: 1. *Shows patterns and/or relationships of data at a point in time*; 2. *Shows patterns and/or relationships of data over time*; 3. *Shows how/where things are distributed or located*; 4. *Relates time and activities*; 5. *Shows how things are organised, arranged, interconnected or interrelated*; 6. *Represents a procedure or algorithm*; 7. *Shows how things proceed, evolve or work*; 8. *Shows how to do things, instructions*; 9. *Language-based (linguistic) communication*; 10. *Formally expresses a mathematical, logical or scientific concept*; 11. *Aid to memory*, and, 12. *Symbolizes an idea or concept.* The functional categories were derived from a sourcebook [30] and literature on the functional roles of diagrams [17, 33].

Production - assessed via a naming task One NPV item per trial was presented. Subjects were prompted "*What is it?*". Participants responded by choosing one name from 77 specific names for each of the 90 NPV items. The names were organised by category (maps, plans, notations, set diagrams, etc) in the response interface. Participants were instructed to select one name by clicking a response button adjacent to it. Examples of names provided include: 'Venn diagram', 'music', 'simple list', 'timetable', 'truth table', 'box plot', 'scatterplot', 'Gantt chart', 'entity relation (ER) diagram', 'decision tree', 'syllogism', 'logo', 'anatomical illustration', 'logic', 'pictogram', 'engineering drawing', 'assembly instructions', 'floor plan'...etc. Participants were provided with 90 specific names and were constrained to one selection per trial.

The tasks were administered online and the order of presentation of NPVs was randomised across subjects. The first session (decision task) was of approximately 30 minutes duration. Participants performed the categorisation, functional knowledge and naming tasks in a second session of approximately 90 minutes duration. In each task, subjects were presented with the NPVs sequentially, one at a time. Each participant performed the four tasks in the order 1. decision, 2. categorisation, 3. functional knowledge, and 4. naming. The decision responses were scored by assigning 1 for correctly identified items (maximum score 112). For the categorisation, functional knowledge and naming tasks each correct response scored 1 (maximum 90). Proportion correct scores for each subject on each item on each task were computed. Mean proportion correct (mpc) across all NPV items within each task (recognition, semantic categorisation, semantic functional knowledge and naming) were computed for each subject (4 task mpc's per subject, n=54).

3 Results and Discussion

Models of picture processing predict that to name an object an individual requires some conceptual understanding of the object - i.e. its category membership and meaning. To test this prediction with respect to NPVs, an initial analysis was conducted to determine the number of items correct in naming and in semantic categorisation of the NPVs. Table 1 presents the mpc scores for all subjects on each task. It was determined that the proportion scores were normally distributed with skewness within acceptable limits (+/- 2). Means rather than medians were therefore used.

Table 1. Mean proportion correct (mpc) scores for each task (n=54)

Task	min	max	**mpc**	sd	skewness
Recognition	.36	.93	**.77**	.10	-1.856
SemCat	.42	.82	**.66**	.09	-.175
SemFunc	.32	.74	**.50**	.10	.202
Naming	.30	.78	**.57**	.09	-.371

Table 2. Between-task correlations, (n=54, ** = $p < 0.01$, 2-tailed)

Task	SemCat	SemFunc	Naming
Recognition	.25	.38**	.14
SemCat		.53**	.74**
SemFunc			.55**

The mpc was greater for semantic categorisation than for naming (.66 vs .57, Table 1). This difference was statistically significant (t(related)=9.40, df=53, p<.001). The correlation between naming and semantic categorisation was positive and significant (ρ = .74, Table 2). Analyses of individual subjects' scores revealed that 50 out of 54 achieved scores higher in category knowledge relative to their knowledge of NPV names. This very consistent pattern provided good support for the model's prediction of a strong association between semantic knowledge representation and naming ability. It should be noted too that in the case of the four exceptions - although they failed to show the predicted pattern, their category semantic scores were nevertheless very close to their naming performance scores (differences were small - range 0.01 to 0.03).

A second analysis addressed whether semantic *functional* knowledge showed a similar relationship with naming performance. In this case the relationship of mpc scores for semantic knowledge and naming were reversed with naming performance being superior to functional knowledge (.57 vs .50, Table 1). This difference was also significant (t(related)=5.55, df=53, p<.001). The correlation between functional knowledge and naming was positive and moderate ($\rho = .55$).

The finding that naming performance was superior to functional knowledge occurred in 48 out of 54 subjects and demonstrated that NPV naming can occur in the absence of functional knowledge.

A third analysis examined the relationship between semantic knowledge and perceptual recognition of the NPVs. The mpc for the decision (recognition) task was .77. NPV recognition was not highly correlated with either semantic (category) knowledge ($\rho = .25$) or functional knowledge ($\rho = .38$). This finding is aligned with the process model's prediction that individuals can be proficient in recognising NPVs despite the absence of semantic knowledge about NPVs. The correlation of naming with categorisation ($\rho = .74$) was much larger than between naming and functional knowledge ($\rho = .55$, Table 2), the relationship between the two types of semantic knowledge was modestly positive ($\rho = .53$). The mpc on the semantic categorisation task was also significantly greater than semantic functional knowledge (.66 vs .50; t(related)=12.00, df=53, p<.001, Table 1). This supports the claim that people can categorise NPVs without necessarily knowing their function and suggests that the two semantic processes are less tightly coupled in the NPV domain than in the object picture domain. Proficiency in NPV recognition can be present in the absence of semantic knowledge of category or function. Analyses of individuals' data showed that a majority of subjects achieved higher recognition (decision) scores in recognising NPVs relative to their comprehension of these forms. Subjects who show proficiency in semantic and functional knowledge of NPVs showed comparable or better recognition performance.

3.1 Age of Acquisition Effects Upon NPV Naming

As mentioned in the introduction, Garcia Garcia & Cox [10] used the UK National Curriculum to identify the forms of representation that are introduced to students at ages 5 to 9 years and forms that are introduced later (9-14 years). The earlier forms include illustrations, tables, lists, bar charts, maps, flow charts and Venn diagrams. Later-introduced forms include entity-relation diagrams, network diagrams, tree diagrams, pie charts and Cartesian coordinate forms such as line graphs. The NPV corpus was therefore partitioned into 'early' and 'late' representations. The mpc for the naming task for early items was .64 and for late items it was .50. A related samples t-test showed that the difference was significant (t=7.48, df=53, p<.001). Age of acquisition effects are therefore seen in NPV naming. In fact, there were significant differences between performance on 'early' acquired and 'late' acquired items for all 4 tasks. Recognition task means were: early NPV items=.86, late=.76; categorisation task, early=.66, late=.58; functional knowledge, early=.52, late=.39. Taken together, these results provide support for the idea that the processing of NPVs shares characteristics and processes similar to those observed in object picture processing studies.

3.2 Functional Knowledge and Naming of NPV Item Categories

The NPV corpus contained 12 categories of item. Figure 3 shows the mpc scores for each category of NPV item, ordered by naming score. In general the pattern reflects the ubiquity of the representations in school subjects, the media and everyday life (text, lists, notational systems such as music, maps, scientific illustration, plans, graphs and charts). These representations are somewhat generic and are used in a wide variety of contexts. Maps, for example, are topographically isomorphic with the real-world and commonly used for navigation in everyday life. More specialist, domain-specific forms (network or node & arrow diagrams, set diagrams) seem to be less well understood and named. The representations on the right-hand side of Figure 3 tend to be less pictorial and textual and more purely diagrammatic in that they relate two or more dimensions by segmentation (tables) or metric axes (graphs). Functional knowledge of tables, trees, and node and arrow (network) diagrams (matrices, networks and hierarchies) are related to each other as forms of representation but they differ subtly in terms of global structure, their basic features (cells, nodes/edges and constraints on mappings between representation and domain represented [17]. For example, understanding set diagrams requires an understanding of their metaphor i.e. that container-contained, overlap, spatial disjunction, shading conventions and so on graphically represent the relationships between sets.

Respondents' functional knowledge of tables and naming accuracy for tables was rather low despite their ubiquity in everyday life. Examining individual items, 96% of respondents correctly named a timetable and a logic truth table was correctly named by 70% of respondents. A bidirectional table was correctly named by 50% and a feature table (e.g. a table with ticks in cells showing features present/absent) by 54%, a multi-way table was only correctly named by 13% of respondents.

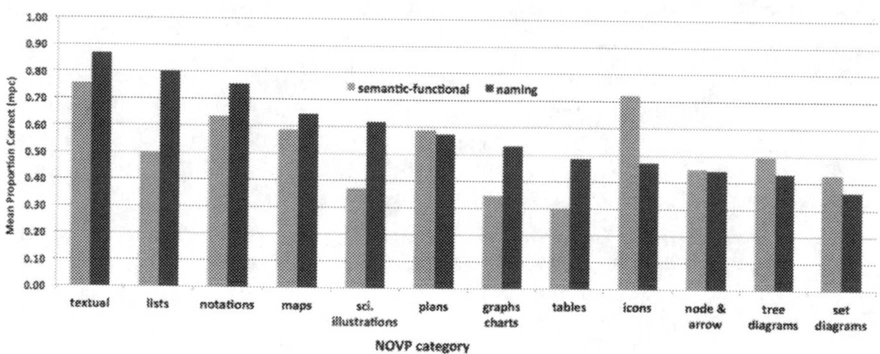

Fig. 3. mpc scores for NPV item categories, ordered by naming score

The general pattern of results reflect the age of acquisition effects mentioned in the previous section with the exception perhaps of set diagrams for which naming mpc was lowest and functional knowledge was also low. However, the category of set diagrams in the NPV corpus also contained Euler's circles and a figure of the Gergonne relations. Two of the NPV items were Venn diagrams and they were correctly named by 67% and 54% of respondents, respectively. On the Venn diagram items the erroneous responses were most frequently "Euler diagram" and (to a lesser extent) "geometry figure". For the two Euler's circle items the error rate was high (74% and 78%) with "Venn diagram" accounting for 66% of the errors. This suggests that "Venn diagram" is synonymous with "set diagram" for many respondents. In the case of tree diagrams the best-named was a grammar parse tree (76%) followed by a decision tree (54%). A dendrogram and a branching probability tree were not accurately named (26%, 17%). The table and tree diagram items in the corpus were therefore quite heterogeneous, with subjects naming common examples relatively well but performing less well on more esoteric forms.

The largest difference between functional knowledge and naming scores was observed for lists and scientific illustration (functional < naming). Participants easily and correctly named most of the examples of lists (a shopping list, a recipe, medication times, a directory,..). In terms of functional knowledge of lists 44% of respondents correctly nominated "memory aid" as the function of a shopping list. 44% correctly gave "how to do things/instructions" as the function of a list of medications and when to take them (the most frequent error was "memory aid"). 78% stated that a recipe's function is to show "how to do things, instructions"[5].

Scientific illustrations were relatively accurately named (geometry figures, anatomical illustration, a physics illustration, a mechanical diagram (pulley system)). Perhaps subjects readily recognised the domain from which the representation originated and that gave them an indication of its name (e.g. anatomical diagram)? However for functional knowledge, subjects found it difficult to discriminate between NPVs that 'show where objects are distributed or located', 'show patterns or relationships in data at a single point in time', and 'show patterns or relationships in data over time'. Assigning the correct function for this class of NPVs requires quite specialised knowledge and knowledge of domain-specific graphical conventions.

The iconic NPVs showed greater functional knowledge scores than naming scores. This may have been because for 9 of the 10 examples the function was simply "symbolise a concept". An exception was an example of a Chinese character for which the correct name ("pictogram") was only given by 9% of respondents (the most frequent error was "handwriting"). However 57% respondents correctly gave "communication" as the character's function, with 43% opting for "symbolise a concept".

[5] Though two subjects nominated "algorithm" which perhaps should have been scored as correct...

Conclusions NPVs are processed similarly to object pictures in that recognition performance can be observed in the absence of semantic and naming knowledge. Effects akin to the age of acquisition effect were observed on recognition, categorisation, naming, and functional knowledge tasks. NPVs differ from visual objects in that functional knowledge is frequently seen to be lower than naming accuracy. However, the relationship between semantic and naming performance varies across sub-categories of NPV. Response patterns like those seen for object pictures (naming > semantic knowledge) are observed for more textual forms, maps and illustrations. For more purely 'diagrammatic' NPV forms (network diagrams, tree diagrams, set diagrams) semantic knowledge tends to be greater than naming accuracy. This difference between the more textual and graphical information representation modalities probably reflects the interaction of several factors - familiarity, ubiquity in the real-world, and age-of-acquisition effects. In the case of network, tree and set diagrams there is a tendency for them to be known by different names across different domain contexts - computer scientists tend to refer to network diagrams as 'graphs', for example. In everyday contexts, 'Venn diagrams' tends to be used as the category term for set diagrams, tree diagrams can be referred to as directed acyclic graphs or hierarchies and column graphs are routinely termed 'bar charts'. In this study, correct NPV responses often required specific and precise naming whereas in object naming studies using the Snodgrass stimuli, category names are often accepted in lieu of precise names (e.g. 'saw' for a depiction of what is actually a panel saw, 'hat' for trilby hat). Another difference between the NPV study reported here and typical picture recognition/naming studies is that in the latter the dependent measure tends to be response latency rather than response accuracy because many people perform at ceiling on naming accuracy tasks when pictorial items such as those in the Snodgrass and Vanderwart corpus are used. It would be useful to investigate response latency in addition to accuracy in future NPV studies. In this study non-pictorial images such as diagrams appear to be processed similarly to pictures of objects in that people tend to be able to recognise and categorise them better than they can name them or identify their function. Non-pictures such as diagrams differ from pictures of objects in that the age at which knowledge of them was acquired, or their commonplaceness, affects performance not just on naming task but also on recognition and functional knowledge. To conclude, examining NPVs from an information processing perspective has the potential to inform how we teach people to use diagrams and provides a basis for understanding data on human performance with external representations. Information processing approaches that are commonly used in studies of picture recognition and naming provide a useful theoretical framework for understanding how graphs and diagrams are cognitively processed.

Acknowledgements. Thanks to the reviewers, to the study participants, to Jonathan Kilgour for programming assistance and to Beate Grawemeyer, Grecia Garcia Garcia and Pablo Romero for assistance with data collection.

References

1. Humphreys, G.W., Riddoch, M.J.: Visual object processing: A cognitive neuropsychological approach. Lawrence Erlbaum Associates, Hillsdale (1987)
2. Shah, P., Freedman, E.G., Vekiri, I.: The comprehension of quantitative information in graphical displays. In: Shah, P., Miyake, A. (eds.) The Cambridge Handbook of Visuospatial Thinking, pp. 426–476. Cambridge University Press (2005)
3. Pinker, S.: A theory of graph comprehension. In: Freedle, R. (ed.) Artificial Intelligence and the Future of Testing, pp. 73–126. Lawrence Erlbaum Associates, Hillsdale (1990)
4. Cripwell, K.R.: Non-picture visuals for communication in health learning manuals. Health Education Research 4(3), 297–304 (1989)
5. Baddeley, A.: Working memory, thought, and action. Oxford University Press (2007)
6. Snodgrass, J.G., Vanderwart, M.: A standardized set of 260 pictures: Norms for name agreement, image agreement, familiarity, and visual complexity. Journal of Experimental Psychology: Human Learning and Memory 6, 174–215 (1980)
7. Johnston, R.A., Dent, K., Humphreys, G., Barry, C.: British-English norms and naming times for a set of 539 pictures: The role of age of acquisition. Behavior Research Methods 42(2), 461–469 (2010)
8. Morrison, C., Hirsh, K.W., Duggan, G.B.: Age of acquisition, ageing, and verb production: Normative and experimental data. The Quarterly Journal of Experimental Psychology 56(4), 705–730 (2003)
9. Johnston, R.A., Barry, C.: Age of acquisition and lexical processing. Visual Cognition 13(7/8), 789–845 (2006)
10. Garcia Garcia, G., Cox, R.: Diagrams in the UK national school curriculum. In: Stapleton, G., Howse, J., Lee, J. (eds.) Diagrams 2008. LNCS (LNAI), vol. 5223, pp. 360–363. Springer, Heidelberg (2008)
11. Cleveland, W.S., McGill, R.: Graphical perception: Theory, experimentation, and application to the development of graphical methods. Journal of the American Statistical Association 79(387), 531–554 (1984)
12. Canham, M., Hegarty, M.: Effects of knowledge and display design on comprehension of complex graphics. Learning & Instruction 20(2), 155–166 (2010)
13. Shah, P., Freedman, E.G.: Bar and line graph comprehension: An interaction of top-down and bottom-up processes. Topics in Cognitive Science, 1–19 (2009)
14. Shah, P., Hoeffner, J.: Review of graph comprehension research: Implications for instruction. Educational Psychology Review 14(1), 47–69 (2002)
15. Lohse, J.: A cognitive model for the perception and understanding of graphs. Human-Computer Interaction 8, 353–388 (1993)
16. Trafton, J.G., Trickett, S.B.: A new model of graph and visualization usage. In: Moore, J.D., Stenning, K. (eds.) Proceedings of the 23rd Annual Conference of the Cognitive Science Society, pp. 1048–1053. Lawrence Erlbaum Associates, Mahweh (2001)
17. Novick, L.R., Hurley, S.M.: To matrix, network or hierarchy: That is the question. Cognitive Psychology 42, 158–216 (2001)
18. Tversky, B.: Spatial schemas in depictions. In: Gattis, M. (ed.) Spatial Schemas in Abstract Thought, pp. 79–111. MIT Press (2001)
19. Hollands, J.G., Spence, I.: The discrimination of graphical elements. Applied Cognitive Psychology 15, 413–431 (2001)

20. Trickett, S.B., Trafton, J.G.: Toward a comprehensive model of graph comprehension: Making the case for spatial cognition. In: Barker-Plummer, D., Cox, R., Swoboda, N. (eds.) Diagrams 2006. LNCS (LNAI), vol. 4045, pp. 286–300. Springer, Heidelberg (2006)
21. Tabachneck-Schijf, H.J.M., Leonardo, A.M., Simon, H.A.: CaMeRa: A computational model of multiple representations. Cognitive Science 21(3), 305–350 (1998)
22. Leinhardt, G., Zaslavsky, O., Stein, M.K.: Functions, graphs, and graphing: Tasks, learning, and teaching. Review of Educational Research 60(1), 1–64 (1990)
23. Garcia Garcia, G., Cox, R.: "Graph-as-picture" misconceptions in young students. In: Goel, A.K., Jamnik, M., Narayanan, N.H. (eds.) Diagrams 2010. LNCS (LNAI), vol. 6170, pp. 310–312. Springer, Heidelberg (2010)
24. Cox, R., Romero, P., du Boulay, B., Lutz, R.: A cognitive processing perspective on student programmers' 'Graphicacy'. In: Blackwell, A.F., Marriott, K., Shimojima, A. (eds.) Diagrams 2004. LNCS (LNAI), vol. 2980, pp. 344–346. Springer, Heidelberg (2004)
25. Grawemeyer, B., Cox, R.: The effects of users' background diagram knowledge and task characteristics upon information display selection. In: Stapleton, G., Howse, J., Lee, J. (eds.) Diagrams 2008. LNCS (LNAI), vol. 5223, pp. 321–334. Springer, Heidelberg (2008)
26. Lohse, G.L., Biolsi, K., Walker, N., Rueter, H.: A classification of visual representations. Communications of the ACM 37(12), 36–49 (1994)
27. Cox, R.: Representation construction, externalised cognition and individual differences. Learning and Instruction 9, 343–363 (1999)
28. Cox, R., Grawemeyer, B.: The mental organisation of external representations. In: Schmalhofer, F., Young, R., Katz, G. (eds.) Proceedings of the 1st European Cognitive Science Conference (EuroCogSci 2003), Lawrence Erlbaum Associates, Osnabrück (2003)
29. Twyman, M.: A schema for the study of graphical language. In: Kolers, P.A., Wrolstad, M.E., Bouma, H. (eds.) Processing of Visible Language, vol. 1, pp. 117–150. Plenum Press, New York (1979)
30. Harris, R.L.: Information graphics: A comprehensive illustrated reference. OUP, Oxford (1999)
31. Lockwood, A.: A visual survey of graphs, maps, charts and diagrams for the graphic designer. Studio Vista, London (1969)
32. Magnié, M.N., Besson, M., Poncet, M., Dolisi, C.: The Snodgrass and Vanderwart Set Revisited: Norms for Object Manipulability and for Pictorial Ambiguity of Objects, Chimeric Objects, and Non-objects. Journal of Clinical & Experimental Neuropsychology 25(4), 521–560 (2003)
33. Cheng, P.C.-H.: Functional roles for the cognitive analysis of diagrams in problem solving. In: Cottrell, G.W. (ed.) Proceedings of the Eighteenth Annual Conference of the Cognitive Science Society, pp. 207–212. Lawrence Erlbaum Associates, Mahweh (1996)

Exploring the Effects of Colouring Graph Diagrams on People of Various Backgrounds

Beryl Plimmer[1], Ann Morrison[2], and Hendrik Knoche[2]

[1] Dept Computer Science University of Auckland
beryl@cs.auckland.ac.nz
[2] Department of Architecture, Design and Media Technology
{morrison,hk}@create.aau.dk

Abstract. Colour is one of the primary aesthetic elements of a visualization. It is often used successfully to encode information such as the importance of a particular part of the diagram or the relationship between two parts. Even so, there are few investigations into the human reading of colour coding on diagrams from the scientific community. In this paper we report on an experiment with graph diagrams comparing a black and white composition with two other colour treatments. We drew our subjects from engineering, art, visual design, physical education, tourism, psychology and social science disciplines. We found that colouring the nodes of interest reduced the time taken to find the shortest path between the two nodes for all subjects. Engineers, tourism and social scientists proved significantly faster with artist/designers just below the overall average speed. From this study, we contribute that adding particular colour treatments to diagrams increases legibility. In addition, preliminary work investigating colour treatments and schemes indicates potential for future gains.

Keywords: Colour encoding of diagrams, Graph diagram aesthetics, Subjects Disciplines' Effects.

1 Introduction

The aesthetics of a diagram is acknowledged as an important factor in the ease to which a person can read and reason with the diagram. There has been considerable research into diagram layout and there are now well understood principles as to what layout strategies make a graph diagram easy (or hard) for people to interpret. These layout principles can be encoded into automatic layout algorithms so that the algorithms produce optimum layouts for human understanding.

There are many other elements of a diagram where aesthetic treatments can automatically be applied such as colour, fonts, line treatments, node shapes and sizes and iconic representations. There are bodies of knowledge about these aesthetic elements which could also inform diagram aesthetics. Attending to such aesthetics has been shown to be useful for user interface design [1, 2] reporting increased tolerance towards errors of 'beautiful' websites [2], and a strong correlation between the user's perception of a website's visual aesthetics and their perception of its credibility [3, 4] and trustworthiness [5].

T. Dwyer et al. (Eds.): Diagrams 2014, LNAI 8578, pp. 176–189, 2014.

Colour is a major component of aesthetics. It is often used in graph diagrams to encode information (e.g. Fig. 1) or simply to make the diagram look more pleasing. Yet, to our knowledge, there have been no experiments on the effects on human interpretation of different colour treatments of graph diagrams.

In this paper we report on an empirical shortest path experiment with three colour treatments applied to nodes of non-directional graphs. We selected people from a broad mix of disciplines as the subjects to investigate whether prior training or knowledge impacted on the experience. For example, we sought to determine if engineers would make the same errors, draw the same paths and/or prefer similar colour schemes as artist/designers, or people who specialise in physical education.

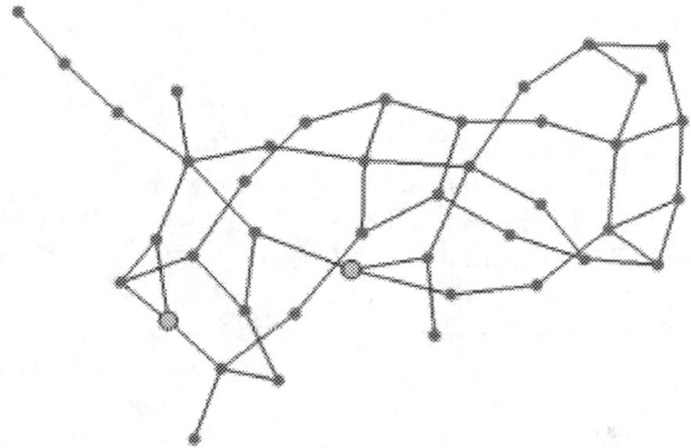

Fig. 1. Diagram used by [6] to explore effects of path straightness

Important, as background for this study, we note that Gestalt theory and psychology argues that we visually identify features and complete shapes rather than seeing an ad-hoc assortment of simple lines, edges and bends as part of the laws of perceptual organization [7]. This 'seeing the whole' enables faster decision-making and uses less cognitive load by calling up information and understandings already gained as prior knowledge [8 ,9]. Furthering this work, Hatfield & Epstein [10] argue the perceptual system reduces its processing to the lowest possible effort by describing the outer world we encounter in the simplest means. Pomerantz [11], adds that colour perception acts in a similar holistic manner to shape perception as understood by Gestalt philosophy. This theory is well-placed to help us identify the role of colour in visual search tasks.

While Art and Design disciplines are clear in their use of colour [12], operating from a strong set of established design and colour guidelines that evolve as the formats change with time, artist/designers do not work from studies that 'prove' their theorems. Rather, they work from established conventions that derive from practice. While there are clearly successful results, this 'experience evidence' and the language that supports it, does not translate easily for more scientific disciplines who require contributions in the form of testing for evidence, findings and conclusions. There is considerable work to be done in the passing on of this knowledge from the art and design communities' perspectives to the diagrams community, with no easy way for direct translation of the accumulated knowledge.

In this paper we explore effects of both colour *treatments* and colour *schemes*. We use the term *treatments* as the semantic of the colour application – for example, colour to particular types of nodes. And colour *schemes* to be the colours applied – for example Fig. 1 the *treatment* is to colour the nodes for their shortest path task differently to all the other nodes on the diagram, the *scheme* is red/cyan.

The structure of this paper is as follows. In section 2 we review the work on diagram aesthetics and colouring of visualizations. Section 3 describes the experimental stimulus used including our design choices. Section 4 describes the methodology for the experiments followed by the analysis and results in Sections 5 and 6. Finally in Section 7 we discuss our findings, the implications of these findings and suggest future research.

2 Related Work

The work of Purchase [13] on the aesthetics of diagram layout has been the basis of many investigations into the aesthetics of graph diagrams. Purchase showed that there are a number of layout principles that can be systematically applied to a diagram that make the diagram easier for people to understand. The most important of these principles is to decrease edge crossings. Purchase and others have expanded on these findings in various contexts and with dynamic graphs [e.g. 14, 15]. Purchase reported that the most significant layout aesthetics are to minimise edge crossings and edge bends and maximise symmetry. These findings have stood the test of time and are now regularly prioritised in layout algorithms [e.g. 16]. Without exception these studies have been conducted without regard to colour. Most studies have used black diagrams (nodes, edges and text) on a white background.

Colour is another fundamental aspect of aesthetics and visualizations. The perceptual and cognitive effects of colour have been studied for over 100 years. For more information we refer the reader to [17] that provides an excellent recent survey on the use of colour in visualizations and an overview of colour theory. It is notable that [17] has no human studies examples on the effectiveness of various colouring strategies from the field of abstract diagrams using graphs or set diagrams.

There have been numerous studies and algorithms proposed for automatic computer generation of colour schemes for particular contexts. These include [18] early work on contrasting and harmonious colour schemes for maps, and [19] for graph diagrams. These colouring generators are based on the theories of colour perception (see [20] as an example). However human experiments on the interpretation of the resultant colouring schemes are not reported. Others [21, 6] suggest that human experiments are necessary to validate such automated application of aesthetic elements, such as Ware's experiments for geographical map colouring [22].

Tsonos & Kouroupetroglou [23] investigate font size and styles and font/background colour combinations and found that in terms of colour combinations, black text on white background, white on blue, and green on yellow combinations were rated as the most pleasant.

Art and Design disciplines offer a variety of perspectives on the use of colour that fall beyond the scope of this study, however practical concepts such as colour relativity, intensity and temperature impact directly on the design considerations [24], with placement of colours to cause vibrations and the illusion of depth set up as background considerations.

Many abstract diagrams such as variants of graph diagrams and set diagrams use colour. There are a variety of ways colour is employed. Some use colour as a part of the syntax, for example [25, 6] use colour to differentiate nodes of interest. Others use colour as an extra level of information, for example [26] suggests the use of colour for alerting readers to important aspects of a diagram and [27] uses colour to indicate the degree of constraint satisfaction. In other cases, colour is used as eye-candy rather than for information for example [28]. None of these papers discuss the colour choices or reasons for them. So far as we are aware the only empirical studies into the effects of different colouring strategies are on concrete diagrams such as geospatial maps. We make an attempt with this initial study, to see if design understandings of colour can be translated into scientific evidence.

3 Experiment Design

Clearly there is unlimited space for experimentation when applying colour to graph diagrams, with an infinite number of colours and colour combinations– at the extreme every node, edge and character could be a different colour. Choices must be made in order to produce empirical results. In this section we describe the experimental stimulus that we designed together with our reasoning for particular decisions.

Given that there is no prior work with abstract diagrams we chose to limit the experiment to fundamentals. We selected to experiment on fundamental graph diagrams – those with just nodes and edges. Based on Purchase's [13] graphs we devised three diagrams with equal numbers of nodes (16) and edges (28) and a similar distribution of edges per node, in addition each has an identifiable sub graph (see Fig. 2). We applied Purchase's [13] layout principles to the graphs so that there were no crossing edges, all edges were straight and that the diagrams were generally symmetric. This was a manual process as the layout algorithms in the software we used gave unsatisfactory results. We then flipped each diagram horizontally, vertically and horizontally and vertically resulting in 12 variants of the diagrams.

We then chose 3 different node pairs on each diagram for our shortest path questions. The paths varied in length from 2-5 edges. We visually identified these nodes in each diagram by allocating the Font Arial Rounded MT Bold, 18pt to the nodes that needed to be easily identified for answering the questions, whereas all other nodes were Helvetica Regular 10. We did this for all graphs regardless of colour treatments and schemes. The flipped graphs also had their label's font correspondingly flipped for easy legibility, that is, so the font was the correct way up.

We ran an informal experiment with 6 subjects and black & white rendering of the flipped diagrams to check whether people would recognize that there were only 3 diagrams. None of the subjects noticed this and all were surprised when we pointed it out to them.

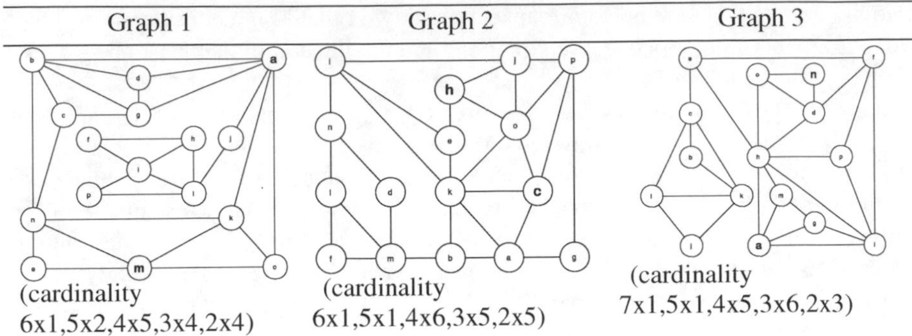

Fig. 2. Three graphs used in experiment with black and white treatment. Nodes of interest have larger font and bolded labels. Cardinality refers to number of joining lines to one node. We refer to the black and white treatment as Set 1.

A major decision was on which colour treatments and schemes to apply. Black and white with the nodes of interest bolded as described above is retained as the default case. We considered a wide range of colouring treatments including: colouring all nodes the same, colouring nodes depending on their degree of connectiveness (cardinality), colouring nodes of interest as a highlighting strategy (design), colouring to highlight sub-regions differently (gestalt). From these we chose the two later treatments: nodes of interest (Set 2) and gestalt (Set 3).

The next decision was the colour schemes – that is what colours to use. We devised 3 colour schemes each consisting of two web safe colours. The colours we chose can be easily differentiated by the majority of the sighted public (taking into account various colour blindness conditions). The colours also largely cover the red-green-blue, red-yellow-blue primary colours spectrum. The colours for each graph were: Graph 1: D2FFFF (Pale Blue) and A5D5D5 (Grey_Green); Graph 2: FF0000 (Red) and FF6600 (Orange); and Graph 3: FF2425 (Scarlet) and A9ACD2 (Lilac) (see Fig. 3). A different colour scheme was applied to each graph.

We chose the colour combinations to create distinctions [12, 24] so information would sit on different levels, and to use primary (basic, pure), secondary (mix of two primaries) and tertiary (mix of primary and secondary) colours from intense saturated hues to luminance light-reflecting tints across the entire colour spectrum (designing around colour blindness). In the choices we made, we layered and separated out the nodes from the white background and the black text by using:

- Graph 1: Grey_Green/Pale Blue - cold, dull-hue, analogous (close to each other in the colour wheel) passive (visually receding) tints (colour added to white) with a high luminance (light reflective) value. We refer to this scheme as *calm*.
- Graph 2: Red/Orange – warm, active (visually appearing to advance to the foreground), highly-saturated (pure, undiluted by white), bright, intense analogous colours. We refer to this scheme as *bright*.

- Graph 3: Scarlet/Lilac - contrast between a warm, active, intense, saturated secondary colour (scarlet) and a cold, passive, dull, high luminance tertiary tint (lilac). We refer to this scheme as *contrast*.

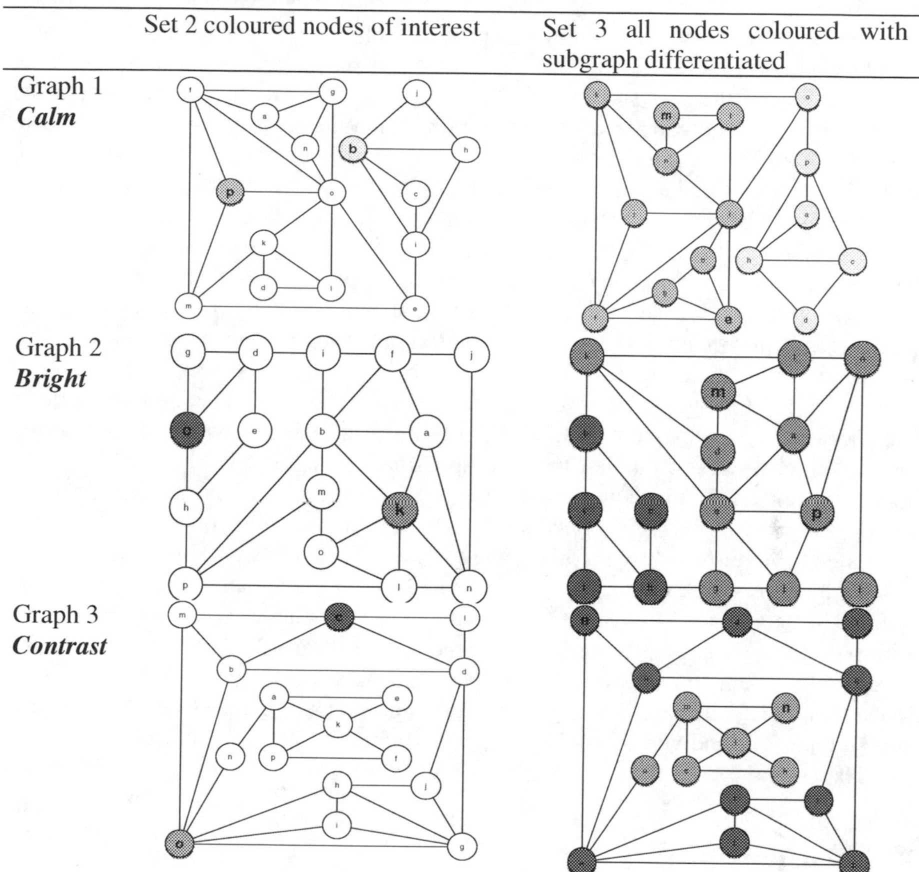

Fig. 3. Two sets of colouring treatments with three colour schemes

4 Methodology

The task selected was *shortest path* where the subjects were asked to mark the path that traversed the least number of edges between the two marked nodes. From the graph sets described in section 3 we randomly selected 9 graphs from each set (B&W, coloured nodes, coloured subgroups). For sets 2 and 3 the random selection was to randomly take 3 from each colour scheme. The selected graphs were added to a set of Powerpoint slides. Before these experimental graphs there were 3 training sections. The first was to familiarize the subjects with drawing on Powerpoint with a stylus. They were asked to draw a house and then on the following slide a person and any other object. This was followed by an explanation with examples of how to identify a shortest path. There were then 6 simple black and white graphs with a pair of bolded nodes that they completed with the facilitator. Following this they were told that the real experiment was beginning, however there were a further 6 training slides with increasingly complex graphs. The experimental graphs were then presented in the order black and white, coloured nodes and coloured subgroups. We chose not to counterbalance the sets to simplify the experiment, reasoning that with sufficient training (12 puzzle slides plus learning drawing with 3 slides before the experiment began), there should be no learning effect, and indeed the analysis of results indicated that this was true. In addition, based on pre-tests of fully randomized presentation in which the subjects found it difficult to identify the different colour schemes and even colour treatments, we created separate sets for each colour treatment, where for set 2 and 3 the colour schemes were in a fixed random order. Usually we would counterbalance the sets for order effects but it was more important that our subjects were able to reliably identify their preferences which they were not able to achieve in our randomized pilot study.

The subjects were asked to mark the shortest path on the Powerpoint slide and were given a maximum of 10 seconds per diagram to complete the task. The sessions were videoed and the time spent on each diagram and errors were later extracted from the video footage. After finishing all the tasks, the subjects completed a questionnaire on their perception of the effect of colouring the graphs and ended with a verbal debrief with the facilitator to elicit more qualitative information.

In selecting subjects we aimed for a range of disciplines from more visual (artistic) to more engineering styles of thinkers with a mix in between this polarized spectrum. To this end, we purposefully recruited a broad mix of 48 subjects. This mix was comprised of subjects with ages ranging from 20-38 and approximately even gender mix with 19 females and 29 males. Of these their background was: 18 artist/designers (8 female, 10 male); 8 male engineers; 8 mixed skill set of (2 more visually oriented females and 3 female and 3 males from engineering psychology); 6 physical education students (1 female and 5 male); and 8 tourism or social science students (5 female and 3 male). The individual sessions were facilitated by 28 students from a Masters User Experience Design class as part of their course work.

5 Analysis

From our 48 subjects, each solving 3x9 diagrams, we ended up with a total of 1296 solved diagrams. We investigated *completion time* and *errors* as dependent variables. *Completion time* was measured from the point at which the diagram appeared on the screen until the subjects had completed drawing the path from start to finish. This could include repairs in case a subject discovered having drawn a path that was too long and then drew another that was the shortest. Completion time was recorded with an accuracy of one second derived from the start and end time readings as identified and extracted from the video recordings. An *error* was scored when the subjects drew and settled on a path that was not the shortest.

Our independent variable was the

- Three different *sets* (*black and white, coloured nodes* and *coloured subgroups*) and

For control variables we had

- The three different *diagrams* (*graphs 1, 2 and 3*) paired with three *colour schemes* (*calm, bright and contrast*) and black and white
- The number of times the subjects had seen the diagram before during the experiment (*diagram repetition*),
- The number of times the subjects had seen the diagram with the same start and end points (*puzzle repetition*),
- Length of the solution counted in nodes between the start and the end node excluding these terminal nodes (*solution length*),
- The five different *disciplines* (*artist/designers, engineers, mixed skill set, physical education, tourism/social science).*
- Age and
- Gender.

Before introducing the control variables as predictors of *completion time* in a multiple regression we tested them individually. We checked for effects of *gender* (t-test), *discipline* (ANOVA) and *age* (single factor regression) on the subjects' averaged *completion time* and for effects of *diagrams, diagram repetition, puzzle repetition and solution length* (single factor regressions) on *completion time. Discipline* ($F(4,43)=4.25$, $p<.01$) and *solution length* ($b=1.33$, $t(1)=19.5$, $p<.0001$), were significant in predicting *completion time* as single factors and included in the subsequent analysis. We found no significant differences for *gender* $t(1)=-0.78$, $p=.439$, *age* $t(1)=3.94$, $p=.763$, *diagrams* $t(2)=0.32$, $p=.725$, *puzzle repetition* ($b=0.35$, $t(1)=3.94$, $p=.763$) or *diagram repetition* $t(1)=0.6$, $p=.549$ and excluded these from the subsequent multiple regression analysis.

We introduced *sets, demographics* and *solution length* as predictors for *completion time* into a multiple regression. Since we had this range of potential explanatory variables and could not refer to existing theory on which to base the model selection we carried out a stepwise linear regression. At each step the variable with the largest predictive power from the set with at least .05 significance was chosen and included in the model. Previously chosen predictors were removed from this set if their significance level went above a .10 cutoff.

In total our 48 subjects made 51 errors solving the diagrams. We used a logistic regression to analyze the independent and control variables as predictors for *error*.

6 Results

6.1 Completion Time

Controlling for all significant variables the regression showed that the subjects completed Set 2 with the coloured nodes of interest faster than the other two sets ($b=-.32$, $t(2)=-3.60$, $p<.001$) on average it took them 0.5 seconds less. *Solution length* significantly predicted *completion time* ($b=1.29$, $t(1)=378.2$, $p<.0001$) - the longer the solution path was the more time it took the subjects to complete the diagram (this is consistent with [6]). Our Set 2 had on average shorter solutions (2.9) than Set 1 (3) and Set 3 (3.4) but this was controlled for in the regression and the result that Set 2 was faster than Set 1 and 3 as described above still holds true.

Discipline (Fig. 4) was another variable that significantly predicted *completion time* (t(4)=31.6, p<.001). *Engineers* and *tourism/SocSci* were faster and both the *physical education* and *mixed skill set* were slower than the average (see Eq (1) below for the amounts of time according to the regression model). It should be noted that the data from subjects from each discipline were collected with the same procedure but different researchers facilitated the experiments.

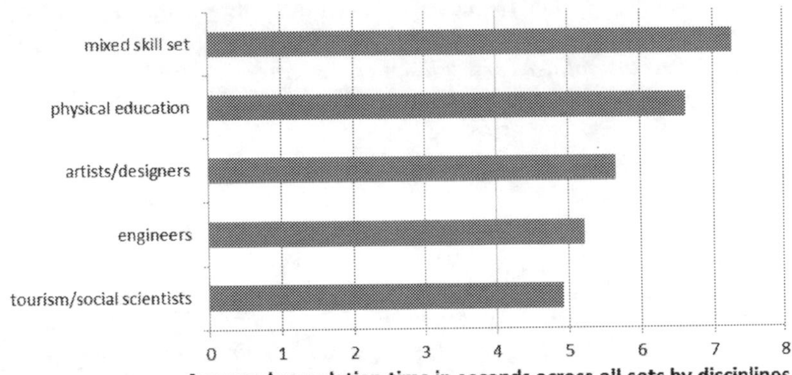

Fig. 4. Completion times by disciplines

$$ct = 3.1 + 1.3 \times sl + \begin{cases} art, design: -0.3 \\ engineers: -0.7 \\ mixed\ skill\ set: 1.3 \\ physical\ educ: 0.7 \\ tourism, SocSci: -1 \end{cases} + \begin{cases} set1: +0.1 \\ set2: -0.3 \\ set3: +0.2 \end{cases} \quad (1)$$

We tested for an interaction effect between *disciplines* and *sets* to see whether a given discipline benefited more or less from the different *sets*. The regression showed no evidence for an interaction indicating that our discipline groups were not differently affected by the sets. We also tested for an interaction effect of *solution length* and *sets* to check whether the subjects' performance for a given *solution length* was affected differently in the three *sets*. The regression yielded no significant effect.

The model to predict *completion time* (*ct*) in seconds of the experiment overall depicted in (1) was based on the three significant variables *solution length* (*sl*), *discipline* and *sets*. This model had a fit of R^2=.32, which is not very high but since our goal was not to model completion time but to understand whether it was significantly moderated by the predictor variables this was of no concern.

In order to be able to analyse the contribution of the different colour schemes without the confounding influence of the diagrams they appeared with, we needed to remove the diagrams as a factor. To this end we employed Set 1(black and white) as a baseline to establish the difficulty of the diagrams. We obtained an estimate of diagram difficulty by using a regression on the Set 1 data only with *solution length* and *disciplines* as variables alongside *diagrams*. All three variables were significant in predicting *completion time*. We then introduced the obtained values as a new variable *diagram difficulty* into our analysis and ran two regressions that compared Set 2 and Set 3 separately with Set 1 with *solution length, disciplines, diagram difficulty* and *colour scheme* as predictor variables. The first comparison (Set 1 vs. 2) looked at the advantage of colour schemes marking start and end nodes distinctly from the other nodes and the second (Set 1 vs. 3) tested how different colour schemes aided partitioning the Graph visually into sub-graphs. We did not include *sets* as a predictor since it could be expressed as linear combinations from *colour schemes*. In both regressions all the predictor variables that had been significant before (*solution length, disciplines,* and *diagram difficulty*) remained significant predictors.

In the Set 1 vs. 2 comparison we found a significant effect for colour combination (p=.004) as a whole. The brighter the intensity and the stronger the saturation of the colours between the background (white) and the colour scheme, the faster the subjects were: with black and white being significantly slower than all other colour schemes (*b*=.32, t(1)=2.91, p=.004), then calm (*b*=.22, t(1)=1.42, p=.15), contrast (*b*=-.27, t(1)=-1.92, p=.19) and bright – the fastest (*b*=-.28, t(1)=2.91, p=.13). However, the differences between the three colour schemes were not significant. For the second comparison of Set 3 and Set 1 we did not find a significant overall effect for *colour scheme* (F(3,847)=1.73, p=.16) despite two of the categorical levels being significant – specifically the bright condition being slowest (*b*=.39, t(1)=2.07, p=.039) and the contrast condition fastest (*b*=-.44, t(1)=-2.05, p=.04).

6.2 Error Rate

Whether or not our subjects made errors in the whole experiment was influenced by *gender*, *age* and *solution length* - they all contributed significantly to the variance of the data in the logistic regression. Errors were more likely to occur for women than for men ($\chi^2(1)=13.81$, p<.001), for younger than for older subjects ($\chi^2(1)=18.5$, p<.0001) and the longer the solution was ($\chi^2(1)=11.3$, p<.001). However, taken together these three variables only explained 13% of the variance of the data and we therefore do not report changes in odd ratios. Errors did not depend on the *diagram repetition, puzzle repetition, diagrams* or *sets*.

6.3 Self-reported Results

The subjects were asked questions about the colour treatments and experience. They were asked to rank the colour treatments by preference (Fig. 5). The majority preferred Set 2 with Set 1 second favourite and just 5 subjects (4f, 1m) preferred Set 3. Interestingly, these 5 were significantly faster in finishing Set 3 than the other subjects and solved both Set 3 and Set 1 faster than Set 2. Secondly, the subjects ranked the colour treatments according to the confidence they had in completing a graph correctly in that condition. These results were almost identical to the preference ranking.

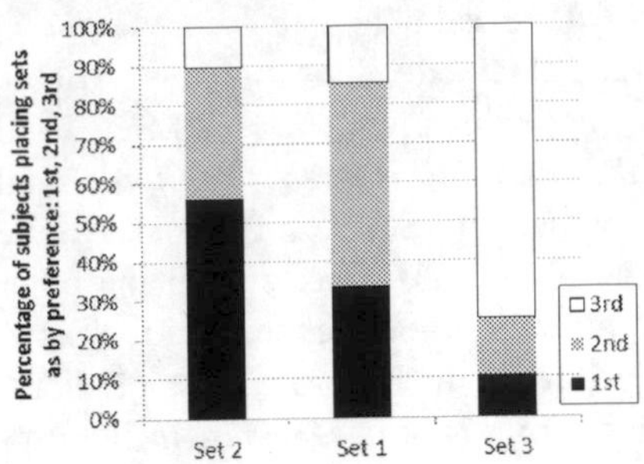

Fig. 5. Subjects rank preference for colour treatments

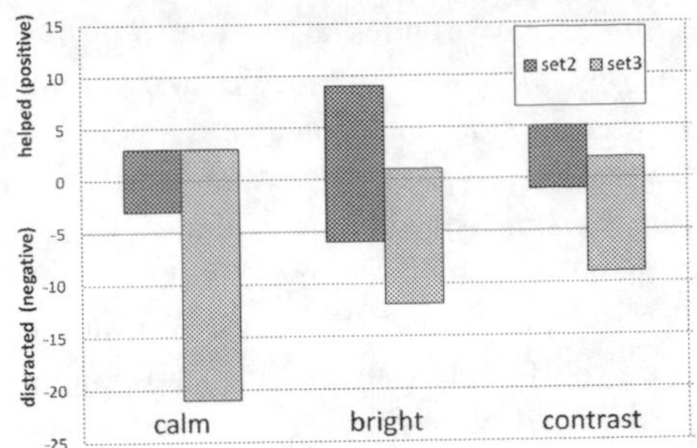

Fig. 6. Number of subjects commenting that the colour schemes helped or distracted in Set 2 and Set 3

In the verbal debrief the subjects commented on their experiences solving the graphs in the different sets. We counted occurrences in which the subjects made references to the colour schemes being helpful in or whether they distracted from solving the graphs. The results are summarized in Fig. 6. For Set 2, the subjects reported that overall the colour schemes overall helped more than they distracted. This is in line with the performance measures. The subjects (9) most often mentioned the *bright* colour scheme as helpful, ahead of *contrast* (5) and *calm* (3). This tendency matches the performance results when comparing Set 1 to Set 2 where *bright* was fastest, *contrast* next, then *calm* with *black and white* being slowest. For Set 3 overall (where subjects were generally slower) the subjects reported the colours as being distracting - with *calm* (21) being worst. In contrast to Set 2 where *bright* was reported as more helpful ahead of *contrast,* in Set 3 more subjects (12) reported *bright* as being distracting than *contrast* (9).

7 Discussion and Conclusions

Our goal with this work is to begin an investigation into the effect of different colouring strategies on graph drawing's legibility. This is an initial investigation to see if results 'understood' in the art and design community and, in the case of our Highlighted Nodes of Interest, often used in the diagrams community, can be proven and useful to the diagrams community. To this end we devised two colour treatments and three colour schemes which we then applied to graph diagrams. Participants from a range of disciplines completed shortest path tasks on the diagrams.

There are a number of limitations of the study and as pointed out in [29] these can affect the results. Clearly we could only choose a limited number of treatments and schemes. Our treatment selections are based on common strategies applied to graph diagrams and gestalt theories of visual grouping. The devised colour schemes are based on visual design practice and perception theory. The execution of the study also poses potential threats to validity. The facilitators were Masters students groups who worked with different discipline groups. While there were comprehensive instructions and training there is always the possibility of varying facilitator bias. Further we chose to fix the order of presentation of the diagrams. The diagrams were clustered into treatments, as pilot studies suggested that randomizing the treatments confused subjects. Likewise we found that fixing the order sets were presented in, assisted subjects to identify the colour treatments and to counter this we included sufficient pre-training examples to negate learning effects.

Of the colour treatments trialled, Set 2 with Highlighted Nodes of Interest, is an undeniable winner for the task tested. Subjects performed significantly better and most preferred it. There are many variations on this theme, such as Fig. 1, that could be further investigated. Certainly it would be straight forward to apply this treatment to interactive diagrams. It could be applied automatically to indicate, for example, faulty nodes in a network. Or as a response to user interaction, for example the user could colour a node of interest to mark it while searching the graph for other nodes.

Set 3, with the sub-graphs differentiated (gestalt), scored similarly to Set 1 (black & white) in performance but Set 1 was preferred by more. There was an interesting effect with 10% (5) subjects preferring Set 3 (gestalt) and preforming better with it: there is an obvious correlation between performance and preference. Why these subjects were different in preference and performance requires more data and investigation. The gestalt colour treatment enables subjects to more easily identify the connection paths between the sub-graphs. This set is interesting from another perspective. As the treatment does not directly relate to the task, the treatment could confound rather than assist subjects: however the visual grouping (seeing the whole gestalt) makes it easier to navigate.

Our colour schemes were based on design theory and visual perception. Using these theories, colour is used to effectively separate and layer information in a manner that results in intuitive correct interpretation of the data. We trialled three schemes each of which was designed to generate a layered effect on the 2-dimensional plane and create some kind of visual impact for the subjects. The calm colours pull the eyes into the background, we wanted to see if these colours would even be noticeable. The bright colours stood out from the background as we wanted to make sure people noticed there was colour. The contrast set created a 3-dimensionsal effect on the 2D plane with the soft cold colour drawing the eye back and the strong bright colour standing out from the background. While the number of data-points we have in each treatment/scheme is quite small, there are statistically significant results. When this is combined with subjects' comments on the effectiveness of different colour schemes it is a clear indication that colour choice matters. Interestingly, the bright colours that stood strongly out from the background were most helpful for Highlighted Nodes of Interest.

We have just scratched the surface of colour application to graph diagrams - we have applied two colour treatments and three schemes to only the nodes. Graph diagrams consist of nodes (that have borders and fill), edges which can also be coloured and have line treatments applied to them (heavier, lighter, dashed etc) and labels which have fonts, colours and weight.

There is an interesting dichotomy in the literature when one reviews the use of colours in graph interpretation studies. Most have been conducted on black and white diagrams similar to our Set 1. Those that have used colour, such as [25, 6] have done so without detailing the reasoning of their choices. Our results show that colour treatment and schemes effect subject performance, so careful consideration for colour choices should be taken. It would be interesting to test whether there are colour treatments that hinder interpretation. For example, if some nodes on a graph are highlighted with colour but these are *not* the nodes of interest for a particular problem, does this slow interpretation? Or do the coloured nodes provide spatial points of reference regardless?

Performance varied significantly in the different disciplines. However, we found colour treatments and schemes had a similar effect across the disciplines. Colours have cultural connotations, but whether these connotations affect the legibility of diagrams remains an open question.

Outside the diagrams community it is unusual to see a plain black and white diagram. General use, together with our results, suggest that colouring of the nodes of interest improves the legibility of the diagram for path-finding. Furthermore our model indicates a trend towards more efficiency with bright and/or contrasting colour schemes in comparsion to duller (calm) colours. Partitioning a diagram using colour, means the colour treatments heighten the underlying graph structure rather than highlighting the start and end points. The results for this treatment are mixed, suggesting further investigation is required. This first investigation into the effects of colouring graphs provides foundational evidence that colouring does effect legibility. More material from design knowledge needs to be translated and tested before a comprehensive set of colour guidelines could be defined and converted to algorithms for automatic application to diagrams.

References

1. Lindgaard, G., Fernandes, G., Dudek, C., Brown, J.: Attention Web Designers: You Have 50 Milliseconds to Make a Good First Impression! Behaviour & Information Technology 25(2), 115–126 (2006)
2. Hartmann, J., Sutcliffe, A., Angeli, A.D.: Towards a Theory of User Judgment of Aesthetics and User Interface Quality. ACM Transactions on Computer-Human Interaction 15(4) (2008)
3. Robins, D., Holmes, J.: Aesthetics and credibility in web site design. Information Processing & Management 44, 386–399 (2008)
4. Alsudani, F., Casey, M.: The effect of aesthetics on web credibility. In: BCS People and Computers, pp. 512–519 (2009)
5. Lindgaard, G., Dudek, C., Sen, D., Sumegi, L., Noonan, P.: An Exploration of Relations Between Visual Appeal, Trustworthiness and Perceived Usability of Homepages. ACM Transactions on Computer-Human Interaction 18(1) (2011)
6. Ware, C., Purchase, H., Colpoys, L., McGill, M.: Cognitive measurements of graph aesthetics. Information Visualization 1(2), 103–110 (2002)
7. Köhler, W.: Gestalt Psychology. Liveright, New York (1929)
8. Köhler, W.: Zum Problem der Regulation (On the problem of regulation). In: Henle, M. (ed.) The Selected Papers of Wolfgang Köhler, pp. 305–326. Liveright, New York (1927)
9. Tufte, E.R.: Envisioning Information, 2nd edn. Graphics Press USA (1990)

10. Hatfield, G., Epstein, W.: The status of minimum principle in the theoretical analysis of visual perception. Psychological Bulletin 97(20), 155–186 (1985)
11. Pomerantz, J.R.: Colour as a Gestalt: Pop out with basic features and with conjunctions. Visual Cognition 14(4-8), 619–628 (2006)
12. Albers, J.: Interaction of Color: 50th Anniversary Edition. Yale University Press (2013)
13. Purchase, H.: Which aesthetic has the greatest effect on human understanding? In: DiBattista, G. (ed.) GD 1997. LNCS, vol. 1353, pp. 248–261. Springer, Heidelberg (1997)
14. Purchase, H.C., Samra, A.: Extremes Are Better: Investigating Mental Map Preservation in Dynamic Graphs. In: Stapleton, G., Howse, J., Lee, J. (eds.) Diagrams 2008. LNCS (LNAI), vol. 5223, pp. 60–73. Springer, Heidelberg (2008)
15. Schneider, E., Boucheix, J.-M.: On Line Elaboration of a Mental Model During the Understanding of an Animation. In: Barker-Plummer, D., Cox, R., Swoboda, N. (eds.) Diagrams 2006. LNCS (LNAI), vol. 4045, pp. 40–54. Springer, Heidelberg (2006)
16. Wybrow, M., Marriott, K., Stuckey, P.J.: Orthogonal Hyperedge Routing. In: Cox, P., Plimmer, B., Rodgers, P. (eds.) Diagrams 2012. LNCS (LNAI), vol. 7352, pp. 51–64. Springer, Heidelberg (2012)
17. Silva, S., Sousa Santos, B., Madeira, J.: Using color in visualization: A survey. Computers & Graphics 35(2), 320–333 (2011)
18. Della Ventura, A., Schettini, R.: Computer aided color coding. In: Communicating with Virtual Worlds, pp. 62–75. Springer Japan (1993)
19. Dillencourt, M.B., Eppstein, D., Goodrich, M.T.: Choosing colors for geometric graphs via color space embeddings. In: Kaufmann, M., Wagner, D. (eds.) GD 2006. LNCS, vol. 4372, pp. 294–305. Springer, Heidelberg (2007)
20. Ware, C.: Information visualization: perception for design. Elsevier (2012)
21. Kosara, R., Healey, C.G., Interrante, V., Laidlaw, D.H., Ware, C.: Visualization viewpoints. IEEE Computer Graphics and Applications 23(4), 20–25 (2003)
22. Ware, C.: Color sequences for univariate maps: Theory, experiments and principles. IEEE Computer Graphics and Applications 8(5), 41–49 (1988)
23. Tsonos, D., Kouroupetroglou, G.: Modeling reader's emotional state response on document's typographic elements. Adv. in Hum.-Comp. Int. 2 (January 2011)
24. Morten, J.L.: Color Logic, Colorcom (2008)
25. Burns, K.: Bar-Gain Boxes: An Informative Illustration of the Pairing Problem. In: Blackwell, A.F., Marriott, K., Shimojima, A. (eds.) Diagrams 2004. LNCS (LNAI), vol. 2980, pp. 379–381. Springer, Heidelberg (2004)
26. Lowe, R., Boucheix, J.-M.: Supporting Relational Processing in Complex Animated Diagrams. In: Stapleton, G., Howse, J., Lee, J. (eds.) Diagrams 2008. LNCS (LNAI), vol. 5223, pp. 391–394. Springer, Heidelberg (2008)
27. Cheng, P.C.H., Barone, R.: Representing Rosters: Conceptual Integration Counteracts Visual Complexity. In: Blackwell, A.F., Marriott, K., Shimojima, A. (eds.) Diagrams 2004. LNCS (LNAI), vol. 2980, pp. 385–387. Springer, Heidelberg (2004)
28. Burns, K.: Bayesian Boxes: A Colored Calculator for Picturing Posteriors. In: Blackwell, A.F., Marriott, K., Shimojima, A. (eds.) Diagrams 2004. LNCS (LNAI), vol. 2980, pp. 382–384. Springer, Heidelberg (2004)
29. Purchase, H.C.: Sketched Graph Drawing: A Lesson in Empirical Studies. In: Wismath, S., Wolff, A. (eds.) GD 2013. LNCS, vol. 8242, pp. 232–243. Springer, Heidelberg (2013)

An Empirical Study of Diagrammatic Inference Process by Recording the Moving Operation of Diagrams

Yuri Sato, Yuichiro Wajima, and Kazuhiro Ueda

Interfaculty Initiative in Information Studies, The University of Tokyo
{sato,ywajima}@iii.u-tokyo.ac.jp, ueda@gregorio.c.u-tokyo.ac.jp

Abstract. In this study, we investigate how people manipulate diagrams in logical reasoning, especially no valid conclusion (NVC) tasks. In NVC tasks, premises are given and people are asked to judge whether "no consequence can be drawn from the premises." Here, we introduce a method of asking participants to directly manipulate *instances* of diagrammatic objects as a component of inferential processes. We observed how participants move Euler diagrams, presented on a PC monitor, to solve syllogisms with universally quantified sentences. In the NVC tasks, 88.6% of our participants chose to use an enumeration strategy with multiple configurations of conclusion diagrams and/or a partial-overlapping strategy of placing two circles. Our results provide evidence that NVC judgment for tasks with diagrams can be reached using an efficient way of counter-example construction.

Keywords: inference process, NVC task, Euler diagram.

1 Introduction

The efficacy of diagrammatic representations in logical reasoning has been discussed in a number of theoretical studies (e.g., [9]). An important assumption among the literatures is that diagrammatic inference is achieved by manipulating diagrams; namely, diagrams are syntactic objects that are manipulated along with inference processes. However, the empirical plausibility of this assumption is unclear.

To provide evidence for the above assumption, Sato et al. [5] compared the performance of three groups: a *Linguistic group* using linguistic syllogistic tasks, a *Venn group* using syllogistic tasks with Venn diagrams, and a *Euler group* using syllogistic tasks with Euler diagrams. The Venn and Euler groups received information regarding the meaning of the diagrams, but were not given instructions about how to manipulate the diagrams. The study reported that the performance accuracy of the Euler group was significantly higher than that of the Venn group, and the performance of the Venn group was significantly higher than that of the Linguistic group. In the experimental setting, the participants in the Venn and Euler groups both had access to semantic information regarding diagrammatic

T. Dwyer et al. (Eds.): Diagrams 2014, LNAI 8578, pp. 190–197, 2014.

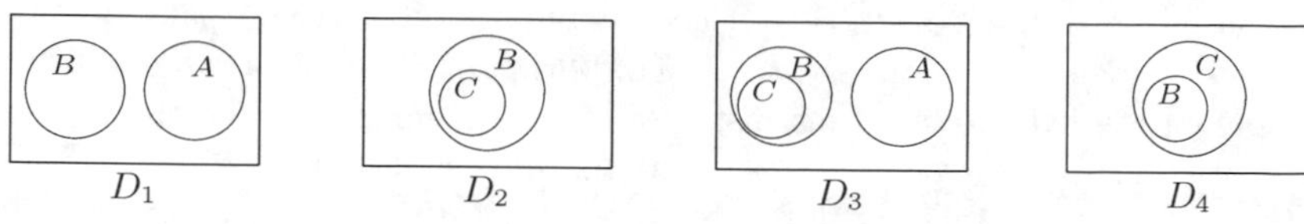

Fig. 1. Euler diagrams for syllogistic reasoning

representations. Therefore, the differences between the Venn and Euler groups appear to be related to additional factors, such as those related to the effect of diagram manipulation for each type of diagram. It can be explained that Euler diagrams are manipulated to solve syllogisms but Venn diagrams are not [6,7].

Based on the above findings, we investigated how diagrams are manipulated during logical reasoning, particularly in no valid conclusion (NVC) tasks. In NVC tasks, premises are given and participants are asked to judge whether "no valid conclusion or no consequence can be drawn from the premises." Take a non-problematic task: i.e., a non-NVC task. Consider a valid syllogism *No B are A, All C are B; therefore, No C are A*. The first premise is represented by D_1 in Fig. 1, and the second premise is represented by D_2. By unifying diagrams D_1 and D_2, one can obtain diagram D_3. Here, circle C is excluded from circle A, corresponding to the correct conclusion *No C are A*. Thus, the process of unifying premise diagrams resulted in a single determinate configuration of circles.

Conversely, NVC tasks are relatively complex [4]. Consider a syllogism with the premises *No B are A* and *All B are C*. The first premise is represented by D_1 in Fig. 1, and the second premise is represented by D_4. Here the process of unifying premise diagrams gives rise to indeterminacy, i.e., there are three possible configurations of circles C and A (C excludes A; C includes A; C and A partially overlap each other). A possible strategy for making judgments in NVC tasks is to enumerate possible configurations of conclusion diagrams. Eventually, this will result in the realization that there is no relationship between C and A holding in all the conclusion diagrams. An alternative strategy is to place circles C and A such that they partially overlap. If such a circle configuration holds, then both *No C are A* and *All C are A* are invalid conclusions. Additionally, when (1) a syllogisms consists of universally quantified sentences or (2) Euler diagrams do not hold the existential assumption for minimal regions, one can answer NVC using the partial-overlapping strategy.

Thus, there are two strategies for making judgments in NVC tasks: the enumeration strategy and the partial-overlapping strategy. However, previous empirical studies have not provided positive support for either strategy. Several investigations of *protocol analysis* have focused on the cognitive processes involved in logical reasoning with diagrams. For instance, Ford [2] collected participants' verbal and written protocols for solving (sentential) syllogistic tasks. They found that half the participants independently constructed Euler diagrams and used them to solve the tasks. However, NVC tasks, which are the focus of the current study, were not assessed in Ford's experiment. A subsequent study by Bucciarelli and Johnson-Laird [1] conducted the syllogistic tasks including NVC tasks by means of a similar protocol experiment. They reported that "they [participants] tended to construct just a single diagram, even for premises that could in principle be represented by several distinct diagrams (p. 265)." This suggests that few participants actually use the enumeration strategy with multiple configurations of conclusion diagrams when completing the NVC tasks.

To improve the situation revealed in the above protocol studies, our experiment controlled for the following factors: (i) We used the logical reasoning tasks with *externally* provided diagrams, since we focus on the inference (manipulation) processes rather than self-construction processes of diagrams[1] (ii) We asked participants to manipulate *instances* of diagrammatic objects[2]. We expected that if the participants directly moved the diagrams they would be less likely to consider information that was external to relative positional relationships when making judgments. In these settings, we examined whether participants used the enumeration strategy and/or the partial-overlapping strategy of diagrammatic inference during NVC tasks.

2 Experiment

We observed the way in which participants moved Euler diagrams presented on a PC monitor when solving syllogisms with universally quantified sentences. In our task, the Euler diagrams had specific sizes (dimensions), and the participants were not able to change the circle size from the default. The default size was determined such that the participants would be able to construct each of the three circle-configurations of conclusion diagrams (i.e., inclusion, exclusion, and partial-overlap) during the NVC tasks.

2.1 Methods

Participants. Twenty-four undergraduate and graduate students from the University of Tokyo (mean age 20.17 ± 1.20 SD) were recruited by means of a poster placed on the campus. All participants provided informed consent and were paid for their participation. All procedures in this experiment were approved by the Ethics Committee of the University of Tokyo. The participants were Japanese-speaking students, and the sentences and instructions were given in Japanese. Two participants were excluded–one for a computer malfunction, and one for response times more than two SD from the mean.

[1] As emphasized in [11], cognitive studies of logical reasoning should distinguish between two types of reasoning: reasoning *toward* an interpretation of premises (i.e., interpretation) and reasoning *from* a fixed interpretation of premises (inference). In reasoning tasks with diagrams, externally provided diagrams strongly impose a particular interpretation of premise sentences, and can selectively contribute to revealing psychologically plausible processes of inferences under the controlled conditions.

[2] For a discussion of diagram *instances* or *tokens*, see [3].

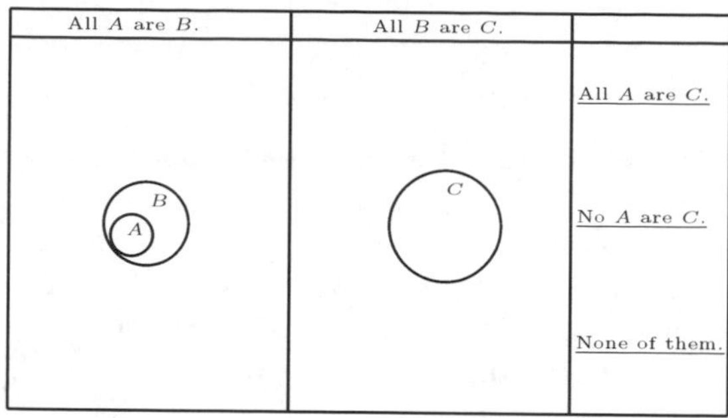

Fig. 2. An example of syllogistic reasoning task with direct manipulation of Euler diagram; the correct answer is *All A are C*

Materials. We displayed syllogisms with Euler diagrams (such as the one in Fig. 2) on a PC monitor. The participants were given two premises and asked to choose the sentence corresponding to the correct conclusion, from a list of three possibilities. The list consisted of the following: *All C are A*, *No C are A*, and *None of them*. Only in the case of task example, as shown in Fig. 2, the term order was presented as *AC*. During the task, participants were also asked to use diagrams, which were shown at the center of the monitor (the lower side of the premise sentences), and to move circle *C* to the left side diagram by drag operation using a mouse (the left side diagram cannot be moved). Based on the coordinate values of the images, we defined a configuration that remained for more than 0.3 seconds in the conclusion diagram as a unit of circle-configuration. The following cases were not considered to be units of circle configuration: (1) the default configuration, and (2) unintentional configurations produced by the mouse movements that were too fast to correctly drag the pictures (screencasts were used to identify these exceptional cases). The task was first displayed such that the premise sentences and diagrams were surrounded by frames, as shown in Fig. 2. These frames were removed after five seconds, and the participants were instructed to start changing the position of circle *C* to solve the syllogistic reasoning tasks. There was no time limit for solving the reasoning tasks.

For simplicity, the premises and conclusions of the syllogisms used in this experiment were universally quantified sentences either of the form *All A are B* or *No A are B*. All the syllogisms used in this experiment are shown in Table 1.

Table 1. Syllogisms used in the experiment

Syllogisms having no-valid conclusion		Syllogisms having valid conclusion	
AE1	*All B are A, No C are B.*	AE2	*All A are B, No C are B.*
AE3	*All B are A, No B are C.*	AE4	*All A are B, No B are C.*
EA3	*No B are A, All B are C.*	EA1	*No B are A, All C are B.*
EA4	*No A are B, All B are C.*	EA2	*No A are B, All C are B.*
		AA1	*All B are A, All C are B.*

We used nine different types of syllogisms in total, out of which 4 syllogisms had no-valid conclusion ("NVC") and 5 syllogisms had a valid conclusion. The valid conclusion of AE2, AE4, EA1, and EA2 syllogisms was *No C are A* ("VC-*no*"), and that of AA1 is *All C are A* ("VC-*all*").

Procedure. The experiment was conducted individually. First, the participants were provided with a one-page document outlining the meaning of Euler diagrams. We then conducted a pretest to determine whether the participants understood the instructions, which focused on the correspondence between universally quantified (affirmative and negative) sentences and Euler diagrams.

Before attempting the reasoning tasks, the participants were given oral instructions regarding the reasoning task and familiarized with the mouse operations required for circle manipulation. Note that the experimenter did not provide any specific instructions about how to manipulate the circle C. One task example was displayed on the PC monitor (Fig. 2). The nine different reasoning tasks were then presented in random order (one of three patterns).

2.2 Predictions

In the syllogisms with NVC, we expected the participants to choose either the enumeration strategy with multiple configurations of conclusion diagrams (inclusion & exclusion & overlap; inclusion & exclusion), or the partial-overlapping strategy of placing the two circles (overlap) (the two strategies were considered to be confounded in inclusion & overlap and exclusion & overlap). In the syllogisms with VC-*all* and VC-*no*, we did not expect the participants to choose the enumeration strategy or the partial-overlapping strategy.

2.3 Result and Discussion

In the Euler diagram pretest, all of the participants chose the correct answer for each item. In the syllogistic reasoning tasks with direct manipulation of Euler diagrams, the accuracy rates for the three task types were substantially high: 92.0% for NVC, 100% for VC-*all*, and 96.6% for VC-*no*. There was no significant difference among them.

Table 2 shows the appearance frequencies of (combinations of) configurations of circles C and A in diagram manipulation for solving syllogisms. The bold numbers refer to the frequencies of circle-configurations, which were consistent with our predictions. The subscript **c** denotes the correctly answered items. In the NVC tasks, 88.6% of the participants chose the expected strategies (the sum of 3, 4, 5, 6, and 7): 35.2% of the participants chose the partial-overlapping strategy (3), 26.1% of the participants chose the enumeration strategy (the sum of 4 and 7), and 27.3% of the participants chose a confounding strategy (the sum of 5 and 6). When we excluded incorrectly answered items, we obtained similar results: 91.1% for the expected strategies (36.6% for the partial-overlapping strategy; 25.2% for the enumeration strategy; 29.3% for a confounding strategy). Furthermore, in the VC-*all* tasks, 100% of the participants constructed a circle

Table 2. The appearance frequencies of circle-configurations in direct manipulation of Euler diagrams for solving syllogistic tasks

	NVC	VC-*all*	VC-*no*	NVC^c	VC-*no*^{c}
1. inclusion	4.5	**100.0**	0.0	3.6	0.0
2. exclusion	6.8	0.0	**96.6**	5.3	**97.7**
3. overlap	**35.2**	0.0	0.0	**36.6**	0.0
4. inclusion & exclusion	**15.9**	0.0	2.3	**13.6**	1.1
5. inclusion & overlap	**11.4**	0.0	0.0	**12.5**	0.0
6. exclusion & overlap	**15.9**	0.0	1.1	**16.8**	1.1
7. inclusion & exclusion & overlap	**10.2**	0.0	0.0	**11.5**	0.0
The sum of 3, 4, 5, 6, and 7	**88.6**	0.0	3.4	**91.1**	2.2
The sum of 4 and 7	**26.1**	0.0	2.3	**25.2**	1.1
The sum of 5 and 6	**27.3**	0.0	1.1	**29.3**	1.1

configuration such that C is included in A. In the VC-*no* tasks, 96.6% of the participants constructed a circle configuration such that C is excluded from A.

The average response time was significantly longer for the NVC tasks (13.098 sec) than for the other two task types: 7.448 sec for VC-*all* tasks and 8.229 sec for VC-*no* tasks. The comparison tests (Bonferroni, Analysis of variance) revealed significant differences of $p < 0.01$ between NVC and VC-*all* (F (1, 21) = 7.070) and between NVC and VC-*no* (F (1, 21) = 6.093). Note that when we exclude the incorrectly answered items, we obtained similar results for each comparison (12.960 sec for NVC; 7.448 sec for VC-*all*; 8.070 sec for VC-*no*). The response time distribution suggests that NVC judgment involves complex processes such as enumeration and partial-overlapping strategies.

3 General Discussion

In this experiment we observed processes of diagram manipulation in NVC tasks. We found evidence for (i) the enumeration strategy with multiple configurations of conclusion diagrams and (ii) the partial-overlapping strategy of placing two circles. In contrast, for syllogisms with a valid conclusion, the participants chose neither the enumeration strategy nor the partial-overlapping strategy. From a theoretical viewpoint, there appears to be a commonality underlying the use of the two strategies observed in this study. Generally, the judgment of NVC or non-consequence in reasoning is thought to be performed by showing the existence of counter-examples to an argument. This process requires not only object-level information in which argument is expressed, but also meta-level information concerning semantic (truth) values. Such meta-level processes involve a high degree of effort. Thus, it may be relatively difficult for untrained users to reason by counter-example construction. In fact, in one cognitive theory (i.e., mental model theory [1,12]), tendencies of human performance errors in (sentential) reasoning tasks are explained by the difficulties encountered when constructing counter-example. In diagrammatic inference, abstract processes involved in counter-example construction can be replaced by concrete manipulation of diagrams to show the indeterminacy on circle-configuration. More specifically, the manipulation of diagrams is performed mainly as processes associated with external (visible) objects in the enumeration strategy, and relative internal processes in the partial-overlapping strategy.

Thus, based on our findings, we propose a role for the cognitive processes implicated in *indeterminacy* or *indefiniteness* in integrating diagrams, especially for NVC tasks. Clarifying these processes has been difficult using the traditional protocol analysis method, as mentioned before, but also in terms of recent developments produced by an eye-tracking study. Indeed, Shimojima and Katagiri [10] studied the exploitations of spatial constraints of position diagrams in relational inference tasks by analyzing participants' eye-movements. In their experimental setting, a diagram (of two terms) corresponding to a first premise was prepared and then supplemented with audio information regarding further premises. The position of a new term (object) in a diagrammatic representation can be set such that it is not uniquely determined in the case of indefiniteness (when solving four-term relational inference tasks). The authors found that the participants' gaze points were varied in location, supporting the exploitations of the spatial constraints. However, the following questions, which were examined in this study, were left unanswered: how are indefinite objects integrated to other diagrams? Do participants construct multiple configurations of such diagrams?

A potential limitation of our experiment is that the sizes (instances) of the diagrams were specified in spite of "Euler diagrammatic reasoning"; the configuration of the circles should not be a geometrical relationship but be a topological relationship. To address this concern, we [8] plan to conduct an experiment with diagrams that have less-specific sizes (instances) by adding the condition that participants change the circle sizes.

References

1. Bucciarelli, M., Johnson-Laird, P.N.: Strategies in syllogistic reasoning. Cognitive Science 23, 247–303 (1999)
2. Ford, M.: Two modes of mental representation and problem solution in syllogistic reasoning. Cognition 54, 1–71 (1994)
3. Howse, J., Molina, F., Shin, S.-J., Taylor, J.: On diagram tokens and types. In: Hegarty, M., Meyer, B., Narayanan, N.H. (eds.) Diagrams 2002. LNCS (LNAI), vol. 2317, pp. 146–160. Springer, Heidelberg (2002)
4. Mineshima, K., Sato, Y., Takemura, R., Okada, M.: Towards explaining the cognitive efficacy of Euler diagrams in syllogistic reasoning: A relational perspective. Journal of Visual Languages and Computing 25, 156–169 (2014)
5. Sato, Y., Mineshima, K., Takemura, R.: The efficacy of Euler and Venn diagrams in deductive reasoning: Empirical findings. In: Goel, A.K., Jamnik, M., Narayanan, N.H. (eds.) Diagrams 2010. LNCS, vol. 6170, pp. 6–22. Springer, Heidelberg (2010)
6. Sato, Y., Mineshima, K., Takemura, R.: Constructing internal diagrammatic proofs from external logic diagrams. In: Proceedings of 32nd Annual Conference of the Cognitive Science Society, pp. 2668–2673 (2010)

7. Sato, Y., Mineshima, K.: The efficacy of diagrams in syllogistic reasoning: A case of linear diagrams. In: Cox, P., Plimmer, B., Rodgers, P. (eds.) Diagrams 2012. LNCS (LNAI), vol. 7352, pp. 352–355. Springer, Heidelberg (2012)
8. Sato, Y., Wajima, T., Ueda, K.: Visual bias of diagram in logical reasoning. Accepted for publication in Proceedings of 36th Annual Conference of the Cognitive Science Society (2014)
9. Shimojima, A.: Inferential and expressive capacities of graphical representations: Survey and some generalizations. In: Blackwell, A.F., Marriott, K., Shimojima, A. (eds.) Diagrams 2004. LNCS (LNAI), vol. 2980, pp. 18–21. Springer, Heidelberg (2004)
10. Shimojima, A., Katagiri, Y.: An eye-tracking study of exploitations of spatial constraints in diagrammatic reasoning. Cognitive Science 37, 211–254 (2013)
11. Stenning, K., van Lambalgen, M.: A little logic goes a long way: basing experiment on semantic theory in the cognitive science of conditional reasoning. Cognitive Science 28, 481–529 (2004)
12. Sugimoto, Y., Sato, Y., Nakayama, S.: Towards a formalization of mental model reasoning for syllogistic fragments. In: AIC 2013. CEUR, vol. 1100, pp. 140–145. CEUR-WS.org, Aachen (2013)

Neural Mechanisms of Global Reading

Takeshi Sugio

Doshisha University, 1-3 Tatara-Miyakodani, Kyotanabe, Kyoto, Japan
tsugio@mail.doshisha.ac.jp

Abstract. The neural mechanisms underlying global reading on tabular representations were investigated using the task-switching paradigm in event-related functional magnetic resonance imaging. Participants were required to make an appropriate response based on the latest task cue using stimuli tabulated in five rows and five columns with labels. The task was either local or global, and critical events included both cue and target events, which enabled separate analyses of the preparation and execution stages of each type of reading process. Neuroimaging results revealed differential activations between local and global tasks in both preparation and execution stages. For the preparation stage, global cues led to larger activation in the extrastriate cortex, which has been shown as the neural basis of selective attention in the literature. For the execution stage, the left middle temporal gyrus and inferior parietal lobule were more activated in the local task. These areas comprise an object-based attentional selection network, which serves to attend to a particular element in the table that changed with each event. For the global task, the left inferior frontal junction showed high activation, suggesting that the task demanded more cognitive control. The implications of these findings are discussed with respect to the characteristics of global reading.

Keywords: global reading, fMRI, cognitive control.

1 Introduction

1.1 Global Reading of Diagrams

Reading a diagram is a mundane activity in daily life. For example, nutrition labels are often in bar graph format, and we can easily determine whether vitamin C is contained in a packaged leafy vegetable. In some situations, we might want to know whether the nutritional balance is appropriate for a meal that we are considering. In that case, we need to figure out how the lengths of the bar elements in the graph compare to each other. In order to accomplish this task, we must organize the elements into a perceptual whole, and then make a decision as to whether this global shape is balanced or not.

In the case of a table, we may experience something similar. From a timetable at a bus stop, we can obtain information about the times that buses to the city will come. At the same time, we can also see whether there are fewer buses on that route in the morning compared to those in the afternoon. Putting it all together, we not only allocate our attention to a particular element in the graph or table, but also integrate multiple elements and make various types of judgments on the integrated whole [1]. Such diagram reading is termed "global reading", and may be the basis for the flexibility of the human information acquisition process.

T. Dwyer et al. (Eds.): Diagrams 2014, LNAI 8578, pp. 198–212, 2014.

In order to read a certain diagram globally, we must first determine the hierarchical structure of information units of the diagram. The hierarchical structure can be revealed via perceptual organization, and surface characteristics such as the homogeneity of the properties of closed regions play a functional role in the success of such processes [2]. Watson and Kramer found that both surface and geometric characteristics define the candidate objects available for attentional selection [3]. Such psychological evidence suggests that our ability to selectively attend to a particular region within a diagram lies in how the diagram is perceptually organized. For example, coloring cells within the same row of a table may well facilitate the acquisition of information on the global properties of the colored row.

However, such explicit cues to grouping perceptual elements in a diagram might distract the flexibility of the reading process in some situations. Diagrams that we normally see in daily life can be read in many different ways in accordance with our behavioral goals. We can determine whether there are fewer buses to the city than to the airport, or whether there are more buses available from 07:00 to 08:00 in the morning than from 08:00 to 09:00. Using a spatial cueing paradigm, Sugio and colleagues found that we can mentally segregate a table into information units without any perceptually defined boundaries [4].

1.2 Global Reading and Cognitive Control

Such processes of mental segregation call for a high-load cognitive operation that relies on working memory, as we need to maintain the virtual boundaries of a diagram without the help of any salient visual features. To cope with such a difficult situation, cognitive control is necessary. The key aspects of cognitive control are how the task is represented and how such representation directs subsequent processing to accomplish the goal [5]. To read a table globally in an effective manner, we need to keep in mind how we select elements (cells) and integrate them as an information unit. The purpose of this study was to reveal differences in the involvement of cognitive control between two different types of table reading processes.

There are some other processes that modulate subsequent processing to achieve a particular goal in global reading. For example, switching attention from one information unit to another involves cognitive control and is essential to obtain information from the entire table. In some cases, we might need to ignore or inhibit information that has already been attended, particularly when the information leads to conflict or inappropriate responses. Scheduling a sequence of selective attention processes might also be necessary. Shomstein and Yantis suggested that object-based selection may reflect an object-specific attentional prioritization strategy [6]. Such strategies can be influenced by various factors, including cultural (e.g., attending from left to right when reading a text written with an alphabet) and perceptual factors (e.g., attending to the most salient information unit first, then to the next and so forth). Besides these processes, monitoring performance is undertaken to evaluate the current reading strategy.

Psychological processes involved in cognitive control have been studied using various experimental paradigms. A typical experiment proceeds as follows: participants perform a task on the first trial, and another task on the next trial based on the type of a cue presented between trials. In some trials, participants continue to perform the same task on consecutive trials, and switch between two different tasks in other trials. Rogers and Monsell has employed digit and letter decision tasks to explore the cost of this alternation of tasks [7]. All trials contained a digit and a letter, and participants had to decide whether the digit was above or below five for digit decision trials, while they had to decide whether the letter was a vowel or consonant for letter decision trials. Two types of blocks were used: alternating blocks, in which the task was switched; and pure blocks, in which only a single task was applied. Participants were found to require more time to respond on alternating blocks, referred to as the switching cost. In order to accomplish the switching task effectively, participants had to keep in mind the current requirement of the reading task and at the same time attend to information that signalled the task switch.

1.3 The Need for Neuroimaging Studies

Although a task-switching paradigm is potent in investigating the nature of the cognitive control process, psychological experiments are often insufficient to differentiate between two cognitive processes, because response times and error rates are not sensitive enough to identify the difference in strategies that lead to similar levels of performance. Meanwhile, recent advances in neuroimaging studies have revealed different neural systems engaged in seemingly similar cognitive tasks.

Chiu and Yantis used rapid event-related functional magnetic resonance imaging (fMRI) to reveal whether different domains of cognitive control are associated with separate mechanisms or a common source of control initiates switching in all domains [8]. In their experiment, eight letters (including one digit in a target event trial) were displayed at each trial, and participants were asked to maintain fixation at a central fixation point. At the beginning of the session, an instruction was presented on the screen and specified the initial task (either to attend at the left or right of the fixation point and which of the two categorization rules was to be performed). Two categorization rules were applied to a digit presented at the attended location (either left or right). One was a magnitude categorization, and the other was a parity judgment. Sustained and transient effects were modeled separately. Increased activation in the extrastriate cortex and posterior intraparietal sulcus contralateral to the locus of spatial attention were found for the sustained effect. As for the transient effect, median superior parietal lobule showed activation for two different domains of control, suggesting the existence of a single domain-independent control mechanism. However, Osaka and his colleagues found that differences in working memory capacity result in the differential activation in regions related to executive function [9]. Whether the differences in cognitive control are reflected in distinct neural mechanisms for a specific task (e.g. table reading) remains to be elucidated.

The neuroimaging literature on the neural mechanisms underlying subitizing and counting also suggests the differential activation between local and global reading. Demeyere and her colleagues have shown that separate attentional mechanisms are involved in subitizing and counting [10][11][12]. Counting requires more focused attention compared to subitizing, which leads to the activation in the parietal areas while subitizing relys on the visual processing at the posterior occipital cortex. The result from the MEG-recordings also supports this distinction [13]. For successful global reading, spatially distributed attention is required to gather necessary information from the entire stimuli. From these evidences, it can be presumed that the underlying neural mechanism of global reading may also reflect a difference in attentional strategies.

In the present study, the neural mechanism engaged in global reading was investigated by rapid event-related fMRI. The sort of cognitive control undertaken in global reading can be inferred from the regions showing larger activation in global event trials. The task-switching paradigm was employed across eight blocks, and cue trials within each block directed the task of the subsequent target trials that participants were asked to read and respond to appropriately.

2 fMRI Experiment

2.1 Method

Participants. Fifteen right-handed neurologically healthy adults (four females; 20-27 years old) participated in the study. The study protocol was approved by the Doshisha University Institutional Review Board, and informed consent was obtained from all participants before experiment. Data from two participants were not included in the following analysis due to performance (percentage correct was under <90%).

Stimuli and apparatus. All stimuli were rendered in white on a black background, with the exception of colored row and column labels (red and green). The software package Presentation (Neurobehavioral Systems, Inc.) was used for stimulus presentation, image pulse acquisition, and recordings of participant responses during tasks. Stimuli were projected onto a screen mounted to the top of the magnet bore behind the head of the participant. Participants viewed the screen reflected from a mirror at a distance of 70 cm. The size of the screen was 37 cm in width and 28 cm in height (resolution: 1024 pixels × 768 pixels). The size of a table-like stimulus including labels was 250 pixels in both width and height (approximately 7.33 ° × 7.33 ° in visual angle). Table-like stimuli consists of five rows and five columns, with both rows and columns labeled (from "1" to "5" for rows and from "A" to "E" for columns). One of the row labels was colored red, and one of the column labels was colored green. The selection was pseudo-random in that when both the selected row and column labels were the center ones, the selection process was redone. The element of a cell was either "O", "X", or blank (Figure 1). The cell at the center of the table-like stimuli (central row and central column) was centered with respect to the screen. Participants held an MR-compatible response box with their right index and middle fingers places on one of two buttons.

Procedure. Participants were instructed to maintain fixation at a fixation cross at the beginning of each run. For each run, after the presentation of a fixation cross for 10 s, event trials began. The first trial for each run was always a cue event. At the cue event, either "L," red "G," or green "G" appeared at the center of the table-like stimulus (Figure 2). Participants were required to press an assigned button as fast as possible to indicate that they had seen the cue. They had to maintain the current task designated by the preceding cue and when the target stimuli was presented, they performed the corresponding task.

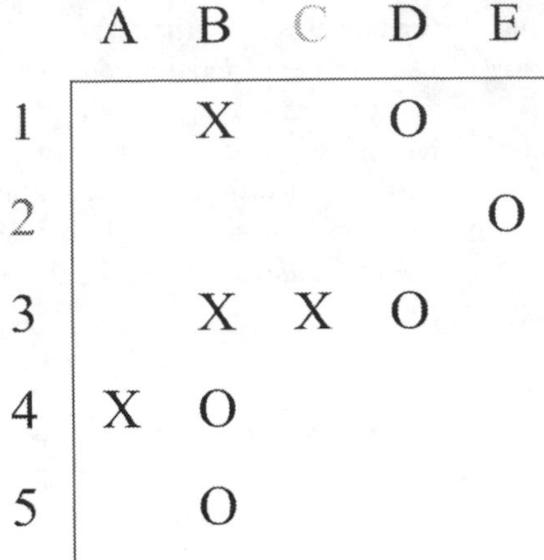

Fig. 1. An example of the table stimuli used in the experiment

When the cue was "L", participants were required to identify the element of the cell at the intersection between the colored row and column in the subsequent target trials. When the element was either "O" or "X", they pressed the assigned button as fast as possible without error. When the element was blank, participants had to withhold the response until the next trial began. The response assignments of the buttons were counterbalanced across participants.

When the cue "G" was presented, participants were required to perform a global reading task with the subsequent table-like stimuli. Participants had to judge whether there are more than two identical elements in the colored row or column. The color of the cue "G" was either red or green, determining the direction to attend (row for the red cue and column for the green cue). There were five cells in either row or column direction and participants pressed the assigned button as fast as possible without error. Response was one of the following: "O", "X", both "O" and "X", and none. When both "O" and "X" was less than two in the assigned direction, participants had to withhold the response until the next trial. Response assignments of buttons were also counterbalanced across participants. Figure 3 illustrates the virtual path that attention is assumed to follow.

One of the advantages of such global task was the use of the same stimulus set throughout the experiment. Participants had to maintain the current task in the working memory because it was impossible to determine which task was to be performed from a target stimuli. This was intended to facilitate task preparation at the cue event. Another feature of the global task was that the part of the response set was identical between the tasks. This was also important in that it reduced the cognitive load of reestablishing the relation between the stimulus and the response. Furthermore, the difficulty of the task was comparable between the two tasks, which was confirmed in a pilot study.

On target events, the table-like stimuli were presented for 2 s, and participants had to make an appropriate response to the stimuli based on the current task. Cue and target events were separated in time by 1-3 s, with an average of 2 s between them. During the interval, a fixation cross was presented and participants were required to

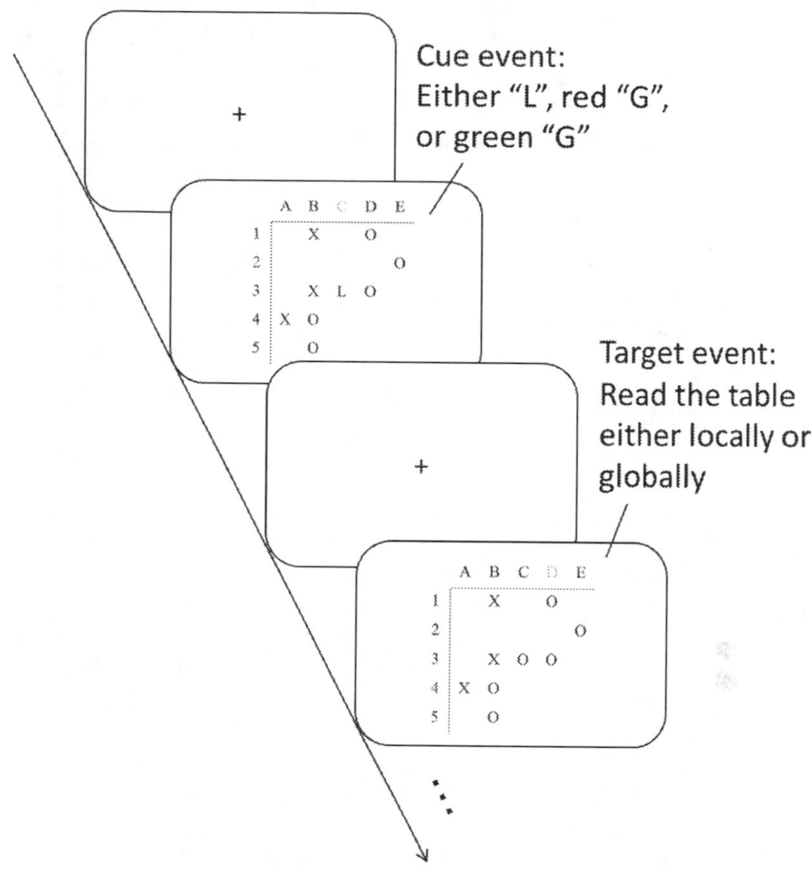

Fig. 2. Cue and target events in the experiment

maintain their fixation. Forty-eight events were randomly intermixed and one-half of the events were cues and the other half were targets. As for cue events, the number of local cues and global cues were equal. There were eight runs in total, and participants took a brief rest between the fourth and fifth runs. Participants made a practice outside the scanner for a single run to ensure that participants understood the task.

For both cue and target events, response time and accuracy were recorded. When the participant did not make a response within 2 s, the trial was coded as an error. No feedback was given to the participant. In cases where a participant missed a cue, they were told to wait for the next cue. When the overall accuracy was below 90%, the data were not included in subsequent analyses. Each run in the scanner lasted for 212 s with an additional 10 s for dummy scans.

Image acquisition and processing. MRI was performed using a 3-T Siemens Trio scanner in the Kokoro Research Center at Kyoto University. Anatomical images were acquired using an MP-RAGE T1-weighted sequence that yielded images with a 1-mm

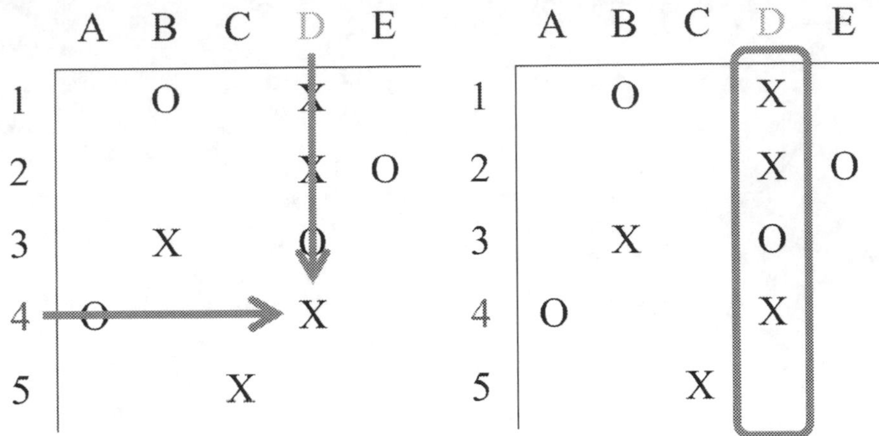

Fig. 3. Virtual attentional paths and boundaries formed by participants in the experiment for local and global tasks

isotropic voxel resolution [repetition time (TR), 2250 ms; echo time (TE), 3.51 ms; flip angle, 9 deg; inversion time, 900 ms]. Such high-resolution anatomical images can be used to compensate for the loss of spatial information due to the maximization of temporal resolution [14]. Whole-brain echo planar functional images (EPIs) were acquired with a twelve channel parallel-imaging (GRAPPA) head coil in 39 transverse slices (TR, 2000 ms; TE, 25 ms; flip angle, 75 °; matrix, 64 × 64; field of view, 224 mm; slice thickness, 3.5 mm, no gap), yielding 3.5 mm isotropic voxel resolution. Among these parameters, the repetition time (TR) and voxel size are important for the analysis of fMRI data [14]. The TR is the time between successive whole-brain scans and affects temporal resolution. On the other hand, voxel size determines the spatial resolution of the data. 3.5 mm isotropic voxel resolution was within a typical range [14].

Imaging data were analyzed using SPM8 software (Wellcome Department of Imaging Neuroscience, London, UK [15]). First, slice timings of images were corrected with reference to the middle slice. Images were then realigned to correct for movement and anatomical images were coregistered to the mean functional image. After realignment, images were normalized to the MNI (Montreal Neurological Institute) EPI template available in SPM8 and smoothed with a FWHM (full width at half maximum) of 8 mm. The FWHM values represent the amount of smoothing, which was widely used in the analysis of fMRI data [14].

The data were modeled for each voxel using a general linear model (GLM) that included regressors obtained by convolving each event-related unit impulse with a canonical hemodynamic function. Duration was set zero for each function. This first-level analysis included four types of events (local cue, global cue, local target, and global target) with correct responses. By such modeling, two cue regressors are presumed to account for two possible preparatory states while two target regressors account for execution-dependent aspects of the table reading task.

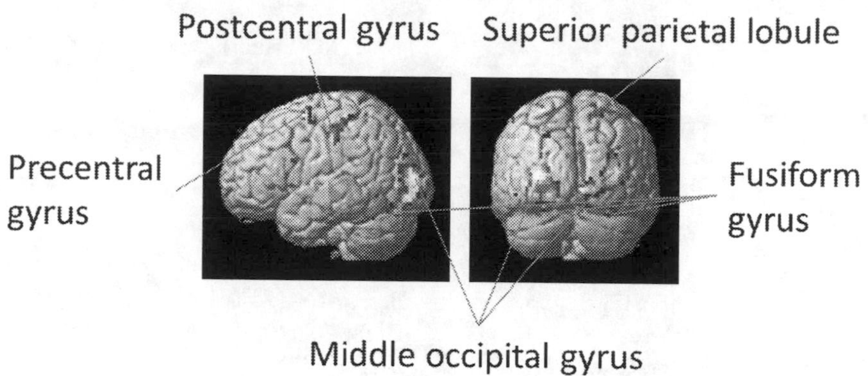

Fig. 4. The regions activated in the conunction analysis

Mean images from the first-level analysis were entered into a random effects analysis accounting for the between participant variance. The significance threshold was set at a voxel-level inference of $p < 0.001$ (uncorrected). Anatomical locations of the peak voxels were labelled using the SPM Anatomy Toolbox version 1.8 [16] and xjView 8 (http://www.alivelearn.net/xjview/).

2.2 Results

Average overall accuracy across participants was 96.6%. The change in accuracy levels across eight runs showed that accuracy for the first run was 92.3%, and above 95% from the second to the eighth runs.

Activated regions were quite similar between different event types. A conjunction analysis was performed to determine the brain regions that were activated in the same manner across these event types. Figure 4 shows the regions that showed activation across the four event types. These included bilateral posterior occipital cortex (including middle occipital gyrus, fusiform gyrus, and primary visual cortex), right superior parietal lobule, and left precental and postcentral gyri. These commonly activated regions were assumed to be the elements of the current task. The visual processing of the stimuli was an essential component of the task, and it has been claimed that the superior parietal lobule plays an important role in the top-down attentional orienting [17][18]. Participants were asked to press an appropriate button with their right hand, which lead to the activation in the left motor-related area (including precentral gyrus).

Differences in activation were revealed when contrasted between event types for both cue and target events. Each of the eight runs in the experiment was treated as a sequence of variable-length epochs during which participants were prepared for either local or global reading task. Contrasts between these two preparatory states revealed larger activation in extrastriate cortex for global cue (Table 1). The MNI coordinates, which is a standard stereotaxic space that are widely used in the neuroimaging studies, of the peak were reported for each activated cluster. Figure 5 shows the regions exhibiting a higher activation for global cue events. The preparatory state for local task showed larger

Table 1. Brain regions involved in the preparation of local and global reading

Area	Side	MNI (mm)
Local Cue > Global Cue		
Cerebellum	Right	12 -43 -38
Middle frontal gyrus	Right	36 50 13
Global Cue > Local Cue		
Fusiform gyrus	Right	36 -67 -17
hOC4v (V4)	Right	33 -70 -14
Fusiform gyrus	Left	-36 -58 -17
Area 17	Right	12 -67 10
hOC4v (V4)	Left	-33 -82 -17

activation in right cerebellum and middle frontal gyrus, although the sizes of clusters were rather small (less than 10 voxels).

Contrasts between two execution states also showed differential activations (Table 2). The left middle temporal gyrus and inferior parietal lobule (supramarginal gyrus and angular gyrus) showed activations larger for local target events, while frontal regions including the superior precentral gyrus, supplementary motor area (SMA), and inferior frontal junction were more activated for global target events. Lateral views of activations for both contrasts are shown in Figure 6.

Table 2. Brain regions involved in the execution of local and global reading

Area	Side	MNI (mm)
Local Target > Global Target		
Middle temporal gyrus	Left	-63 -10 -11
Supramarginal gyrus (PGa)	Left	-48 -49 25
Angular gyrus	Left	-45 -61 25
Cerebellum	Right	30 -76 -38
Global Target > Local Target		
Postcentral gyrus (Area 3b)	Left	-39 -28 52
Precentral gyrus (Area 6)	Left	-39 -4 64
Inferior frontal gyrus	Left	-39 11 28
Putamen	Left	-18 8 4
Thalamus	Left	-9 -13 7
Cerebellum	Right	9 -46 -11
SMA (Area 6)	Left	-3 20 58
Middle frontal gyrus	Right	45 38 19
Inferior frontal gyrus	Right	42 5 22

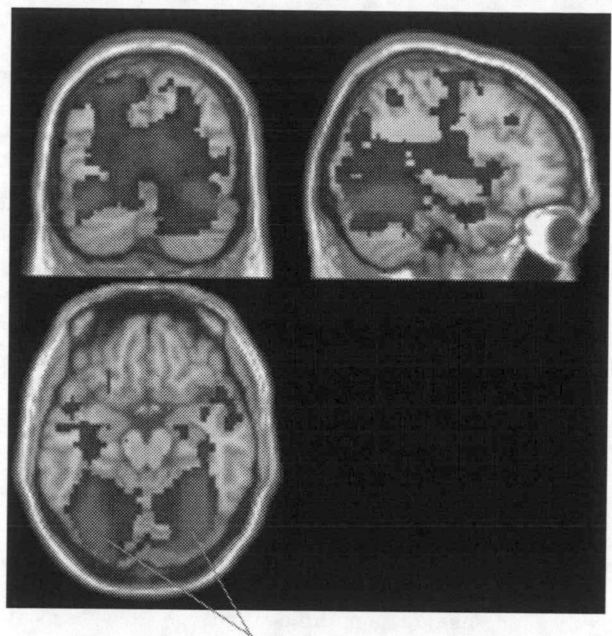

Fig. 5. The region in the extrastriate cortex that showed higher activation in the global cue event

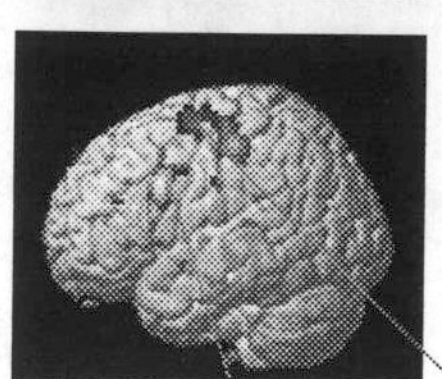

Fig. 6. Differential activations between execution stages of local and global events

2.3 Discussion

There is consistent evidence that the extrastriate cortex modulates attention by filtering out irrelevant information in visual scenes. When multiple stimuli are presented in the visual field, activation in the extrastriate cortex is reduced in general due to interactions among the stimuli. Spatial attention is assumed to reduce these interactions at its directed region [19][20]. The result of larger activation in the extrastriate cortex for global cue events compared to local cue events suggests that the preparation for global reading involves an enhancement of particular region within the table by partially canceling out suppressive effects from surrounding irrelevant regions. Interestingly, no significant differences in activation in the extrastriate cortex were seen between execution stages of local and global reading. This might suggest that once the information unit necessary for global reading is fixed, the fixed region is temporarily stored in the working memory. The larger activation in the right middle frontal gyrus (Brodmann area 46) shown in the global target events supports this hypothesis [21]. However, the possibility remains that the diversity of color in the global task cues evoked activation of the extrastriate cortex. Multiple functional regions including V4 are encompassed in the extrastriate cortex. Many studies have revealed the essential role of V4 in color processing [22]. The current global task required the discrimination of the cue color in order to determine the reading direction of the stimuli. Further research is required to clarify the exact role of the extrastriate cortex in the task preparation.

During the execution stage, the left middle temporal gyrus and inferior parietal regions showed larger activations in the local reading task. Arrington and colleagues had shown a left-lateralized network for attentional selection to a region bounded by an object [23]. In the current local reading task, participants had to detect the colored row and column labels and then identify the crossed element (cell) in the table. To identify a single cell in the table, attentional selection based on an object is necessary. The right temporo-parietal junction has been claimed to represent a critical part of the attentional control network [17]. Its primary role is purportedly to reorient attention toward an unexpected, task-relevant object. However, Geng and Vossel raised questions about lateralization and the perceptual role of this region [24]. They concluded that substantial evidence supports the idea that the signal from this temporo-parietal region is involved in contextual updating and is not necessarily right-lateralized. Participants in the present study had to spatially attend to a different element (cell) on each local target event, which demands updating of internal task-dependent representation of the table for the purpose of making appropriate responses. In essence, both the left middle temporal gyrus and inferior parietal lobule (supramarginal and angular gyri) serve to maintain the optimal representation of the table in order to locally identify its content.

The superior precentral gyrus has been implicated as the human homolog of the frontal eye field [25]. Several studies has revealed that this region is involved in covert shifts of spatial attention [26][27]. However, because we did not measure eye movements, the possibility remains that differential activation in the superior precentral gyrus is due to differences in the amount of eye movements.

The inferior frontal junction, located in the region near the junction of the inferior frontal sulcus and the inferior precentral gyrus, showed larger activation in the global target events. The functional role of the inferior frontal junction was studied mainly by those tasks that required cognitive control. Derrfuss and colleagues claimed that the region is involved in the updating of task representations [28]. For example, in the Stroop task, participants have to enforce the relevant but nondominant action against a dominant one. In the present global reading task, participants may need to inhibit the dominant response resulting from the local reading of the cells within the attended region. Evidently, such a task requires a high degree of cognitive control, which resulted in larger activation in the inferior frontal junction. Prior research has suggested that the global feature of an image is identified more readily than the local features [29]. Although present findings may seem to contradict with this global precedence effect, it is possible that both behavioral goals and the association between stimuli and response might effect the occurrence of this effect. These conditions should be studied further.

3 Conclusion

The main purpose of this study was to examine whether the global reading process for tables involved higher cognitive control in comparison to local reading. Using a task-switching paradigm, the neural mechanisms of both preparing and executing the table-reading task were investigated in event-related fMRI.

The main findings of the study showed differential activations between local and global tasks in both preparation and execution stages. For the preparation stage, global cues led to a larger activation in the extrastriate cortex, which has been shown as the neural basis of selective attention in the literature. For the execution stage, the left middle temporal gyrus and inferior parietal lobule were more activated in the local task. These areas comprise an object-based attentional selection network, which serves to attend to a particular element in a table that changed with each event. For the global task, the left inferior frontal junction showed high activation, suggesting that the task demanded more cognitive control. For example, to accomplish the global task, participants had to inhibit an irrelevant but dominant response based on a local element of the table. Figure 7 shows the proposed cognitive mechanism of diagram reading. The perceptual organization process operated on a diagram enables the reader to access the hierarchical structure of information in the diagram [2]. Local reading can be achieved through the access to the appropriate element in the structure. The attentional control network including the temporo-parietal junction is assumed to serve this purpose. Cognitive control, such as attentional switching and response inhibition, is exerted to perform global reading in an appropriate manner.

For both local and global reading tasks, it was necessary to update the current task representation to make an appropriate response. While selection based on a localized and meaningful unit is necessary for local tasks, interactive suppressions among units must be reduced and an irrelevant but dominant responses need to be inhibited for the global task. Future research may need to focus on the more detailed conditions that affect the degree of cognitive control. These conditions might be useful for determining the layout of diagrams that are suitable for global reading. One of the advantages of global reading is that we can set an object on a diagram in a flexible manner. By reducing the degree of congitive control due to inappropriate responses evoked by local elements, we might be able to design a diagram that permit the efficient usage of such process.

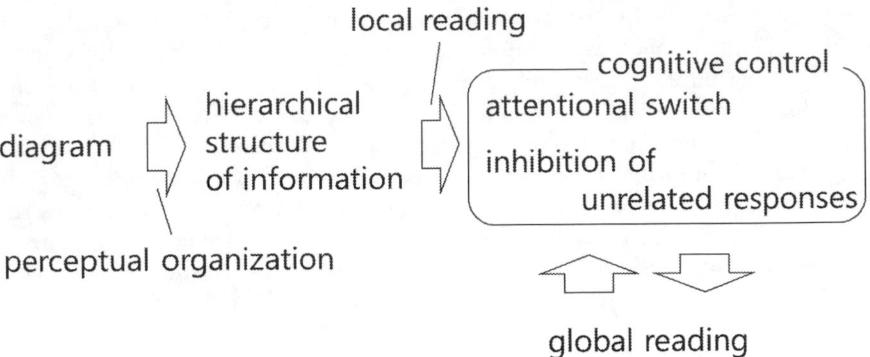

Fig. 7. Cognitive mechanism of diagram reading

Furthermore, it might be possible that some types of perceptual grouping cues are more efficient for global reading than others in terms of generating more dominant responses. For example, it might be helpful to avoid using local elements with low associative value. The association between elements might not be limited to the same modality. Recent psychological studies have shown that crossmodal correspondences exist between all possible pairings of modalities [30]. If a certain local element leads to automatic perceptual experiences in the different modalities, such "strong" element can intercept the global reading process. The shape congruence between the local element and the global object might also lead to the cognitive conflict that need to be reduced. Becker and her colleagues found that the perceptual intensity of the element and congruence among elements of the package design may impact product evaluations [31]. Such result might suggest that how we construct and evaluate a global object depends on the perceptual and cognitive properties of local elements. There are certainly some other possibilties and how such cognitive mechanims contribute to designing the layout of diagrams needs to be investigated much more throughly.

Acknowledgement. I would like to thank Dr. Atsushi Shimojima for his fruitful comments on the manuscript. This work was supported by JSPS KAKENHI Grant Numbers 24240041 and 23300101.

References

1. Tufte, E.R.: Envisioning Information. Graphics Press, Cheshire (1990)
2. Palmer, S., Rock, I.: Rethinking perceptual organization: The role of uniform connectedness. Psychonomic Bulletin & Review 1(1), 29–55 (1994)
3. Watson, S.E., Kramer, A.F.: Object-based visual selective attention and perceptual organization. Perception & Psychophysics 61(1), 31–49 (1999)
4. Sugio, T., Shimojima, A., Katagiri, Y.: Psychological evidence of mental segmentation in table reading. In: Cox, P., Plimmer, B., Rodgers, P. (eds.) Diagrams 2012. LNCS (LNAI), vol. 7352, pp. 124–131. Springer, Heidelberg (2012)

5. Miller, E.K., Cohen, J.D.: An integrative theory of prefrontal cortex function. Annual Review of Neuroscience 24, 167–202 (2001)
6. Shomstein, S., Yantis, S.: Object-based attention: Sensory modulation or pripríty setting? Perception & Psychophysics 64(1), 41–51 (2002)
7. Rogers, R.D., Monsell, S.: Costs of a predictable switch between simple cognitive tasks. Journal of Experimental Psychology: General 124(2), 207–231 (1995)
8. Chiu, Y.C., Yantis, S.: A domain-independent source of cognitive control for task sets: Shifting attention and switching categorization rules. The Journal of Neuroscience 29(12), 3930–3938 (2009)
9. Osaka, N., Osaka, M., Kondo, H., Morishita, M., Fukuyama, H., Shibasaki, H.: The neural basis of executive function in working memory: an fMRI study based on individual differences. NeuroImage 21, 623–631 (2004)
10. Demeyere, N., Humphreys, G.W.: Distributed and focused attention: neuropsychological evidence for separate attentional mechanisms when counting and estimating. Journal of Experimental Psychology: Human Perception and Performance 33(5), 1076–1088 (2007)
11. Demeyere, N., Lestou, V., Humphreys, G.W.: Neuropsychological evidence for a dissociation in counting and subitizing. Neurocase 16(3), 219–237 (2010)
12. Demeyere, N., Rotshtein, P., Humphreys, G.W.: The neuroanatomy of visual enumeration: Differentiating necessary neural correlates for subitizing versus counting in a neuropsychological voxel-based morphometry study. Journal of Cognitive Neuroscience 24(4), 948–964 (2012)
13. Vuokko, E., Niemivirta, M., Helenius, P.: Cortical activation patterns during subitizing and counting. Brain Research 1497, 40–52 (2013)
14. Ashby, F.G.: Statistical Analysis of fMRI Data. The MIT Press, Cambridge (2011)
15. Friston, K.J., Frith, C.D., Turner, R., Frackowiak, R.S.: Characterizing evoked hemodynamics with fMRI. NeuroImage 2, 157–165 (1995)
16. Eickhoff, S.B., Stephan, K.E., Mohlberg, H., Grefkes, C., Fink, G.R., Amunts, K., Zilles, K.: A new SPM toolbox for combining probabilistic cytoarchitectonic maps and functional imaging data. NeuroImage 25, 1325–1335 (2005)
17. Corbetta, M., Shulman, G.L.: Control of goal-directed and stimulus-driven attention in the brain. Nature Reviews Neuroscience 3, 201–215 (2002)
18. Shomstein, S.: Cognitive functions of the posterior parietal cortex: top-down and bottom-up attentional control. Frontiers in Integrative Neuroscience 6, 1–7 (2012)
19. Kastner, S., De Weerd, P., Desimone, R., Ungerleider, L.G.: Mechanisms of directed attention in the human extrastriate cortex as revealed by functional MRI. Science 282, 108–111 (1998)
20. Rao, A.W., Chelazzi, L., Connor, C.E., Conway, B.R., Fujita, I., Gallant, J.L., Lu, H., Vanduffel, W.: Toward a unified theory of visual area V4. Neuron 74, 12–29 (2012)
21. D'Esposito, M., Aguirre, G.K., Zarahn, E., Ballard, D., Shin, R.K., Lease, J.: Functional MRI studies of spatial and nonspatial working memory. Cognitive Brain Research 7, 1–13 (1998)
22. Wade, A., Augath, M., Logothetis, N., Wandell, B.: fMRI measurements of color in macaque and human. Journal of Vision 8(10), 1–19 (2008)
23. Arrington, C.M., Carr, T.H., Mayer, A.R., Rao, S.M.: Neural mechanisms of visual attention: Object-based selection of a region in space. Journal of Cognitive Neuroscience 12, 106–117 (2000)
24. Geng, J.J., Vossel, S.: Re-evaluating the role of TPJ in attentional control: Contextual updating? Neuroscience and Biobehavioral Reviews 37, 2608–2620 (2013)
25. Paus, T.: Location and function of the human frontal eye-field: A selective review. Neuropsychologia 34(6), 475–483 (1996)
26. Serences, J.T., Yantis, S.: Spatially selective representations of voluntary and stimulus-driven attentional priority in human occipital, parietal, and frontal cortex. Cerebral Cortex 17, 284–293 (2007)

27. Kelley, T.A., Serences, J.T., Giesbrecht, B., Yantis, S.: Cortical mechanisms for shifting and holding visuospatial attention. Cerebral Cortex 18, 114–125 (2008)
28. Derrfuss, J., Brass, M., Neumann, J., von Cramon, D.Y.: Involvement of the inferior frontal junction in cognitive control: Meta-analyses of switching and stroop studies. Human Brain Mapping 25, 22–34 (2005)
29. Navon, D.: Forest before trees: The precedence of global features in visual perception. Cognitive Psychology 9(3), 353–383 (1977)
30. Spence, C.: Crossmodal correspondences: A tutorial review. Attention, Perception, & Psychophysics 73(4), 971–995 (2011)
31. Becker, L., van Rompay, T.J.L., Schifferstein, H.N.J., Galetzka, M.: Tough package, strong taste: The influence of packaging design on taste impressions and product evaluations. Food Quality and Preference 22, 17–23 (2011)

The Relationship between Aristotelian and Hasse Diagrams

Lorenz Demey[1] and Hans Smessaert[2]

[1] Center for Logic and Analytic Philosophy, KU Leuven
Lorenz.Demey@hiw.kuleuven.be
[2] Department of Linguistics, KU Leuven
Hans.Smessaert@arts.kuleuven.be

Abstract. The aim of this paper is to study the relationship between two important families of diagrams that are used in logic, viz. Aristotelian diagrams (such as the well-known 'square of oppositions') and Hasse diagrams. We discuss some obvious similarities and dissimilarities between both types of diagrams, and argue that they are in line with general cognitive principles of diagram design. Next, we show that a much deeper connection can be established for Aristotelian/Hasse diagrams that are closed under the Boolean operators. We consider the Boolean algebra $\mathbb{B}_n$ with 2^n elements, whose Hasse diagram can be drawn as an n-dimensional hypercube. Both the Aristotelian *and* the Hasse diagram for $\mathbb{B}_n$ can be seen as $(n-1)$-dimensional vertex-first projections of this hypercube; whether the diagram is Aristotelian or Hasse depends on the projection axis. We show how this account provides a unified explanation of the (dis)similarities between both types of diagrams, and illustrate it with some well-known Aristotelian/Hasse diagrams for $\mathbb{B}_3$ and $\mathbb{B}_4$.

Keywords: Aristotelian diagram, Hasse diagram, square of oppositions, logical geometry, hexagon, rhombic dodecahedron, hypercube.

1 Introduction

Logicians make use of several kinds of diagrams for a variety of purposes, such as obtaining new results and communicating their findings more effectively. Roughly, the diagrams used in logic can be divided into two broad classes. On the one hand, there are diagrams that visualize formulas from some given logical system; typical examples include Euler diagrams, Venn diagrams, spider diagrams, Peirce's existential graphs, etc. [1–5]. In these cases, one diagram visualizes a single formula, and visual operations on the diagram correspond to logical operations on that formula. Diagrammatic reasoning thus consists in a sequence of operations that gradually transforms an initial diagram into another diagram. On the other hand, logicians also use diagrams to visualize certain *relations* between formulas from some given logical system. Typical examples include Aristotelian diagrams, Hasse diagrams and duality diagrams [6–13]. In these cases, one diagram contains several formulas, and diagrammatic reasoning

T. Dwyer et al. (Eds.): Diagrams 2014, LNAI 8578, pp. 213–227, 2014.

consists in 'traversing' the diagram by making use of the relations between the formulas.[1] Since the focus is on the relations between the formulas, the formulas themselves are usually simply visualized as symbolic labels attached to the vertices in the diagram.[2]

The aim of this paper is to study the relationship between two main types of diagrams of the second class, viz. Aristotelian and Hasse diagrams. On the one hand, there are some obvious similarities between these two types; for example, the relation of logical implication (also called subalternation or entailment) is visualized in Aristotelian as well as Hasse diagrams. On the other hand, there are also some equally obvious dissimilarities; for example, in Hasse diagrams, the implications all go in the same general direction (viz. upwards), but in Aristotelian diagrams, they tend to go in a wide variety of directions. Because of this equivocal evidence, the overall picture of the relationship between Aristotelian and Hasse diagrams has remained unclear up till now.

However, in this paper, we show that there exists a deep connection between these two types of diagrams. On a *visual-cognitive* level, we argue that their dissimilarities can perfectly be explained in terms of general principles of diagram design and information visualization. Next, on a more abstract *geometrical* level, we show that if we restrict ourselves to Boolean closed diagrams, Aristotelian and Hasse diagrams can be seen as different vertex-first projections of one and the same hypercube (whether the resulting diagram is Aristotelian or Hasse depends on the projection axis). This account naturally yields a unified explanation of the obvious similarities and dissimilarities mentioned above. Finally, these results are illustrated by means of some well-known Aristotelian and Hasse diagrams, such as the hexagon and the rhombic dodecahedron (RDH).

The paper is organized as follows. Section 2 formally introduces Aristotelian and Hasse diagrams, and briefly discusses their importance and usage in logic. Next, Section 3 examines some obvious similarities and dissimilarities between both kinds of diagrams, and relates them to general principles of diagram design. Section 4 contains the more technical results of this paper: it shows how Aristotelian and Hasse diagrams can be seen as vertex-first projections of hypercubes, and discusses how this leads to a unified explanation of the (dis)similarities between both types of diagrams. The next two sections illustrate these results by applying them to some well-known Aristotelian and Hasse diagrams, viz. the hexagon (Section 5) and the RDH (Section 6). Finally, Section 7 wraps things up, and mentions some questions that are left for further research.

---

[1] For example, most readers will be familiar with the reasoning task in which one is presented with a square of oppositions and the truth value of (the formula in) one of the square's corners, and is then asked to determine the truth values of the other corners by making use of the Aristotelian relations [14, Exercise 4.5.I].

[2] However, there also exist Aristotelian/Hasse diagrams in which the formulas themselves are visualized as diagrams as well. For example, Bernhard [15] discusses a multi-layered square of opposition in which the four formulas are visualized as small Euler/Venn diagrams that are embedded inside a large Aristotelian diagram.

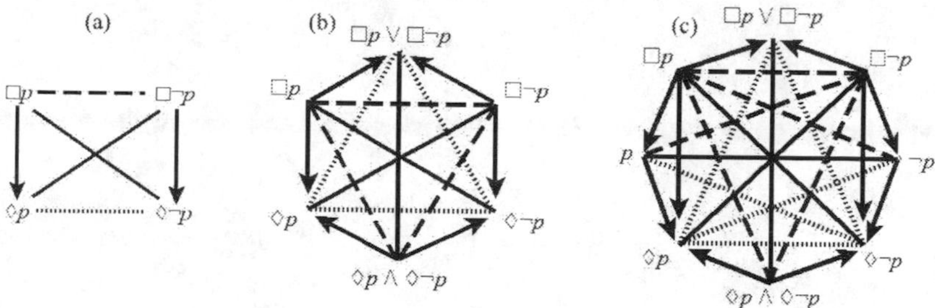

Fig. 1. Aristotelian (a) square, (b) hexagon and (c) octagon for the modal logic S5

2 Aristotelian and Hasse Diagrams

An Aristotelian diagram visualizes a set of logical formulas and the Aristotelian relations between them. These relations are defined as follows (relative to some given logical system S, which is supposed to have the usual Boolean connectives): the formulas φ and ψ are said to be

contradictory	iff	$\mathsf{S} \models \neg(\varphi \wedge \psi)$	and	$\mathsf{S} \models \neg(\neg\varphi \wedge \neg\psi)$,
contrary	iff	$\mathsf{S} \models \neg(\varphi \wedge \psi)$	and	$\mathsf{S} \not\models \neg(\neg\varphi \wedge \neg\psi)$,
subcontrary	iff	$\mathsf{S} \not\models \neg(\varphi \wedge \psi)$	and	$\mathsf{S} \models \neg(\neg\varphi \wedge \neg\psi)$,
in subalternation	iff	$\mathsf{S} \models \varphi \rightarrow \psi$	and	$\mathsf{S} \not\models \psi \rightarrow \varphi$.

Furthermore, almost all Aristotelian diagrams that have appeared in the literature impose the following additional constraints on the formulas that are visualized: these formulas are (i) contingent and (ii) pairwise non-equivalent, and (iii) they come in contradictory pairs (i.e. for a given formula φ, the diagram contains both φ and $\neg\varphi$). Finally, almost all Aristotelian diagrams are centrally symmetric, with all contradictory pairs ordered around the center of symmetry (so that φ is diametrically opposed to $\neg\varphi$).

The most widely known Aristotelian diagram is of course the so-called 'square of oppositions'. This diagram has a rich tradition [16, Chapter 5], but it is also widely used by contemporary logicians to visualize interesting fragments of systems such as modal logic [17], (dynamic) epistemic logic [8, 18] and deontic logic [19, 20]. There is also a vast literature on Aristotelian diagrams other than the traditional square. The most widely known among these is probably the hexagon proposed by Jacoby, Sesmat and Blanché [21–23], but several other hexagons, octagons, etc. have been studied in detail [10, 16, 24]. Figure 1 shows an Aristotelian square, hexagon and octagon for the modal logic S5.[3]

We now turn to the second type of diagrams, viz. Hasse diagrams. In general, these are used to visualize partially ordered sets (posets). A poset consists of

[3] There is no universally accepted way of visualizing the Aristotelian relations. We here follow the convention introduced in [12], and visualize the relations of contradiction, contrariety, subcontrariety and subalternation by means of a full line (—), a dashed line (- - -), a dotted line $(\cdots)$ and an arrow $(\longrightarrow)$, respectively.

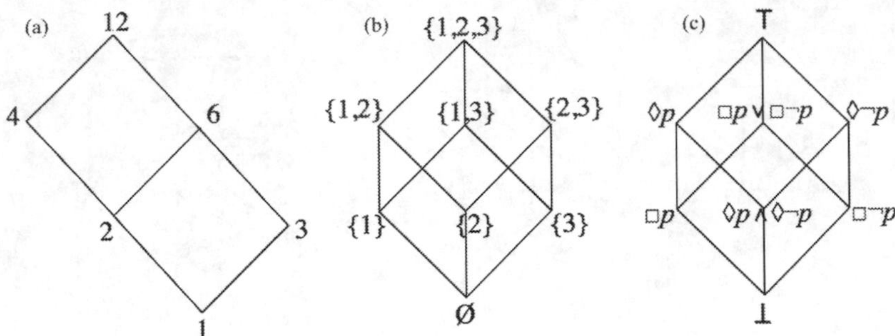

Fig. 2. Hasse diagrams for (a) the divisors of 12, (b) the Boolean algebra $\wp(\{1,2,3\})$, and (c) a Boolean algebra of formulas from the modal logic S5

a set P and a partial ordering $\leq$, i.e. a binary relation on P that is reflexive, transitive, and antisymmetric. If $x \leq y$ and not $y \leq x$, we say that $x < y$. If $x < y$ and there is no $z \in P$ such that $x < z < y$, we say that $x \lhd y$. A Hasse diagram visualizes the poset $(P, \leq)$ in such a way that if $x \lhd y$, the point representing x is connected by a line segment to the point representing y (furthermore, y should be 'above' x, so that the line segment from x to y runs upwards) [6].

Hasse diagrams have a wide variety of applications. Typical examples from mathematics include divisibility posets (in which $x \leq y$ iff x divides y) and subgroup lattices [6]; more practical applications come from formal concept analysis [25]. In this paper, however, we will focus on their applications in logic, and thus assume that the underlying poset (with $\leq$ being logical entailment) is a Boolean algebra, i.e. has top and bottom elements, and meet, join and complementation operations. It is well-known that a Boolean algebra can always be visualized by means of a Hasse diagram that is centrally symmetric (with all complementary pairs of elements ordered around the center of symmetry) [6]. Furthermore, a finite Boolean algebra can be partitioned into 'levels' $L_0, L_1, L_2, \ldots$, which are recursively defined as follows: $L_0 = \{\bot\}$, and $L_{k+1} = \{x \mid \exists y \in L_k \colon y \lhd x\}$. Figure 2 shows Hasse diagrams for the divisors of 12, the Boolean algebra $\wp(\{1,2,3\})$, and a Boolean algebra of formulas from the modal logic S5.

To fully appreciate the results that will be presented in Section 4, it should be realized that although most Aristotelian and Hasse diagrams are two-dimensional, this restriction is certainly not essential. In recent years, several three-dimensional Aristotelian diagrams have been studied, such as octahedrons, cubes, RDHs, etc. [8, 9, 26–28]. Similarly, there have also been studies on three-dimensional Hasse diagrams, such as (hyper)cubes and RDHs [13, 29–31].[4] Figure 3 shows an

[4] Recalling the typology of logic diagrams presented in Section 1, it should be noted that the trend towards three-dimensional diagrams includes not only Aristotelian and Hasse diagrams, but also other diagrams that visualize relations between formulas (e.g. duality diagrams [7]), and even diagrams that visualize single formulas (e.g. Euler and Venn diagrams [32]).

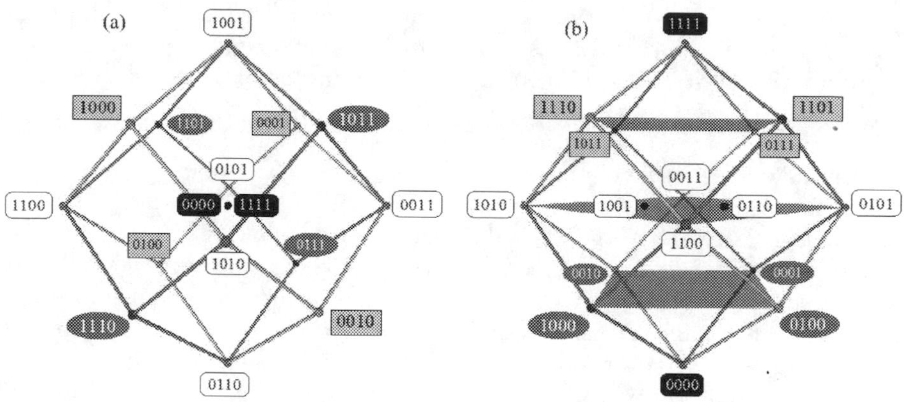

Fig. 3. (a) Aristotelian and (b) Hasse rhombic dodecahedron for CPL

Aristotelian and a Hasse RDH for the binary, truth-functional connectives of classical propositional logic (CPL); each connective is identified with its truth table (for example, conjunction is written as 1000, disjunction as 1110, etc.).

3 Similarities, Dissimilarities, and Diagram Design

We now discuss some obvious similarities and dissimilarities between Aristotelian and Hasse diagrams, and explain how they relate to general cognitive principles of diagram design and information visualization.

Let's start with the similarities between Aristotelian and Hasse diagrams. First of all, it should be noted that both types of diagrams represent the formulas *up to logical equivalence*, i.e. if φ and ψ are logically equivalent, then they cannot appear as distinct formulas in either type of diagram. In the case of a Hasse diagram for some Boolean logical system S, this is very clear: the poset that is visualized by the Hasse diagram is the Lindenbaum-Tarski algebra of S, whose elements are not individual formulas, but rather equivalence classes of formulas [6, p. 254ff.]. The literature on Aristotelian diagrams is less explicit about this logical equivalence condition (although there are exceptions [8, 12]), but nearly all Aristotelian diagrams that have been studied so far do indeed satisfy it.

A second, more important observation is that there is a large *overlap in the relations* visualized by the two types of diagrams. The Aristotelian relation of subalternation is identical to the notion of 'one-way entailment' $<$ (recall that $x < y$ iff $x \leq y$ and not $y \leq x$), which is itself the transitive closure of the covering relation $\lessdot$ visualized by Hasse diagrams. Hence, if φ and ψ occur in an Aristotelian diagram D_1 and a Hasse diagram D_2, then there is a subalternation arrow from φ to ψ in D_1 iff there is an upward path (i.e. a sequence of $\lessdot$-edges) from φ to ψ in D_2. Since all Aristotelian relations can be reduced to subalternation and contradiction,[5] and Hasse diagrams can represent these two relations (subalternation: just discussed; contradiction: by means of central symmetry), it follows that Hasse diagrams can represent all Aristotelian relations.

We now turn to the dissimilarities between Aristotelian and Hasse diagrams. The first difference concerns the *non-contingent formulas* $\bot$ and $\top$. These constitute the natural begin- and endpoints of the entailment ordering of a given Boolean logic S, and are thus visualized as resp. the lowest and the highest point of the Hasse diagram for S. This is the case regardless of whether this Hasse diagram happens to be two- or three-dimensional; for example, see the Hasse hexagon for S5 in Fig. 2c and the Hasse RDH for CPL in Fig. 3b. In contrast, almost all Aristotelian diagrams that have appeared in the literature so far contain only contingent formulas. One possible explanation for this restriction is that $\bot$ and $\top$ enter into many 'vacuous' Aristotelian relations with contingent formulas,[6] which would only clutter the diagrams [8]. Although most Aristotelian diagrams do not represent the non-contingent formulas at all, some authors [27, 33] prefer to think of them as coinciding in the center of the diagram. From this perspective, $\bot$ and $\top$ still do not occupy any 'real' vertices of the diagram, but they are 'hidden' in its center of symmetry (which is itself *not* a separate vertex of the diagram); for example, see the Aristotelian RDH in Fig. 3a, and the formulas $\bot = 0000$ and $\top = 1111$ coinciding in its center.

A second difference concerns the *general direction of the entailments.* We have seen above that entailment is visualized in both Aristotelian and Hasse diagrams (resp. as subalternation and as the transitive closure of the covering relation $\lhd$). By definition, all individual $\lhd$-edges in a Hasse diagram are directed upwards, and hence, all paths of such edges are also directed upwards. In Hasse diagrams, all entailments thus have the same general direction (viz. upwards). By contrast, in Aristotelian diagrams the subalternation arrows generally do not share a single direction.[7] Consider, for example, the Aristotelian hexagon in Fig. 1b: the subalternations starting in $\Diamond p \wedge \Diamond\neg p$ and those ending in $\Box p \vee \Box\neg p$ all have the same general direction (viz. upwards), but the subalternations from $\Box(\neg)p$ to $\Diamond(\neg)p$ run in the exact opposite direction (viz. downwards).

The third difference is related to the *visualization of the levels* of the Boolean algebra that the formulas are taken from. In a Hasse diagram, the levels L_k are visualized as hyperplanes that are orthogonal to the general entailment direction. Since this entailment direction is vertically upwards (cf. supra), the levels are horizontal hyperplanes. In Aristotelian diagrams, by contrast, there is generally a complete 'mixing' of levels. For example, compare the Aristotelian and Hasse hexagons for S5 in Fig. 1b and 2c, and focus on the L_1-formulas $\Box p$, $\Diamond p \wedge \Diamond\neg p$

[5] Writing CD and SA for contradiction and subalternation, respectively, we have that φ and ψ are contrary iff $SA(\varphi, \neg\psi)$ iff $\exists\theta\big(CD(\psi,\theta)$ and $SA(\varphi,\theta)\big)$, and similarly, φ and ψ are subcontrary iff $SA(\neg\varphi,\psi)$ iff $\exists\theta\big(CD(\varphi,\theta)$ and $SA(\theta,\psi)\big)$ [12, § 3.3.3].

[6] E.g. $\bot$ is contrary to every contingency, and $\top$ is subcontrary to every contingency.

[7] The main exception to this claim is, of course, the Aristotelian square, in which all subalternations are directed downwards; see Fig. 1a.

and $\Box\neg p$. In the Hasse hexagon, these formulas constitute a horizontal line, whereas in the Aristotelian hexagon, they constitute a triangle. Similarly, when comparing the RDHs for CPL, we see that the L_1-formulas 1000, 0100, 0010 and 0001 constitute a horizontal plane in the Hasse RDH in Fig. 3b (visualized in transparent grey), but a tetrahedron in the Aristotelian RDH in Fig. 3a.

The similarities and dissimilarities between Aristotelian and Hasse diagrams discussed above are fully in line with general principles about diagram design, such as *congruity*, *apprehension* and *information selection* [34, 35]. Both types of diagrams are used to visualize certain logical relations between formulas. Hasse diagrams primarily focus on the structure of the entailment ordering $<$, and try to establish a strong congruence between this logical structure and the visual structure of the diagram. For example, the Hasse hexagon in Fig. 2c visualizes the fact that $\Box p$ entails $\Diamond p$ by putting $\Box p$ lower than $\Diamond p$; similarly, it visualizes the fact that the L_1-formulas $\Box p$, $\Diamond p \wedge \Diamond\neg p$ and $\Box\neg p$ do not entail each other (and are thus *independent* with respect to the entailment ordering) by putting them on a line that is *orthogonal* to the direction of the entailment ordering.

From an Aristotelian perspective, however, $\Box p$, $\Diamond p \wedge \Diamond\neg p$ and $\Box\neg p$ are all contrary to each other. Since these formulas lie on a single line in the Hasse hexagon in Fig. 2c, the contrariety edges between them would overlap and would thus not be visually discernible from each other; this is a grave violation of the apprehension principle. The Aristotelian hexagon in Fig. 1c solves this problem by moving $\Diamond p \wedge \Diamond\neg p$ away from the horizontal line between $\Box p$ and $\Box\neg p$: the three contrariety edges now form a triangle and are thus clearly discernible. Of course, the price that has to be paid for achieving this is that the resulting diagram mixes the levels and no longer has a clear entailment direction. However, these properties correspond to the structure of the *entailment ordering*, so by the principle of information selection, it is no problem for an *Aristotelian* diagram to distort them in order to better visualize the *Aristotelian* relations.

4 A Unified Account: Projections of Hypercubes

We now begin with the development of a unified perspective on Aristotelian and Hasse diagrams, by showing how both types of diagrams can be seen as vertex-first projections of n-dimensional hypercubes. This development will be carried out in full generality (for arbitrary n), using some basic tools from linear algebra [36]. Concrete applications to the Aristotelian/Hasse hexagons ($n = 3$) and the Aristotelian/Hasse RDHs ($n = 4$) will be discussed in the next sections.

We will restrict ourselves to Aristotelian diagrams that represent all formulas of a given finite Boolean algebra (except for $\bot$ and $\top$, of course; cf. supra). This restriction is harmless, since every Aristotelian diagram that is not Boolean closed can be embedded inside one that is.[8] It is well-known that every finite Boolean algebra $\mathbb{B}_n$ can be represented as the powerset of a finite set $\{1, \ldots, n\}$,

[8] For example, the Aristotelian square in Fig. 1a can be embedded inside the Aristotelian hexagon in Fig. 1b, and the Aristotelian octagon in Fig. 1c can be embedded inside the Aristotelian RDH in Fig. 3a.

or equivalently, as the set $\{0,1\}^n$ of all *bitstrings* of length n [37, 38]. The latter representation is the most convenient for our purposes, and will thus be used.

It is well-known that the Boolean algebra $\mathbb{B}_n$ can be represented as a hypercube $\mathcal{C}_n$ in n-dimensional Euclidean space $\mathbb{R}^n$ [29, 30]. We now argue that this hypercube is a Hasse diagram for $\mathbb{B}_n$.[9] Consider a coordinate mapping $c\colon \{0,1\}^n \to \mathbb{R}^n$, which maps each bitstring $\varphi \in \{0,1\}^n$ onto its coordinates $c(\varphi) \in \mathbb{R}^n$.[10] It will be convenient to assume that c maps the bits 1 and 0 to the coordinates 1 and -1, respectively—e.g. $c(11010) = (1,1,-1,1,-1)$. The resulting hypercube is centered around the origin $(0,\ldots,0)$ of $\mathbb{R}^n$, with central symmetry representing logical negation: $c(\neg\varphi) = -1 \cdot c(\varphi)$, i.e. $\neg\varphi$ is diametrically opposed to φ. The general direction of entailment runs from $\bot$ to $\top$, and thus corresponds to the vector $c(\top) - c(\bot) = (2,\ldots,2) \sim (1,\ldots,1)$.[11] Any $(n-1)$-dimensional hyperplane orthogonal to this direction has an equation of the form $x_1 + \cdots + x_n = a$. For any bitstring $\varphi \in L_k$ $(0 \leq k \leq n)$, it holds that φ consists of k 1-bits and $n-k$ 0-bits, and hence $\sum_{i=1}^{i=n} c(\varphi)_i = 2k - n$; i.e. $c(\varphi)$ lies on the hyperplane with equation $x_1 + \cdots + x_n = 2k - n$. The level L_k of $\mathbb{B}_n$ thus corresponds to a hyperplane in $\mathbb{R}^n$ that is orthogonal to the entailment direction vector $(1,\ldots,1)$.

The Boolean algebra $\mathbb{B}_n$ has 2^n bitstrings, or equivalently, $\frac{2^n}{2} = 2^{n-1}$ pairs of contradictory bitstrings $(\varphi, \neg\varphi)$. Hence, its hypercube representation $\mathcal{C}_n$ has 2^{n-1} pairs of diametrically opposed vertices $(c(\varphi), c(\neg\varphi))$. Each of these pairs defines a direction vector $\mathbf{d}_{\varphi,\neg\varphi} := c(\varphi) - c(\neg\varphi) = c(\varphi) - (-c(\varphi)) = 2c(\varphi) \sim c(\varphi)$. The vector $\mathbf{d}_{\varphi,\neg\varphi}$ can be taken as the projection axis for a vertex-first projection $\Pi_{\varphi,\neg\varphi}$. We will focus on two such projections. The first one is $\Pi_{\top,\bot}$. The second one is harder to describe in general: it is of the form $\Pi_{\gamma,\neg\gamma}$ for some γ near the 'middle' of $\mathbb{B}_n$ (i.e. in the level L_k where $k = \lceil \frac{n}{2} \rceil$); this will become clearer in the concrete case studies. We will also make use of a linear transformation $T\colon \mathbb{R}^n \to \mathbb{R}^n$ that maps $c(\top)$ onto $c(\gamma)$; the matrix representation of T is a diagonal matrix that has the components of $c(\gamma)$ on its diagonal (i.e. $M_{i,i} = c(\gamma)_i$ and $M_{i,j} = 0$ for $1 \leq i \neq j \leq n$). Note that $T \cdot T = I_n$, and thus $T^{-1} = T$.

The projection axis of $\Pi_{\top,\bot}$ is $\mathbf{d}_{\top,\bot} = (1,\ldots,1)$. This is a projection onto the $(n-1)$-dimensional hyperplane that goes through the origin and is orthogonal to $\mathbf{d}_{\top,\bot}$, which has the equation $x_1 + \cdots + x_n = 0$. It is possible to choose $n-1$

9 Since Hasse diagrams can represent all Aristotelian relations, the hypercube is not only a Hasse diagram, but also an Aristotelian diagram. This dual perspective corresponds exactly to the two projections $\Pi_{\top,\bot}$ and $\Pi_{\gamma,\neg\gamma}$ introduced below.

10 We thus identify the hypercube $\mathcal{C}_n$ with its vertices, and 'ignore' its edges, faces, etc. This is unproblematic, since the latter are linearly generated by the vertices, and all transformations that will be applied to the hypercube are linear transformations. For example, the edge between vertices $\mathbf{x}$ and $\mathbf{y}$ is $E = \{\mathbf{x} + \lambda(\mathbf{y} - \mathbf{x}) \mid \lambda \in [0,1]\}$; for any linear transformation L, it holds that $L[E] = \{L(\mathbf{z}) \mid \mathbf{z} \in E\} = \{L(\mathbf{x}) + \lambda(L(\mathbf{y}) - L(\mathbf{x})) \mid \lambda \in [0,1]\}$, i.e. $L[E]$ is exactly the edge between $L(\mathbf{x})$ and $L(\mathbf{y})$.

11 For direction vectors $\mathbf{x}, \mathbf{y} \in \mathbb{R}^n$, we write $\mathbf{x} \sim \mathbf{y}$ iff $\mathbf{x}$ and $\mathbf{y}$ are identical up to a scalar λ (i.e. $\mathbf{x} = \lambda\mathbf{y}$). After all, we are only interested in the *direction* of these vectors, not in their particular *magnitude*.

pairwise orthogonal vectors $\rho_1, \ldots, \rho_{n-1}$ in this hyperplane.[12] The matrix that contains these vectors as rows is the matrix representation of $\Pi_{\top,\bot}$. It is easy to show that for the matrix representation of $\Pi_{\gamma,\neg\gamma}$, we can take $\Pi_{\gamma,\neg\gamma} := \Pi_{\top,\bot} \cdot T$. Since $T = T^{-1}$, it immediately follows that $\Pi_{\top,\bot} = \Pi_{\gamma,\neg\gamma} \cdot T$.

We now study the result of applying the vertex-first projection $\Pi_{\top,\bot}$ to the hypercube $\mathcal{C}_n$. Since $c(\neg\varphi) = -c(\varphi)$ and $\Pi_{\top,\bot}$ is a linear transformation, it follows that $\Pi_{\top,\bot}(c(\neg\varphi)) = \Pi_{\top,\bot}(-c(\varphi)) = -\Pi_{\top,\bot}(c(\varphi))$, i.e. $\Pi_{\top,\bot}[\mathcal{C}_n]$ is centered around the origin of $\mathbb{R}^{n-1}$, with central symmetry representing negation. Furthermore, since $c(\top)$ and $c(\bot)$ lie along the direction $\mathbf{d}_{\top,\bot}$ of the projection axis, it follows that $\Pi_{\top,\bot}(c(\top)) = \Pi_{\top,\bot}(c(\bot)) = (0, \ldots, 0) \in \mathbb{R}^{n-1}$. Note that this observation perfectly explains the claim that in an Aristotelian diagram, $\top$ and $\bot$ coincide in the diagram's center of symmetry [27, 33]. Next, recall that the hypercube $\mathcal{C}_n$ is a Hasse diagram, with general entailment direction from $c(\bot)$ to $c(\top)$, i.e. $\mathbf{d}_{\top,\bot}$. Since this direction is exactly the projection axis of $\Pi_{\top,\bot}$, it follows that in $\Pi_{\top,\bot}[\mathcal{C}_n]$ the edges no longer share a common entailment direction: the component of the direction vector that they shared has been 'projected away'. To make this more concrete, Fig. 4a provides an example for a projection $\pi\colon \mathbb{R}^2 \to \mathbb{R}^1$. Note that the vectors $\mathbf{a}, \mathbf{b} \in \mathbb{R}^2$ have more or less the same direction, viz. vertically upwards. However, if we define π to be the projection along exactly this vertical direction, then we see that $\pi(\mathbf{a})$ and $\pi(\mathbf{b})$ do not share the same direction at all (the vertical component that they shared has been projected away). Finally, also recall that $\mathcal{C}_n$ represents the levels L_k of $\mathbb{B}_n$ by means of hyperplanes that are orthogonal to $\mathbf{d}_{\top,\bot}$. However, in $\Pi_{\top,\bot}[\mathcal{C}_n]$, the levels will be 'mixed', since the distance (along $d_{\top,\bot}$) that separated them has been 'projected away'. As an illustration, consider Fig. 4b: the points $\mathbf{a}$ and $\mathbf{b}$ belong to some level L_k, and the points $\mathbf{c}$ and $\mathbf{d}$ belong to a different ('higher') level L_m; thus, $\mathbf{a}$ and $\mathbf{b}$ are both 'below' $\mathbf{c}$ and $\mathbf{d}$. However, when we consider their projections, we see that the levels have been completely 'mixed': $\pi(\mathbf{a})$ lies between $\pi(\mathbf{c})$ and $\pi(\mathbf{d})$, while $\pi(\mathbf{d})$ lies between $\pi(\mathbf{a})$ and $\pi(\mathbf{b})$.

We now consider the other vertex-first projection of the hypercube: $\Pi_{\gamma,\neg\gamma}[\mathcal{C}_n]$. Since the direction of the projection axis $\mathbf{d}_{\gamma,\neg\gamma}$ does *not* coincide with the general entailment direction $\mathbf{d}_{\top,\bot}$ of $\mathcal{C}_n$, matters are much simpler in this case. Since $\mathcal{C}_n$ is a Hasse diagram and thus all its edges share a general entailment direction $\mathbf{d}_{\top,\bot}$, it follows that $\Pi_{\gamma,\neg\gamma}[\mathcal{C}_n]$ also has a general entailment direction, viz. $\Pi_{\gamma,\neg\gamma}(\mathbf{d}_{\top,\bot})$, which runs from $\Pi_{\gamma,\neg\gamma}(c(\bot))$ to $\Pi_{\gamma,\neg\gamma}(c(\top))$. However, one problem seems to remain: $\Pi_{\gamma,\neg\gamma}$ maps the formulas γ and $\neg\gamma$ to the origin of $\mathbb{R}^{n-1}$. This problem can easily be solved, but the details are highly dependent on the concrete case (e.g. it matters whether n is odd or even), and will thus be postponed to the concrete case studies in the next sections.

In sum, the discussion above shows that $\Pi_{\top,\bot}[\mathcal{C}_n]$ and $\Pi_{\gamma,\neg\gamma}[\mathcal{C}_n]$ are resp. an Aristotelian diagram and a Hasse diagram of the Boolean algebra $\mathbb{B}_n$. There is thus a deep connection between Aristotelian and Hasse diagrams: both can be seen as vertex-first projections of one and the same hypercube $\mathcal{C}_n$. Additionally,

[12] In the case studies in the next sections, we will show that it is often possible to choose these vectors in particularly elegant ways.

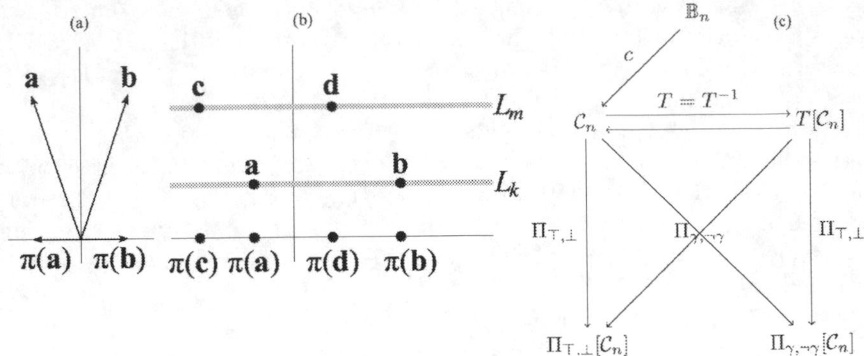

Fig. 4. The effects of projecting along the entailment direction: (a) loss of a shared entailment direction and (b) 'mixing' of levels. Part (c) is a commutative diagram containing the various linear transformations studied in this section.

this viewpoint yields a unified explanation of the various dissimilarities between both types of diagrams (as discussed in Section 3), by showing how they are merely different manifestations of the same underlying process, viz. projecting the Hasse diagram $\mathcal{C}_n$ along its own general entailment direction. Finally, recalling that $\Pi_{\gamma,\neg\gamma} = \Pi_{\top,\bot} \cdot T$ and $\Pi_{\top,\bot} = \Pi_{\gamma,\neg\gamma} \cdot T$, we also have a way to move back and forth between the Aristotelian diagram $\Pi_{\top,\bot}[\mathcal{C}_n]$ and the Hasse diagram $\Pi_{\gamma,\neg\gamma}[\mathcal{C}_n]$, which is summarized by the commutative diagram in Fig. 4c.

5 Case Study I: Hexagons as Projections of the Cube

We will now illustrate the unified account of Aristotelian and Hasse diagrams that was introduced in the previous section, by applying it to some small Boolean algebras. In this section we consider $\mathbb{B}_3$, in Section 6 we will look at $\mathbb{B}_4$.

The Boolean algebra $\mathbb{B}_3$ is represented as a cube $\mathcal{C}_3$ in three-dimensional Euclidean space $\mathbb{R}^3$. This is done by the 'conventional' coordinate mapping c: $\{0,1\}^3 \to \mathbb{R}^3$; see Fig. 5a and the $c(\varphi)$-row in Table 1 at the end of this section. The general entailment direction goes from $c(\bot)$ to $c(\top)$, and is thus $c(\top) - (\bot) = (1,1,1) - (-1,-1,-1) = (2,2,2) \sim (1,1,1)$. We will consider two vertex-first projections of $\mathcal{C}_3$. The first one, $\Pi_{\top,\bot}$, is along the direction $\mathbf{d}_{\top,\bot} = (1,1,1)$, and onto the projection plane $x+y+z=0$ (which is orthogonal to $\mathbf{d}_{\top,\bot}$). For the second one, $\Pi_{\gamma,\neg\gamma}$, we choose $\gamma := 101$; this projection is thus along $\mathbf{d}_{\gamma,\neg\gamma} = c(\gamma) - c(\neg\gamma) = (1,-1,1) - (-1,1,-1) = (2,-2,2) \sim (1,-1,1)$, and onto the projection plane $x-y+z=0$ (which is orthogonal to $\mathbf{d}_{\gamma,\neg\gamma}$). The projection axes and projection planes of $\Pi_{\top,\bot}$ and $\Pi_{\gamma,\neg\gamma}$ are shown in Fig. 5b and Fig. 5c, respectively. The linear transformation T which maps $c(\top) = (1,1,1)$ onto $c(\gamma) = (1,-1,1)$ corresponds to a reflection over the $(x-z)$-plane (compare Fig. 5b and Fig. 5c); it is described by a 3×3 diagonal matrix with 1, -1 and 1 on its diagonal. (The effect of T on all points of $\mathcal{C}_3$ is described in the $T(c(\varphi))$-row of Table 1.) The matrix representations of the vertex-first projections are

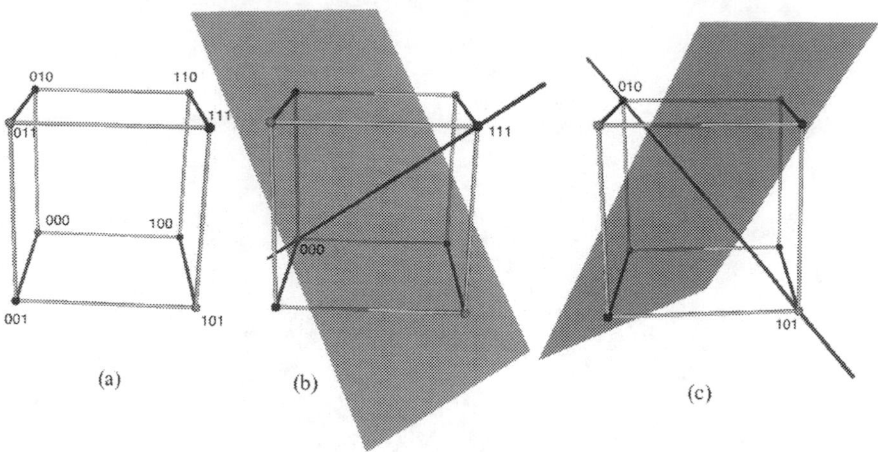

Fig. 5. (a) The cube $\mathcal{C}_3$ with its bitstring decoration; (b) the projection axis and projection plane of $\Pi_{\top,\bot}$; (c) the projection axis and projection plane of $\Pi_{\gamma,\neg\gamma}$

$$\Pi_{\top,\bot} = \begin{pmatrix} \frac{-\sqrt{3}}{4} & 0 & \frac{\sqrt{3}}{4} \\ \frac{1}{4} & \frac{-1}{2} & \frac{1}{4} \end{pmatrix} \quad \text{and} \quad \Pi_{\gamma,\neg\gamma} = \begin{pmatrix} \frac{-\sqrt{3}}{4} & 0 & \frac{\sqrt{3}}{4} \\ \frac{1}{4} & \frac{1}{2} & \frac{1}{4} \end{pmatrix}.$$

Note that the rows of $\Pi_{\top,\bot}$ are two orthogonal vectors that lie in the plane $x+y+z=0$, i.e. the projection plane of $\Pi_{\top,\bot}$. Similarly, the rows of $\Pi_{\gamma,\neg\gamma}$ are two orthogonal vectors that lie in the plane $x-y+z=0$, i.e. the projection plane of $\Pi_{\gamma,\neg\gamma}$. Finally, note that the rows of $\Pi_{\gamma,\neg\gamma}$ are the result of applying the linear transformation T to the corresponding rows of $\Pi_{\top,\bot}$, and thus $\Pi_{\gamma,\neg\gamma} = \Pi_{\top,\bot} \cdot T$.

When the vertex-first projection $\Pi_{\top,\bot}$ is applied to (all points of) the cube $\mathcal{C}_3$, the result is a regular hexagon that is centered around the origin $(0,0)$; this hexagon is shown in Fig. 6a and its concrete coordinates can be found in the $\Pi_{\top,\bot}(c(\varphi))$-row of Table 1. Although the hexagon does not by itself contain any logical relation, it is clear that it is essentially an *Aristotelian diagram* (compare with Fig. 1b). First of all, the non-contingent formulas $\top = 111$ and $\bot = 000$ coincide in the hexagon's center of symmetry. Secondly, the hexagon does not have a single direction of entailment (in contrast, $\mathcal{C}_3$ does have such a direction, viz. $\mathbf{d}_{\top,\bot}$). Finally, the levels L_k have been 'mixed'; for example, the bitstrings of L_1 and those of L_2 form two interlocking triangles (in contrast, in $\mathcal{C}_3$ the levels L_1 and L_2 are represented by means of the planes $x+y+z=-1$ and $x+y+z=1$, respectively).

We now turn to the second projection, viz. $\Pi_{\gamma,\neg\gamma}$. The result is again a regular hexagon that is centered around the origin; this hexagon is shown in Fig. 6b and its concrete coordinates can be found in the $\Pi_{\gamma,\neg\gamma}(c(\varphi))$-row of Table 1. Although this hexagon has $\bot = 000$ at its lowest vertex and $\top = 111$ at its highest, and also has a general direction of entailment—viz. vertically upwards—, it is not a true Hasse diagram, since the bitstrings 101 and 010 coincide in the center, and thus the levels L_1 and L_2 are not represented uniformly.

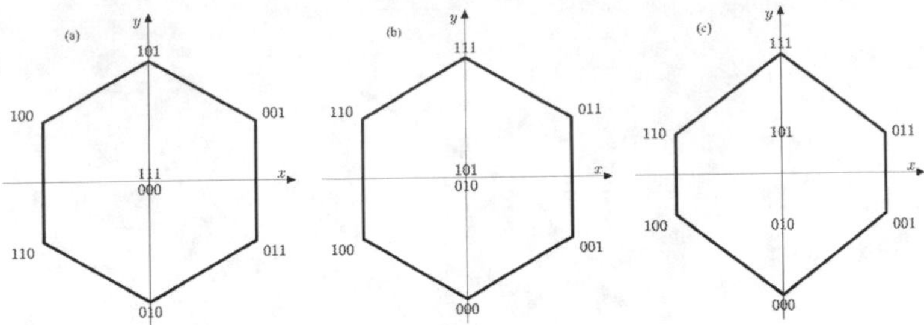

Fig. 6. (a) $\Pi_{\top,\bot}[\mathcal{C}_3]$, (b) $\Pi_{\gamma,\neg\gamma}[\mathcal{C}_3]$, (c) $\Pi^{1/3}_{\gamma,\neg\gamma}[\mathcal{C}_3]$

Table 1. The elements of $\mathbb{B}_3$, $\mathcal{C}_3$, $\Pi_{\top,\bot}[\mathcal{C}_3]$, $\Pi_{\gamma,\neg\gamma}[\mathcal{C}_3]$ and $\Pi^{1/3}_{\gamma,\neg\gamma}[\mathcal{C}_3]$

φ	111	110	101	011	100	010	001	000
$c(\varphi)$	(1,1,1)	(1,1,-1)	(1,-1,1)	(-1,1,1)	(1,-1,-1)	(-1,1,-1)	(-1,-1,1)	(-1,-1,-1)
$T(c(\varphi))$	(1,-1,1)	(1,-1,-1)	(1,1,1)	(-1,-1,1)	(1,1,-1)	(-1,-1,-1)	(-1,1,1)	(-1,1,-1)
$\Pi_{\top,\bot}(c(\varphi))$	(0,0)	$(-\frac{\sqrt{3}}{2},-\frac{1}{2})$	(0,1)	$(\frac{\sqrt{3}}{2},-\frac{1}{2})$	$(-\frac{\sqrt{3}}{2},\frac{1}{2})$	(0,-1)	$(\frac{\sqrt{3}}{2},\frac{1}{2})$	(0,0)
$\Pi_{\gamma,\neg\gamma}(c(\varphi))$	(0,1)	$(-\frac{\sqrt{3}}{2},\frac{1}{2})$	(0,0)	$(\frac{\sqrt{3}}{2},\frac{1}{2})$	$(-\frac{\sqrt{3}}{2},-\frac{1}{2})$	(0,0)	$(\frac{\sqrt{3}}{2},-\frac{1}{2})$	(0,-1)
$\Pi^{1/3}_{\gamma,\neg\gamma}(c(\varphi))$	(0,1)	$(-\frac{\sqrt{3}}{2},\frac{1}{3})$	$(0,\frac{1}{3})$	$(\frac{\sqrt{3}}{2},\frac{1}{3})$	$(-\frac{\sqrt{3}}{2},-\frac{1}{3})$	$(0,-\frac{1}{3})$	$(\frac{\sqrt{3}}{2},-\frac{1}{3})$	(0,-1)

However, this problem can be solved if we introduce a small *perturbation* ε to the vertex-first projection $\Pi_{\gamma,\neg\gamma}$, thus obtaining the 'quasi-projection' $\Pi^{\varepsilon}_{\gamma,\neg\gamma}$. This quasi-projection is defined in such a way that if $\varepsilon = 0$, then $\Pi^{\varepsilon}_{\gamma,\neg\gamma} = \Pi_{\gamma,\neg\gamma}$. For our current purposes, however, we will particularly be interested in the case $\varepsilon = \frac{1}{3}$. Here is the general definition, and the special case $\varepsilon = \frac{1}{3}$:

$$\Pi^{\varepsilon}_{\gamma,\neg\gamma} = \begin{pmatrix} \frac{-\sqrt{3}}{4} & 0 & \frac{\sqrt{3}}{4} \\ \frac{1+\varepsilon}{4} & \frac{1-\varepsilon}{2} & \frac{1+\varepsilon}{4} \end{pmatrix}, \qquad \Pi^{1/3}_{\gamma,\neg\gamma} = \begin{pmatrix} \frac{-\sqrt{3}}{4} & 0 & \frac{\sqrt{3}}{4} \\ \frac{1}{3} & \frac{1}{3} & \frac{1}{3} \end{pmatrix}.$$

Fig. 6c shows the hexagon that results from applying this quasi-projection to the cube $\mathcal{C}_3$; its coordinates can be found in the $\Pi^{1/3}_{\gamma,\neg\gamma}(c(\varphi))$-row of Table 1. Although the hexagon is no longer regular, it can properly be called a *Hasse diagram* (compare with Fig. 2c). It represents $\bot = 000$ and $\top = 111$ at resp. its lowest and highest point. Furthermore, it has a general entailment direction, viz. vertically upward. Finally, the levels are represented by lines that are horizontal (and thus orthogonal to the general entailment direction); for example, L_1 and L_2 correspond to the horizontal lines $y = -\frac{1}{3}$ and $y = \frac{1}{3}$, respectively.

6 Case Study II: RDHs as Projections of the Hypercube

As a second illustration of the unified account described in Section 4, we will now apply it to $\mathbb{B}_4$. This Boolean algebra is represented as a four-dimensional

Table 2. The elements of $\mathbb{B}_4$, $\mathcal{C}_4$, $\Pi_{\top,\bot}[\mathcal{C}_4]$, $\Pi_{\gamma,\neg\gamma}[\mathcal{C}_4]$ and $\Pi^{1/10}_{\gamma,\neg\gamma}[\mathcal{C}_4]$. For reasons of space, the table lists only 8 of the 16 bitstrings of $\mathbb{B}_4$; it can be completed by adding their 8 negations, and recalling that $c(\neg\varphi) = -c(\varphi)$, $T(c(\neg\varphi)) = -T(c(\varphi))$, $\Pi_{\top,\bot}(c(\neg\varphi)) = -\Pi_{\top,\bot}(c(\varphi))$, $\Pi_{\gamma,\neg\gamma}(c(\neg\varphi)) = -\Pi_{\gamma,\neg\gamma}(c(\varphi))$ and $\Pi^{1/10}_{\gamma,\neg\gamma}(c(\neg\varphi)) = -\Pi^{1/10}_{\gamma,\neg\gamma}(c(\varphi))$.

φ	1111	1110	1101	1011	0111	1100	1010	1001
$c(\varphi)$	(1,1,1,1)	(1,1,1,-1)	(1,1,-1,1)	(1,-1,1,1)	(-1,1,1,1)	(1,1,-1,-1)	(1,-1,1,-1)	(1,-1,-1,1)
$T(c(\varphi))$	(1,-1,-1,1)	(1,-1,-1,-1)	(1,-1,1,1)	(1,1,-1,1)	(-1,-1,-1,1)	(1,-1,1,-1)	(1,1,-1,-1)	(1,1,1,1)
$\Pi_{\top,\bot}(c(\varphi))$	(0,0,0)	(-1,-1,1)	(-1,1,-1)	(1,1,1)	(1,-1,-1)	(-2,0,0)	(0,0,2)	(0,2,0)
$\Pi_{\gamma,\neg\gamma}(c(\varphi))$	(0,2,0)	(-1,1,1)	(1,1,1)	(-1,1,-1)	(1,1,-1)	(0,0,2)	(-2,0,0)	(0,0,0)
$\Pi^{1/10}_{\gamma,\neg\gamma}(c(\varphi))$	(0.2,2,0)	(-0.8,1,1)	(1,1,1)	(-1,1,-1)	(1.2,1,-1)	(0,0,2)	(-2,0,0)	(-0.2,0,0)

hypercube $\mathcal{C}_4 \subset \mathbb{R}^4$. We again employ the 'conventional' coordinate mapping $c\colon \{0,1\}^4 \to \mathbb{R}^4$; see the $c(\varphi)$-row in Table 2. The general entailment direction in this hypercube is $(1,1,1,1)$. We consider two vertex-first projections of $\mathcal{C}_4$. The first one, $\Pi_{\top,\bot}$, is along the direction $\mathbf{d}_{\top,\bot} = (1,1,1,1)$, and onto the projection plane $x+y+z+u=0$ (which is orthogonal to $\mathbf{d}_{\top,\bot}$). For the second one, $\Pi_{\gamma,\neg\gamma}$, we choose $\gamma := 1001$; this projection is thus along $\mathbf{d}_{\gamma,\neg\gamma} = (1,-1,-1,1)$, and onto the projection plane $x-y-z+u=0$ (which is orthogonal to $\mathbf{d}_{\gamma,\neg\gamma}$). The linear transformation T which maps $c(\top) = (1,1,1,1)$ onto $c(\gamma) = (1,-1,-1,1)$ is described by a 4×4 diagonal matrix with 1, -1, -1 and 1 on its diagonal; see the $T(c(\varphi))$-row of Table 2. The matrix representations of the projections are

$$\Pi_{\top,\bot} = \begin{pmatrix} -0.5 & -0.5 & 0.5 & 0.5 \\ 0.5 & -0.5 & -0.5 & 0.5 \\ 0.5 & -0.5 & 0.5 & -0.5 \end{pmatrix} \quad \text{and} \quad \Pi_{\gamma,\neg\gamma} = \begin{pmatrix} -0.5 & 0.5 & -0.5 & 0.5 \\ 0.5 & 0.5 & 0.5 & 0.5 \\ 0.5 & 0.5 & -0.5 & -0.5 \end{pmatrix}.$$

The rows of $\Pi_{\top,\bot}$ are pairwise orthogonal vectors that lie in the plane $x+y+z+u=0$, i.e. the projection plane of $\Pi_{\top,\bot}$; similar remarks apply to $\Pi_{\gamma,\neg\gamma}$.

When the vertex-first projection $\Pi_{\top,\bot}$ is applied to the hypercube $\mathcal{C}_4$, the result is an RDH that is centered around the origin $(0,0,0)$. This RDH was already shown in Fig. 3a; its concrete coordinates can be found in the $\Pi_{\top,\bot}(c(\varphi))$-row of Table 2. This RDH is essentially an *Aristotelian diagram*: the non-contingent formulas $\top = 1111$ and $\bot = 0000$ coincide in the RDH's center of symmetry, the RDH does not have a single direction of entailment, and the levels of $\mathbb{B}_4$ have been 'mixed'; e.g. the bitstrings of L_1 and L_3 form two interlocking tetrahedrons.

We now turn to the second projection, viz. $\Pi_{\gamma,\neg\gamma}$. The result is again an RDH that is centered around the origin; its concrete coordinates can be found in the $\Pi_{\gamma,\neg\gamma}(c(\varphi))$-row of Table 2. This RDH is not a true Hasse diagram, since the bitstrings 1001 and 0110 coincide. However, this problem can be solved by introducing a small *perturbation* ε to $\Pi_{\gamma,\neg\gamma}$. The resulting 'quasi-projection' $\Pi^{\varepsilon}_{\gamma,\neg\gamma}$ is defined in such a way that if $\varepsilon = 0$, then $\Pi^{\varepsilon}_{\gamma,\neg\gamma} = \Pi_{\gamma,\neg\gamma}$. Here is the general definition, and the special case $\varepsilon = \frac{1}{10} = 0.1$:

$$\Pi^{\varepsilon}_{\gamma,\neg\gamma} = \begin{pmatrix} -0.5 & 0.5+\varepsilon & -0.5+\varepsilon & 0.5 \\ 0.5 & 0.5 & 0.5 & 0.5 \\ 0.5 & 0.5 & -0.5 & -0.5 \end{pmatrix}, \quad \Pi^{1/10}_{\gamma,\neg\gamma} = \begin{pmatrix} -0.5 & 0.6 & -0.4 & 0.5 \\ 0.5 & 0.5 & 0.5 & 0.5 \\ 0.5 & 0.5 & -0.5 & -0.5 \end{pmatrix}.$$

Fig. 3b shows the RDH that results from applying this quasi-projection to the hypercube C_4; its coordinates are in the $\Pi^{1/10}_{\gamma,\neg\gamma}(c(\varphi))$-row of Table 2. This RDH is a proper *Hasse diagram*: it represents $\bot = 0000$ and $\top = 1111$ at resp. its lowest and highest point, it has a general entailment direction (viz. vertically upward), and finally, the levels are represented by planes that are horizontal (and thus orthogonal to the general entailment direction); for example, L_1, L_2 and L_3 correspond to the horizontal planes $y = -1$, $y = 0$ and $y = 1$, respectively.

7 Conclusion

In this paper we have explored the relationship between two important types of diagrams for representing logical relations between formulas, viz. Aristotelian and Hasse diagrams. After briefly discussing some obvious similarities and dissimilarities, we argued that there exists a deep connection between both types of diagrams. On a *visual-cognitive* level, we showed that their dissimilarities can perfectly be explained in terms of general principles of diagram design, such as congruity, apprehension and information selection. On a more abstract *geometrical* level, we showed that pairs of Boolean closed Aristotelian and Hasse diagrams can be seen as different vertex-first projections of one and the same hypercube, thereby obtaining a unified explanation of their dissimilarities.

In future work, we plan to use this geometrical account in the search for adequate Aristotelian/Hasse diagrams for larger Boolean algebras. For example, although there are currently no satisfactory diagrams for $\mathbb{B}_5$ in 3D, we know that they certainly exist in 4D, viz. as vertex-first projections of the 5D hypercube.

Acknowledgements. The first author is financially supported by a PhD fellowship of the Research Foundation–Flanders (FWO).

References

1. Howse, J., Stapleton, G., Taylor, J.: Spider diagrams. LMS Journal of Computation and Mathematics 8, 145–194 (2005)
2. Rodgers, P.: A survey of Euler diagrams. J. of Visual Lang. & Comp. (in press)
3. Shin, S.J.: The Iconic Logic of Peirce's Graphs. MIT Press (2002)
4. Stapleton, G.: A survey of reasoning systems based on Euler diagrams. Electronic Notes in Theoretical Computer Science 134, 127–151 (2005)
5. Stapleton, G., Howse, J., Rodgers, P.: A graph theoretic approach to general Euler diagram drawing. Theoretical Computer Science 411, 91–112 (2010)
6. Davey, B., Priestley, H.: Introduction to Lattices and Order. Cambridge U. P. (2002)
7. Demey, L.: Algebraic aspects of duality diagrams. In: Cox, P., Plimmer, B., Rodgers, P. (eds.) Diagrams 2012. LNCS, vol. 7352, pp. 300–302. Springer, Heidelberg (2012)

8. Demey, L.: Structures of oppositions for public announcement logic. In: Béziau, J.Y., Jacquette, D. (eds.) Around and Beyond the Square of Opposition, pp. 313–339. Springer (2012)
9. Moretti, A.: The Geometry of Logical Opposition. Ph.D. thesis, Neuchâtel (2009)
10. Smessaert, H.: Boolean differences between two hexagonal extensions of the logical square of oppositions. In: Cox, P., Plimmer, B., Rodgers, P. (eds.) Diagrams 2012. LNCS (LNAI), vol. 7352, pp. 193–199. Springer, Heidelberg (2012)
11. Smessaert, H.: The classical Aristotelian hexagon versus the modern duality hexagon. Logica Universalis 6, 171–199 (2012)
12. Smessaert, H., Demey, L.: Logical geometries and information in the square of oppositions. Forthcoming in Journal of Logic, Language and Information (2014)
13. Zellweger, S.: Untapped potential in Peirce's iconic notation for the sixteen binary connectives. In: Houser, N., Roberts, D.D., Van Evra, J. (eds.) Studies in the Logic of Charles Peirce, pp. 334–386. Indiana University Press (1997)
14. Hurley, P.J.: A Concise Introduction to Logic, 11th edn. Wadsworth (2012)
15. Bernhard, P.: Visualizations of the square of opposition. Log. Univ. 2, 31–41 (2008)
16. Seuren, P.: The Logic of Language. Oxford University Press (2010)
17. Carnielli, W., Pizzi, C.: Modalities and Multimodalities. Springer (2008)
18. Lenzen, W.: How to square knowledge and belief. In: Béziau, J.Y., Jacquette, D. (eds.) Around and Beyond the Square of Opposition, pp. 305–311. Springer (2012)
19. McNamara, P.: Deontic logic. In: Stanford Encyclopedia of Philosophy (2010)
20. Moretti, A.: The geometry of standard deontic logic. Log. Univ. 3, 19–57 (2009)
21. Blanché, R.: Structures Intellectuelles. Vrin (1966)
22. Jacoby, P.: A triangle of opposites for types of propositions in Aristotelian logic. New Scholasticism 24, 32–56 (1950)
23. Sesmat, A.: Logique II. Les Raisonnements. La syllogistique. Hermann (1951)
24. Béziau, J.Y.: New light on the square of oppositions and its nameless corner. Logical Investigations 10, 218–232 (2003)
25. Ganter, B., Stumme, G., Wille, R. (eds.): Formal Concept Analysis. LNCS (LNAI), vol. 3626. Springer, Heidelberg (2005)
26. Chatti, S., Schang, F.: The cube, the square and the problem of existential import. History and Philosophy of Logic 32, 101–132 (2013)
27. Smessaert, H.: On the 3D visualisation of logical relations. Logica Universalis 3, 303–332 (2009)
28. Smessaert, H., Demey, L.: Logical and geometrical complementarities between Aristotelian diagrams. In: Dwyer, T., Purchase, H.C., Delaney, A. (eds.) Diagrams 2014. LNCS (LNAI), vol. 8578, pp. 248–262. Springer, Heidelberg (2014)
29. Foldes, S.: A characterization of hypercubes. Discr. Math. 17, 155–159 (1977)
30. Harary, F., Hayes, J.P., Wu, H.J.: A survey of the theory of hypercube graphs. Computers & Mathematics with Applications 15, 277–289 (1988)
31. Kauffman, L.H.: The mathematics of Charles Sanders Peirce. Cybernetics & Human Knowing 8, 79–110 (2001)
32. Flower, J., Stapleton, G., Rodgers, P.: On the drawability of 3D Venn and Euler diagrams. Journal of Visual Languages & Computing (in press)
33. Sauriol, P.: Remarques sur la théorie de l'hexagone logique de Blanché. Dialogue 7, 374–390 (1968)
34. Tversky, B.: Prolegomenon to scientific visualizations. In: Gilbert, J.K. (ed.) Visualization in Science Education, pp. 29–42. Springer (2005)
35. Tversky, B.: Visualizing thought. Topics in Cognitive Science 3, 499–535 (2011)
36. Strang, G.: Introduction to Linear Algebra. Wellesley-Cambridge Press (2009)
37. Givant, S., Halmos, P.: Introduction to Boolean Algebras. Springer (2009)
38. Smessaert, H., Demey, L.: The unreasonable effectiveness of bitstrings in logical geometry. In: 4th World Congress on the Square of Opposition (2014)

A Graphical Representation of Boolean Logic

Beáta Bojda[1], Katalin Bubnó[2], Benedek Nagy[1,3], and Viktor Takács[4]

[1] Eastern Mediterranean University, Famagusta, North Cyprus, Mersin-10, Turkey
{bojdab,nbenedek.inf}@gmail.com
[2] Institute of Mathematics, University of Debrecen, Debrecen, Hungary
bubno.kati@gmail.com
[3] Faculty of Informatics, University of Debrecen, Debrecen, Hungary
Faculty of Economics, University of Debrecen, Debrecen, Hungary
viktor.takacs@econ.unideb.hu

Abstract. A new graphical representation/model of Boolean logic is presented. We use horizontal and vertical lines to represent the logical constants, serial and parallel circuits for conjunction and disjunction. For negation of a (sub)formula, we use a turn of the circuit by $\frac{\pi}{2}$. A web-based system that is built on the model is also presented.

Keywords: Boolean Logic, visual representation, educational tool.

1 A Diagrammatic Theory of Boolean Algebra

In set theory, there are well-known graphical representation techniques, e.g., Euler-Venn diagrams, intervals [1]. For Boolean logic, there are also various techniques used, e.g., logic gates, pipe systems, but it is not straightforward to understand them; students may have problems, specially with negation. A technique is presented here based on a new idea.

Boolean constants (true and false) are usually denoted by 1, 0, by T and F, or by $\top$ and $\bot$. In our approach we use the notation | and – for the true and false, respectively. We use | for true because it is somehow similar to 1 or I (note that the word true is starting with an 'I' in Hungarian, it is 'Igaz'). The notation – for the false (it is 'Hamis' in Hungarian) also seems to be a good choice, since both the negative sign and the complement can be connected to this sign, and both of them are somehow connected to negativity, falsity. We may also use variables, each of them with its own color different from the others' color. A variable may be represented by a sign † and its rotated form, depending on its actual value at the evaluation of the formula.

As one can build up a Boolean formula (a complex statement) from atomic statements (true and false statements allowing variables and constants), we build our representation analogously in iterative way. We use "intelligent wiring" technique at the base circuits. We define the three basic Boolean operations in our system in the following way (see also Fig. 1):

- Negation: the figure of the subformula is turned by $\frac{\pi}{2}$ (clockwise direction),
- Conjunction: we put the two subformulae one under the other and connect them in a serial way (series circuit);

T. Dwyer et al. (Eds.): Diagrams 2014, LNAI 8578, pp. 228–230, 2014.

- Disjunction: we put the two subformulae next to each other and connect them in a parallel way (parallel circuit).

One can prove by induction (using De Morgan's laws for negation) that our representation is correct. Moreover it is complete.

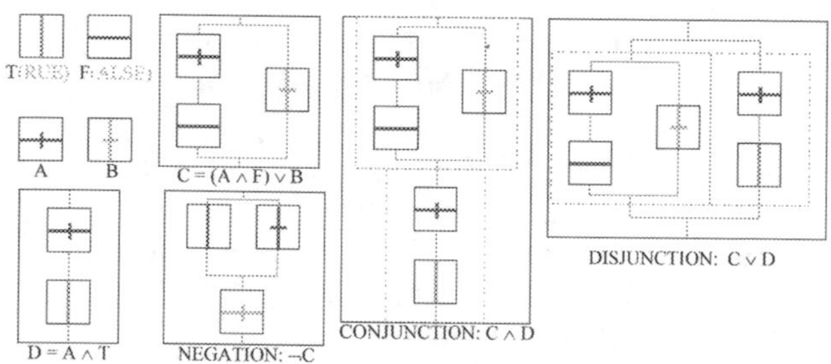

Fig. 1. Representation of the constant truth values, variables and the basic connectives; at negation the wiring automatically changes from serial to parallel and vice versa

Logic is not trivial therefore visual information could be a key point in teaching it. We believe that visual capture is the most important, which gate is next to/under the other, and then, the wiring comes in an easy way. This is the main advantage of this visual/graphical representation, and makes it applicable in, e.g., teaching. We have designed an interactive web-presentation system based on PHP technology (dynamic pages) with CSS3 animations (Web Kit):

2 The Program

The system works well under Chrome, Opera, Safari, Firefox, Maxthon and similar browsers; and that was one of the most important aspects: the presentation is available anywhere for the students, not only in the classroom, but in their home, in their mobile/tablet as well. For sake of a good design we made the blocks with Google Blockly Block Factory tool [2]. A picture about these blocks (that are used through the presentation tool) is shown in Fig. 2. The classic appearance of block languages as MIT Scratch, MIT App Inventor, as well as our newly developed system have a 'puzzle-design'. The connected blocks (elements of the programming language) are similar to the elements of the jigsaw games. The other design element is influenced by the basic evaluation method of variables in Computer Science. In assignment statements we usually declare the variable in the left side of the formula and we give them a value from the expression of the right side. We follow this order in the visual representation, when we put the light to the left side of the circuit we put the circuit representation of the Boolean formula to the right side. The light is on when the evaluated formula has a true value, and off when it has a false value.

2.1 The Elements of the Circuits

In electrical engineering 0 and 1 represent the two different states of one bit in a digital circuit; they are typically high and low voltage, electricity is flowing in the power line or not. The declared (logical) variable is represented by the bulb. The value of this variable could be either 1 or 0 (usually the task is to determine it). Identity (true) and zero (false) elements, as constants of the circuit are represented by the full or empty battery. The value of the full battery is always 1 (true) and the value of the empty battery is always 0 (false). They are useful in teaching of Mathematics and Computer Science, especially in solving logical equations as we have seen in our experiment [3]. The logical variables of the formula, i.e., the parameters are represented by switches (having horizontal or vertical value). See Figure 2. Our BooleImpress system works in user friendly way, and it can be used to teach Boolean logic. We think that the prospective mathematics teachers should know the importance of the applications and interdisciplinary connection of Boolean logic. In the University of Debrecen we did some promising teaching experiment with our tool [3].

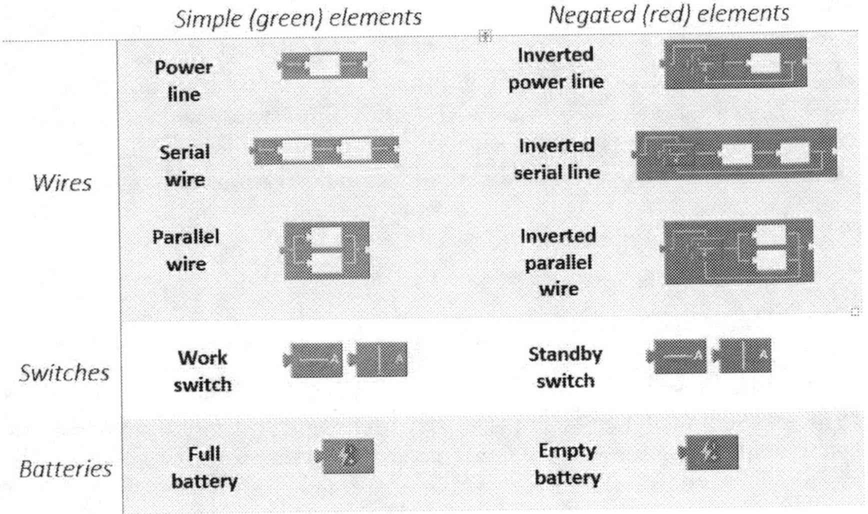

Fig. 2. The basic elements of our system (http://takacs-viktor.info/booleimpress/)

References

1. Nagy, B.: Reasoning by Intervals. In: Barker-Plummer, D., Cox, R., Swoboda, N. (eds.) Diagrams 2006. LNCS (LNAI), vol. 4045, pp. 145–147. Springer, Heidelberg (2006)
2. Google Blockly, `https://blockly-demo.appspot.com/`
3. Bojda, B., Bubnó, K., Nagy, B., Takács, V.: Visualization and efficiency in teaching mathematics. In: de la Puerta, J.G., et al. (eds.) International Joint Conference SOCO'14-CISIS'14-ICEUTE'14. AISC, vol. 299, pp. 541–550. Springer, Heidelberg (2014)

The Barwise-Seligman Model of Representation Systems: A Philosophical Explication

Atsushi Shimojima[1] and Dave Barker-Plummer[2]

[1] Faculty of Culture and Information Science, Doshisha University
1-3 Tatara-Miyakodani, Kyotanabe, 610-0394 Japan
ashimoji@me.com
[2] CSLI/Stanford University, Stanford, CA, 94305, USA
dbp@stanford.edu

Abstract. As an application of their channel theory, Barwise & Seligman sketched a set-theoretic model of representation systems. Their model has the attraction of capturing many important logical properties of diagrams, but few attempts have been made to apply it to actual diagrammatic systems. We attribute this to a lack of precision in their explanation of what their model is about—what a "representation system" is. In this paper, we propose a concept of representation system on the basis of Barwise & Seligman's original ideas, supplemented by Millikan's theory of reproduction. On this conception, a representation system is a family of individual representational acts formed through a repetitive reproduction process that preserves a set of syntactic and semantic constraints. We will show that this concept lets us identify a piece of reality that the Barwise-Seligman model is concerned with, making the model ready for use in the logical analysis of real-world representation systems.

Keywords: representation systems, diagrams, channel theory.

1 Introduction

Channel theory is an attempt to characterize information flows in our environment from a logical point of view. In their book that develops this theory, Barwise and Seligman outlined a general model of representation systems as one of the theory's principal applications [1, Chapter 20]. This model, which we will call "the B&S model," proposes a general framework in which we can investigate logical-semantical properties of a wide range of representation systems, including the systems of spoken language, written language, physical models, and, most importantly for our purpose, diagrammatic representations.

The B&S model is an abstract model of representation systems, and as such it can be used to describe *classes* of representation systems by characterizing abstract properties common to their members. Results proven in the B&S model concerning representation systems with particular properties will apply to all such representation systems, and consequently, verifying that a new representation system has the desired property can be achieved simply by matching the

T. Dwyer et al. (Eds.): Diagrams 2014, LNAI 8578, pp. 231–245, 2014.

system to the model. Barwise and Seligman have shown, for instance, that their model lets us formally characterize some of the fundamental properties of diagrammatic systems, such as free rides, over-specificity, and auto-consistency, that have direct implications on their cognitive efficacy [2,3]. Thus, the B&S model has significant potential as a formal framework for logical study of diagrammatic representations, and when fully developed, will complement the proof-theoretic and model-theoretic framework that have been productively applied to diagrammatic systems [4,5,6, for example].

To our knowledge, however, few attempts have been made to apply the B&S model to actual diagrammatic systems, not even to reveal those fundamental properties it is known to handle well. The mathematical foundation for the B&S model is laid out explicitly by Barwise and Seligman throughout their book, with all its main components derived from standard set theory. We believe that the model has failed to obtain traction, not because of vagueness in the explication of the model, but rather in the lack of specificity concerning the question of how that model is supposed to fit into reality. What aspects of reality are the individual components of the model supposed to capture? Why are those particular mathematical structures required for that purpose? As to these conceptual issues, Barwise and Seligman give only very general clues, leaving some important questions unanswered. In this paper, we lay a conceptual foundation for the avenue of logical study of diagrams that Barwise and Seligman have pointed to.

The first two thirds of this paper are devoted to these specific mathematical questions. After presenting an overview of the structure of the B&S model (Section 2), we will explain the aspects of the B&S model that are relatively easy to interpret (Section 3). We then attack the parts that are less straightforward—the part that develops a rather unusual "two-tier" semantic theory (Section 4).

As the details are filled in, however, it becomes clear that a most fundamental question is yet to be answered. That is, what the B&S model is a model of. Well, it is a model of representation systems, but what is a representation system anyway? Not knowing what the model is about implies not being confident about what the model applies to, and this presents a block to applying the model.

Drawing on the rather scarce clues provided by Barwise and Seligman, we will reconstruct the notion of a representation system that the B&S model apparently presupposes (Section 5). We propose to understand Barwise and Seligman's notion of representation system on the basis of Millikan's theory of reproduction [7,8]. According to this view, a representation system is a family of individual representational acts formed through repetitive reproduction of new representational acts from temporally preceding representational acts. The syntactic and semantic rules associated with a representation system are then explained as the constraints that these individual acts inherit over the reproduction process.

2 The Structure of the B&S Model

In this section, we will present the B&S model in its bare structure. Our purpose here is not to illustrate or explain the model, but to clearly present the mathematical structure posited in the model as preparation for subsequent exposition.

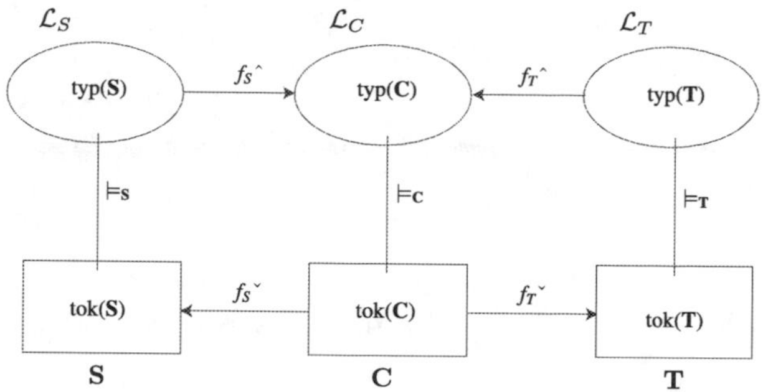

Fig. 1. The Barwise-Seligman Model of Representation Systems

Figure 1 shows the general structure of the model. It is composed of the three *classifications*, three *local logics* on them, and two *infomorphisms* connecting the three classifications. Let us define each component more exactly.

Definition 1 (Classification). *A classification* $\mathbf{A} = \langle tok(\mathbf{A}), typ(\mathbf{A}), \models_{\mathbf{A}} \rangle$ *consists of*

1. *a set, tok*($\mathbf{A}$)*, of objects to be classified, called the* tokens *of* $\mathbf{A}$*,*
2. *a set, typ*($\mathbf{A}$)*, of objects to classify the tokens, called the* types *of* $\mathbf{A}$*, and*
3. *a binary relation,* $\models_{\mathbf{A}}$*, between tok*($\mathbf{A}$) *and typ*($\mathbf{A}$)*.*

Thus, the three classifications involved in the structure in Figure 1 are:

- $\mathbf{S} = \langle \text{tok}(\mathbf{S}), \text{typ}(\mathbf{S}), \models_{\mathbf{S}} \rangle$ (depicted in left),
- $\mathbf{C} = \langle \text{tok}(\mathbf{C}), \text{typ}(\mathbf{C}), \models_{\mathbf{C}} \rangle$ (center), and
- $\mathbf{T} = \langle \text{tok}(\mathbf{T}), \text{typ}(\mathbf{T}), \models_{\mathbf{T}} \rangle$ (right).

We use lowercase Greek letters to refer to types of a classification, and uppercase Greek letters to refer to sets of types. When $a \models_{\mathbf{A}} \alpha$, we say "$a$ is of type α," "a supports α," or "α holds of a."

Given any classification $\mathbf{A}$, we often talk about a pair $\langle \Gamma, \Delta \rangle$ of subsets of typ($\mathbf{A}$). We call such a pair a *sequent* in $\mathbf{A}$. A token a in tok($\mathbf{A}$) is said to *satisfy* a sequent $\langle \Gamma, \Delta \rangle$ if a supports some member of Δ provided a supports all members of Γ. Thus, we are reading Γ conjunctively and Δ disjunctively when we talk about the satisfaction of a sequent.

We use a (subscripted) turnstile $\vdash$ to denote a set of sequents, and write $\Gamma \vdash \Delta$ to mean that $\langle \Gamma, \Delta \rangle$ belongs to $\vdash$. In such a context, we adopt a common abuse of notation for Gentzen sequents. In particular, we omit braces in denoting a unit set (e.g., "$\gamma \vdash \delta$" instead of "$\{\gamma\} \vdash \{\delta\}$") and use a comma to denote the union of sets (e.g., "$\Gamma_1, \Gamma_2 \vdash \Delta_1, \Delta_2$" instead of "$\Gamma_1 \cup \Gamma_2 \vdash \Delta_1 \cup \Delta_2$").

With this preparation, local logics can be defined in the following way:

Definition 2 (Local Logic). *A* local logic $\mathcal{L} = \langle \mathbf{A}, \vdash_{\mathcal{L}}, N_{\mathcal{L}} \rangle$ *on a classification* $\mathbf{A}$ *consists of*

1. *a set* $\vdash_{\mathcal{L}}$ *of sequents in* **A** *satisfying the following closure conditions:*
   Identity : $\alpha \vdash_{\mathcal{L}} \alpha$ *for every* $\alpha \in typ(\mathbf{A})$,
   Weakening : *If* $\Gamma \vdash_{\mathcal{L}} \Delta$, *then* $\Gamma, \Sigma_1 \vdash_{\mathcal{L}} \Delta, \Sigma_2$ *for any* $\Sigma_1, \Sigma_2 \subseteq typ(\mathbf{A})$,
   Global Cut : *If there is a set* $\Sigma \subseteq typ(\mathbf{A})$ *such that* $\Sigma_1, \Gamma \vdash_{\mathcal{L}} \Delta, \Sigma_2$ *for each partition* $\langle \Sigma_1, \Sigma_2 \rangle$ *of* Σ, *then* $\Gamma \vdash_{\mathcal{L}} \Delta$.[1]
2. *a subset* $N_{\mathcal{L}} \subseteq tok(\mathbf{A})$, *called the* normal tokens *of* $\mathcal{L}$, *which satisfy all the sequents of* $\vdash_{\mathcal{L}}$.

A local logic $\mathcal{L} = \langle \mathbf{A}, \vdash_{\mathcal{L}}, N_{\mathcal{L}} \rangle$ is designed to specify a system of *constraints* governing the classification **A**. It does the job by specifying a set $\vdash_{\mathcal{L}}$ of sequents in **A** that all normal tokens in tok(**A**) are supposed to satisfy. Just what token is normal or abnormal is specified by $N_{\mathcal{L}}$, which carves out the members of tok(**A**) that are normal as far as the local logic $\mathcal{L}$ is concerned.[2] We call a sequent in $\vdash_{\mathcal{L}}$ a *constraint in* the local logic $\mathcal{L}$.

The closure conditions in clause 1 of Definition 2 are required if we are to be able to read the sequents in $\vdash_{\mathcal{L}}$ as the constraints satisfied by any set of tokens whatsoever. For example, if the sequent $\langle \Gamma, \Delta \rangle$ belongs to $\vdash_{\mathcal{L}}$, $\langle \Gamma \cup \Sigma_1, \Delta \cup \Sigma_2 \rangle$ necessarily belongs to $\vdash_{\mathcal{L}}$, for if every token supporting all members of Γ supports at least one member of Δ, then every token supporting all members of the superset $\Gamma \cup \Sigma_1$ of Γ supports at least one member of the superset $\Delta \cup \Sigma_2$ of Δ. This means that the set $\vdash_{\mathcal{L}}$ of sequents must satisfy Weakening. Identity and Global Cut are required for similar reasons. [3]

The local logics involved in the structure in Figure 1 are the following:

- $\mathcal{L}_S = \langle \mathbf{S}, \vdash_{\mathcal{L}_S}, N_{\mathcal{L}_S} \rangle$ (placed in upper left)
- $\mathcal{L}_C = \langle \mathbf{C}, \vdash_{\mathcal{L}_C}, N_{\mathcal{L}_C} \rangle$ (upper center)
- $\mathcal{L}_T = \langle \mathbf{T}, \vdash_{\mathcal{L}_T}, N_{\mathcal{L}_T} \rangle$ (upper right)

Definition 3 (Infomorphism). *Given classifications* **A** *and* **B**, *an* infomorphism $f : \mathbf{A} \rightleftarrows \mathbf{B}$ *from* **A** *to* **B** *is a pair of functions* $f = \langle f^{\wedge}, f^{\vee} \rangle$ *such that:*

1. $f^{\wedge} : typ(\mathbf{A}) \rightarrow typ(\mathbf{B})$,
2. $f^{\vee} : tok(\mathbf{B}) \rightarrow tok(\mathbf{A})$, *and*
3. $f^{\vee}(b) \models_{\mathbf{A}} \alpha$ *iff* $b \models_{\mathbf{B}} f^{\wedge}(\alpha)$ *for each token* $b \in tok(\mathbf{B})$ *and each type* $\alpha \in typ(\mathbf{A})$.

Thus, the structure in Figure 1 involves two infomorphisms, $f_S : \mathbf{S} \rightleftarrows \mathbf{C}$ and $f_T : \mathbf{T} \rightleftarrows \mathbf{C}$. One, f_S, consists of a function $f_S{}^{\wedge}$ from typ(**S**) to typ(**C**) and a

[1] $\langle \Sigma_1, \Sigma_2 \rangle$ is a partition of Σ iff $\Sigma_1 \cup \Sigma_2 = \Sigma$ and $\Sigma_1 \cap \Sigma_2 = \emptyset$. Note that this definition allows Σ_1 or Σ_2 in a partition $\langle \Sigma_1, \Sigma_2 \rangle$ to an empty set, unlike the definition of partition adopted in certain contexts.

[2] Just because you specify a subset $N_{\mathcal{L}}$ of tok(**A**), it does not mean that you have specified the reason why some members of tok(**A**) are in $N_{\mathcal{L}}$ while others out. We will come back to this issue later.

[3] In fact, it has been shown that the satisfaction of these closure conditions is also a *sufficient* condition for a set of sequents to be the set of the constraints on a classification. See Section 9.5 of [1] for the more precise formulation of this idea.

function $f_S{}^{\check{}}$ from tok(**C**) to tok(**S**), and the other, f_T, consists of a function $f_T{}^{\hat{}}$ from typ(**T**) to typ(**C**) and a function $f_T{}^{\check{}}$ from tok(**C**) to tok(**T**). The symbol $\rightleftarrows$ suggests the reversed directions of the two functions involved in an infomorphism. Since the two infomorphisms are both connected to the classification **C**, Barwise and Seligman call it the *core* of this structure.

Now, a representation system is just a pair of infomorphisms with a common core, coupled with a local logic on each of the three classifications involved.

Definition 4 (Representation System). *A representation system $\mathcal{R}$ is a quintuple $\langle f_S : \mathbf{S} \rightleftarrows \mathbf{C}, f_T : \mathbf{T} \rightleftarrows \mathbf{C}, \mathcal{L}_S, \mathcal{L}_C, \mathcal{L}_T \rangle$ where f_S and f_T are infomorphisms and $\mathcal{L}_S$, $\mathcal{L}_C$, and $\mathcal{L}_T$ are local logics on the classifications* **S**, **C**, *and* **T**, *respectively.*

3 Easy Part: The Source and Target Logics

Now that we have laid out a mathematical structure, we start our explication of how it is supposed to capture something in real world. Barwise and Seligman intend an individual representation system $\mathcal{R} = \langle f_S : \mathbf{S} \rightleftarrows \mathbf{C}, f_T : \mathbf{T} \rightleftarrows \mathbf{C}, \mathcal{L}_S, \mathcal{L}_C, \mathcal{L}_T \rangle$ to capture a *practice* of producing representations of a particular kind. Such practices include that of producing maps, drawing diagrams, painting pictures, writing sentences, and uttering sentences. How are the individual components of the structure $\mathcal{R}$ fitted to the components of such a practice? We start with the components of $\mathcal{R}$ that are relatively easy to interpret.

3.1 Source

The classification $\mathbf{S} = \langle \text{tok}(\mathbf{S}), \text{typ}(\mathbf{S}), \models_{\mathbf{S}} \rangle$ depicted in the left side of Figure 1 is called the *source* of the representation system, and the members of tok(**S**) are called *representations*. Here, a representation $a \in \text{tok}(\mathbf{S})$ is intended to be such things as an individual diagram drawn on a particular sheet of paper, a map printed on a particular page of a brochure, and a sentence displayed on a particular computer display. As a token, a representation is distinguished from its appearance. So, when one draws exactly the same arrangement of symbols on two different occasions, the result is two different diagram tokens (representations) although they have exactly the same appearance. Similarly, two prints of *Downtown Chicago Map* published by the same map publisher are different representations—map tokens—even though they usually have exactly the same arrangement of symbols and colors.

In contrast, typ(**S**) consist of syntactic properties that classify the representations in tok(**S**) according to what symbols appear in what arrangements. Thus, if tok(**S**) consists of the individual map prints of *Downtown Chicago Map*, typ(**S**) may contain syntactic properties such as the following:

(σ_1) There is a unique street line labeled "E Ontario," drawn left-right.

(σ_2) There are unique street lines labeled "N Rush" and "N Saint Claire," drawn up-down.

(σ_3) A hotel symbol is at the upper-left corner of the intersection of street lines labeled “N Rush” and “E Ontario.”

(σ_4) A hotel symbol is at the lower-right corner of the intersection of street lines labeled “N Saint Claire” and “E Ontario.”

These are types, or properties, in the sense that they can hold of many different map tokens. In fact, they hold of all prints of *Downtown Chicago Map*, with the exception of defective or degenerated copies (changed colors, unprinted symbols, spilled coffee, etc.) These types hold of even *revised* maps of downtown Chicago, as long as they have street lines with appropriate labeling, have a certain line straight and certain lines in parallel, and have hotel symbols in certain positions. In this way, given the sets tok(**S**) and typ(**S**), the relation $\models_{\mathbf{S}}$ is determined by which tokens support which types in reality.

3.2 Target

The classification $\mathbf{T} = \langle \text{tok}(\mathbf{T}), \text{typ}(\mathbf{T}), \models_{\mathbf{T}} \rangle$ depicted in the right side of Figure 1 is called the *target* of the system. Intuitively, a member b of tok(**T**) is something that is actually represented in a representational practice, so in the case of the mapmaking practice for *Downtown Chicago Map*, it is a particular region in Chicago in some period of time. If the company produces revised maps, the same region in a subsequent time period gets represented in this mapmaking practice, so it makes another member of tok(**T**).

Naturally, typ(**T**) consists of types, or properties, that classify these represented objects. In our case of the mapmaking practice, the set consist of types that classify the region of Chicago in different periods of time. It may contain:

(θ_1) There is a unique streets named “E Ontario,” running east-west.

(θ_2) There are unique streets named “N Rush” and “N Saint Claire,” running north-south.

(θ_3) A hotel building is at the north-west corner of the crossing of streets named “North Rush” and “East Ontario.”

(θ_4) A hotel building is at the south-east corner of the crossing of streets named “North Saint Claire” and “East Ontario.”

Note the difference from the types σ_1–σ_4. The types θ_1–θ_4 refer to possible arrangements of streets and buildings on a region in Chicago, rather than possible arrangements of lines and symbols on a sheet of paper.

3.3 Local Logic on the Source

The local logic $\mathcal{L}_S$ shown in the upper left part of Figure 1 is designed to capture a system of constraints governing the source classification **S**. It is the triple $\langle \mathbf{S}, \vdash_{\mathcal{L}_S}, N_{\mathcal{L}_S} \rangle$, and as with every local logic, it does its job by having the second coordinate $\vdash_{\mathcal{L}_S}$ specify a set of sequents that every normal token in tok(**S**) satisfies. In our case of mapmaking practice, an example of such a sequent can be $\langle\{\sigma_1, \sigma_2, \sigma_3, \sigma_4\}, \{\sigma_5\}\rangle$, where σ_1–σ_4 are as above and σ_5 is:

(σ_5) There are at least two hotel symbols on a street line labeled "E Ontario."

This particular constraint $\{\sigma_1, \sigma_2, \sigma_3, \sigma_4\} \vdash_{\mathcal{L}_S} \sigma_5$ is due to a geometrical constraint governing lines and symbols on a plane. The set $\vdash_{\mathcal{L}_S}$ may also contain physical constraints on the coloring of symbols. In addition to these natural constraints, the set typically contains constraints due to the *syntactic stipulations* adopted in the mapmaking practice, regulating such things as what types of building symbols can appear, how labels are placed on them, and what varieties of colors can color regions of a map.

3.4 Local Logic on the Target

The local logic $\mathcal{L}_T$ shown in the upper right part of Figure 1 captures a system of constraints governing the target classification **T**. It is the triple $\langle \mathbf{T}, \vdash_{\mathcal{L}_T}, N_{\mathcal{L}_T}\rangle$, and it works just as the local logic $\mathcal{L}_S$ works on the source classification **S**. Thus, in our case of mapmaking practice, $\vdash_{\mathcal{L}_T}$ contains such sequents as $\langle\{\theta_1, \theta_2, \theta_3, \theta_4\}, \{\theta_5\}\rangle$, where σ_1–σ_4 are as above and σ_5 is:

(θ_5) There are at least two hotel buildings on a street named "East Ontario."

The correspondence between this constraint $\{\theta_1, \theta_2, \theta_3, \theta_4\} \vdash_{\mathcal{L}_S} \theta_5$ on the target classification and the constraint $\{\sigma_1, \sigma_2, \sigma_3, \sigma_4\} \vdash_{\mathcal{L}_S} \sigma_5$ on the source classification is not an accident. Both constraints are based on the same topological law that regulate buildings and streets on a geographical region, as well symbols and lines on a map.

4 Difficult Part: The Core Logic

So far, we have been concerned with the components of the B&S model that are depicted on the left and the right side of Figure 1, and their interpretation was relatively straightforward. What about the components in the middle? What is the center classification **C** supposed to model? What role do the two infomorphisms $f_S = \langle f_S\hat{}, f_S\check{}\rangle$ and $f_T = \langle f_T\hat{}, f_T\check{}\rangle$ play by having **C** as its core?

These medial components are there to define what Barwise and Seligman [1] call the *representation relation* and the *indication relation* supported by the representation system.

Definition 5 (Representation and Indication). *Let* $\mathcal{R} = \langle f_S : \mathbf{S} \rightleftarrows \mathbf{C}, f_T : \mathbf{T} \rightleftarrows \mathbf{C}, \mathcal{L}_S, \mathcal{L}_C, \mathcal{L}_T\rangle$ *be a representation system.*

1. *A token a in tok*(**S**) represents *a token b in tok*(**T**)*, written* $a \rightsquigarrow_{\mathcal{R}} b$*, if there is a token c in tok*(**C**) *such that* $f_S\check{}(c) = a$ *and* $f_T\check{}(c) = b$.
2. *A type* α *in typ*(**S**) indicates *a type* β *in typ*(**T**)*, written* $\alpha \Rightarrow_{\mathcal{R}} \beta$*, if* $f_S\hat{}(\alpha) \vdash_{\mathcal{L}_C} f_T\hat{}(\beta)$.

In this section, we first explicate the two relations $\rightsquigarrow_{\mathcal{R}}$ and $\Rightarrow_{\mathcal{R}}$ defined above, to explain the framework of "two-tier" semantics central to the B&S model.

4.1 Representation Relation

Intuitively, Clause 1 of Definition 5 says that a token in tok(**S**) represents a token in tok(**T**) if they are mediated by a token c in tok(**C**). Barwise & Seligman [1] characterize this c as "the particular spatial-temporal process whereby the representation comes to represent what it does" (p. 236). When a is a map, c is also characterized a "causal link" between a map and what it is a map of (p. 237). In such a case, the classification **C** is said to model the "actual practice" of mapmaking (p. 236), so it is natural to interpret this causal link c as an *individual act* that belongs to this practice of mapmaking.

Thus, in our case of mapmaking practice for downtown Chicago, a token c_1 in tok(**C**) can be an act, conducted mainly by the map publisher, that consists of the sub-acts of assembling relevant information about a particular region in Chicago in a certain period of time, editing the information in the form expressible in a map, and printing it on a particular sheet of paper. When another map is printed, this new printing act is combined with the first two sub-acts of c_1 to make another token c_2 in tok(**C**).

To see this situation more clearly, let e_1 and e_2 be the sub-acts of information collection and information editing, respectively, and p_1 and p_2 be the first and the second printing acts mentioned above. Then, the representational act c_1 is the sequence $e_1 \circ e_2 \circ p_1$ of sub-acts, while the representational act c_2 is the sequence $e_1 \circ e_2 \circ p_2$. Generally, every act p_i of printing a new map based on the information collected and edited in e_1 and e_2 gives rise to a new representational act $c_i = e_1 \circ e_2 \circ p_i$, a new token in tok(**C**).

Conceived in this way, every individual mapmaking act c has a unique map as its product, and a unique region of Chicago in a particular period of time as the object about which information is assembled and edited. We call the former the *representing object* and the latter the *represented object* of the act c.[4]

The function $f_S{}^\vee$ can be interpreted as the assignment of a unique object to the role of representing object in every mapmaking act in tok(**C**). Similarly the function $f_T{}^\vee$ can be interpreted as the assignment of a unique object to the role of represented object in every such mapmaking act.

Combined, the mapmaking act c connects the particular map $f_S{}^\vee(c)$ to the particular region $f_T{}^\vee(c)$ in the particular period of time. This is the situation described by Clause 1 of Definition 5. It is a representational act that connects a token in tok(**S**) to a token in tok(**T**), making the former stand in the representation relation $\rightsquigarrow_{\mathcal{R}}$ to the latter.

[4] When considering the production of representations via acts, the source of the information used in the act is a token in the target classification, and the result of the act is a token in the source classification. It may appear, then, that they have misnamed our classifications. However the names that we have chosen derive from the more common situation where a diagram is the source of information about a target – utilization, rather than production, of the diagram.

4.2 Indication Relation

The notion of role that we have just introduced to explain the functions $f_S\check{}$ and $f_T\check{}$ also helps us to interpret the other functions $f_S\hat{}(c)$ and $f_T\hat{}(c)$, which are depicted in the upper middle part of Figure 1. Given a type α in typ(**S**), we can think of $f_S\hat{}(\alpha)$ as the type that classifies an act in tok(**C**) on the basis of the property of the object playing the role of represented object in it. For example, if a mapmaking act c has produced a map in which there is a unique street labeled "E Ontario," we can classify c as being of the following type:

(ω_1) The object playing the role of representing object is such that there is a unique street line labeled "E Ontario."

Earlier, we considered the following type as a member of typ(**S**):

(σ_1) There is a unique street line labeled "E Ontario."

The types ω_1 and θ_1 are different, but they are closely related in the way the following equivalence holds:

(1) A mapmaking act c is of type ω_1 if and only if the object playing the role of the representing object in c is of type σ_1.

Thus, ω_1 is the property that classifies a mapmaking act c according to whether the object playing the role of representing object in c is of type σ_1.

Generally, for every type σ in typ(**S**), there is a unique type ω in typ(**C**) that classifies a member of tok(**C**) according to whether the object playing the role of representing object in it is of type σ_i. The model uses the function $f_S\hat{}(c)$ to capture this functional relation from typ(**S**) to typ(**C**). In the case of the types ω_1 and σ_1 above, $f_S\hat{}$ assigns ω_1 to σ_1. So, we can paraphrase (1) as:

(2) A mapmaking act c is of type $f_S\hat{}(\sigma_1)$ if and only if the object playing the role of representing object in c is of type σ_1.

Recalling that the object playing the role of the representing object in c is assigned by $f_S\check{}$ to c, this amounts to saying:

(3) A mapmaking act c is of type $f_S\hat{}(\sigma_1)$ if and only if $f_S\check{}(c)$ is of type σ_1.

Here, we see an instance of the equivalence condition stated in the definition of infomorphism (Definition 3). The condition lets us generalize (3) to every token c in tok(**C**) and every type σ in typ(**S**). This way, the infomorphism $f = \langle f_S\hat{}, f_S\check{}\rangle$ captures the partial type-equivalence between an act c and the object that plays the role of representing object in c.

A similar consideration applies to the infomorphism $f = \langle f_T\hat{}, f_T\check{}\rangle$: it captures the partial type-equivalence between an act c and the object that plays the role of represented object in c. For example, consider the type θ_1 in typ(**T**) and the type ω_2 in typ(**C**):

(θ_1) There is a unique street named "E Ontario."

(ω_2) The object playing the role of represented object is such that there is a unique street named "E Ontario."

Here, $f_T\hat{}$ assigns ω_2 to θ_1, so the following equivalence holds:

(4) A mapmaking act c is of type $f_T\hat{}(\theta_1)$ if and only if $f_T\check{}(c)$ is of type θ_1.

Again, the equivalence condition in Definition 3 lets us generalize (4) to every token in c in tok(**C**) and every type θ in typ(**T**).

Now, by the definitions of $f_S\hat{}$ and $f_T\hat{}$, both the image $f_S\hat{}$(typ(**S**)) and the image $f_T\hat{}$(typ(**T**)) are subsets of typ(**C**). The B&S model uses this fact to define the indication relation from types in typ(**S**) to types in typ(**T**). Take the type σ_1 in typ(**S**) and the type θ_1 in typ(**T**) for example. Intuitively, if there is a unique street line labeled "E Ontario" drawn right-left in a downtown Chicago map, it indicates that there is a unique street named "E Ontario" running east-west in the mapped region. That is, the type σ_1 indicates the type θ_1 in this mapmaking practice. Barwise and Seligman models this indication relation as a constraint in the local logic $\mathcal{L}_C$. For them, σ_1 indicates θ_1 just in case there holds the constraint $f_S\hat{}(\sigma_1) \vdash_{\mathcal{L}_C} f_T\hat{}(\theta_1)$.

Remember that $\vdash_{\mathcal{L}_C}$ lists the constraints governing the classification **C** of a set of representational acts. In particular, the constraint $f_S\hat{}(\sigma_1) \vdash_{\mathcal{L}_C} f_T\hat{}(\theta_1)$ states the following:

(5) A representational act in tok(**C**) is such that the representing object (a map, in the present example) is of the type σ_1 only if the represented object (a city region) is of the type θ_1.

This makes it clear that the constraint $f_S\hat{}(\sigma_1) \vdash_{\mathcal{L}_C} f_T\hat{}(\theta_1)$ is something maintained by the effort of people who are involved in the mapmaking acts in tok(**C**). The constraint is essentially *arbitrary* in its origin, but once people start conforming to it and believe that everybody conforms to it, it satisfies significant mutual benefit for them to keep conforming to it. It thus becomes a "self-perpetuating" constraint over the representational acts of a group of people. Lewis [9] has developed a general theory of how such a constraint becomes a stabilized character of human conducts.

The condition $f_S\hat{}(\sigma_1) \vdash_{\mathcal{L}_C} f_T\hat{}(\theta_1)$ is the way Barwise and Seligman capture one of such constraints. Clause 2 of Definition 5 is then a generalization of this strategy of capturing a *semantic constraint* stabilized in a representational practice: it characterizes the indication relation $\Rightarrow_{\mathcal{R}}$ as the relation that holds between a type α in typ(**S**) and a type β in typ(**T**) whenever the constraint of the form $f_S\hat{}(\alpha) \vdash_{\mathcal{L}_C} f_T\hat{}(\beta)$ holds.

Note: Other Constraints on Representational Acts. Typically, our representational acts are constrained not only by these semantic constraints, but also by *syntactic stipulations*, concerning what arrangements of symbols and colors are allowed in representing objects. These stipulations combined with natural (geometrical, topological, and physical) constraints to produce a larger set of constraints on the arrangements of symbols and colors. The B&S model captures their effects as constraints in the local logic $\mathcal{L}_S$ on the source classification, not as constraints in the local logic $\mathcal{L}_C$ on the core classification.

Moreover, our representational acts are typically constrained by *target restrictions*, namely, restrictions on the choice of objects to be represented. For example, only a region of Chicago in some period is chosen as the represented object in the representational practice for *Downtown Chicago Map.* In fact, this is an assumption on which the syntactic stipulations and semantic conventions of this mapmaking practice are stabilized. Totally different stipulations and conventions would be adopted if a region of the Rocky Mountains were the main target of the representational practice. The kind of objects that come in the set tok($\mathbf{T}$) is thus restricted due to target restrictions on our representational acts, and for this reason, a substantial set of constraints hold on the classification $\mathbf{T}$ and get captured in the local logic $\mathcal{L}_T$ on the target classification. The constraint $\{\theta_1, \theta_2, \theta_3, \theta_4\} \vdash_{\mathcal{L}_S} \theta_5$ cited in Section 3.4 is an example of such constraints.

4.3 Two-Tier Semantics

This pair of relations $\rightsquigarrow_{\mathcal{R}}$ and $\Rightarrow_{\mathcal{R}}$ that we have just explained lets us characterize the informational relation between a representation and the represented object in a natural way.

Definition 6 (Representing As). *Let* $\mathcal{R} = \langle f_S : \mathbf{S} \rightleftarrows \mathbf{C}, f_T : \mathbf{T} \rightleftarrows \mathbf{C}, \mathcal{L}_S, \mathcal{L}_C,$ $\mathcal{L}_T\rangle$ *be a representation system. Given a token* a *in tok*($\mathbf{S}$)*, a token* b *in tok*($\mathbf{T}$)*, and a type* β *in typ*($\mathbf{T}$)*,* a represents b as being of type β *if*

1. $a \rightsquigarrow_{\mathcal{R}} b$ *and*
2. *there is a type* α *in typ*($\mathbf{S}$) *such that*
   - $a \models_{\mathbf{S}} \alpha$*, and*
   - $\alpha \Rightarrow_{\mathcal{R}} \beta$.

For example, recall that a map publisher takes an act whose representing object is a particular complete print a of *Downtown Chicago Map* and whose represented object is a particular region b of Chicago in a particular period of time (i.e., $a \rightsquigarrow_{\mathcal{R}} b$), that there is a unique street line labeled "E Ontario" drawn right-left in this particular print a (i.e., $a \models_{\mathbf{S}} \sigma_1$), and that it is a constraint on this mapmaking practice that one tries to produce a map with a unique road line labeled "E Ontario" drawn right-left only if the represented object has a unique street named "E Ontario" running east-west (i.e., $f_S\hat{}(\sigma_1) \vdash_{\mathcal{L}_C} f_T\hat{}(\theta_1)$ and hence $\sigma_1 \Rightarrow_{\mathcal{R}} \theta_1$). Under these conditions, the particular print a of *Downtown Chicago Map* is said to *represent* the region b of Chicago in the particular period of time *as* having a unique road named "E Ontario."

The semantic theory outlined by Definitions 5 and 6 is "two-tier" in that it is formulated with reference to two relations $\rightsquigarrow_{\mathcal{R}}$ and $\Rightarrow_{\mathcal{R}}$. Historically, we can see it as a formal realization of some key ideas of situation semantics [10,11]: (1) It takes the primary carrier of meaning to be a particular object in the world (the token a in Definition 5 and 6) rather than a representation type, (2) it takes a representation as carrying meaning about some particular object in the world (the token b in Definition 5 and 6) rather than an entire (possible) world, and (3) it takes meaning as a special case of information-carrying regularities holding in the environment (Clause 2, Definition 5).

5 Notion of Representational Practice

Our explanations so far have clarified what kinds of objects constitute each of these sets tok($\mathbf{C}$), tok($\mathbf{S}$), and tok($\mathbf{T}$). The set tok($\mathbf{C}$) consists of representational acts that collect information about unique objects to produce unique representations, while tok($\mathbf{S}$) representations produced by the representational acts in tok($\mathbf{C}$) and tok($\mathbf{T}$) represented objects in the representational acts in tok($\mathbf{C}$). As this description makes clear, tok($\mathbf{S}$) and tok($\mathbf{T}$) are defined on the basis of tok($\mathbf{C}$), so they can be defined once tok($\mathbf{C}$) is defined.

The remaining problem is that it is by no means trivial to define tok($\mathbf{C}$). We have seen that Barwise and Seligman conceptualize a representation system $\mathcal{R}$ as a model of a representational *practice*, so that the members of tok($\mathbf{C}$) are individual acts that constitute a representational practice. So the question is what makes an individual act a member of a particular representational practice rather than another. What distinguishes a representational practice from an arbitrarily chosen set of individual acts?

This question is of utmost importance, because what exactly tok($\mathbf{C}$) is determines what constraints hold on the members of tok($\mathbf{C}$), that is, what semantic conventions the relevant representational practice conform to. Furthermore, as tok($\mathbf{S}$) and tok($\mathbf{S}$) are determined on the basis of tok($\mathbf{C}$), the question has ramifications on the set of constraints that the representations in the practice conform to, as well the set of constraints that the objects represented in the practice conform to. All these profoundly affect the effectiveness of the representational practice in question.

Unfortunately, Barwise and Seligman provide no positive clue about this issue. However, it is clear that the satisfaction of all constraints listed in $\vdash_{\mathcal{L}_C}$ *cannot* be the defining character of the set tok($\mathbf{C}$). The definition of a local logic requires all constraints to be satisfied by all members of $N_{\mathcal{L}_C}$, but it does not require them to be satisfied by all members of tok($\mathbf{C}$).

Our proposal is to adopt Millikan's idea of reproduction [7,8] and define tok($\mathbf{C}$) as the class of objects "having been produced from one another or from the same models" [8, p. 20]. A typical example of such a reproductive class is the handshakes occurring in a single culture. Except for a small number of early instances in the history of this culture, individual acts of handshake are not products of somebody's creation, but reproductions of some previous handshakes.

Generally, the members of a reproductive class share certain characters, precisely because they are copies of one another. Thus, handshakes share various physical characters in their movements, and that is because they are reproductions of some previous handshakes or some model handshakes. Following Millikan [7], let us call such a character preserved in a repeated reproduction process a *reproductively established character.* Thanks to humans' ability to copy or repeat previous acts of handshakes, the reproduction process is relatively stable, producing faithful copies over a period of time. However, it does go awry sometimes, resulting in an anomalous handshake that lacks the relevant reproductively established character.

To apply these ideas to characterize the set tok($\mathbf{C}$), let $\mathbb{C}$ be the character of "satisfying a set of syntactic stipulations, semantic constraints, and target restrictions." Then we can see tok($\mathbf{C}$) as a reproductive class having $\mathbb{C}$ as its reproductively established character. On this conception, each representational act in tok($\mathbf{C}$) is a reproduction of previous representational acts in tok($\mathbf{C}$), and precisely because it is a reproduction, it tends to share the character $\mathbb{C}$ with other members of tok($\mathbf{C}$). Again, the process of reproduction is generally stable, but it can go awry, and when it does, there is no guarantee that the resulting act has the character $\mathbb{C}$, that is, it can violate some of the syntactic stipulations, semantic constraints, and target restrictions. We can then interpret $N_{\mathcal{L}_C}$ as those acts for which the reproduction process goes normally and tok($\mathbf{C}$)$-N_{\mathcal{L}_C}$ as those for which the reproduction process goes awry.

Examples

Example 1: System of Downtown Chicago Maps. In the case of the mapmaking practice for downtown Chicago, some people collect information about a region in Chicago in a particular period of time, some designer for the map publisher uses that information to design a map, and the technician prints the first print of the map. This sequence of acts, which we have characterized as $c_1 = e_1 \circ e_2 \circ p_1$ before, is the first token in tok($\mathbf{C}$). This act satisfies a set of syntactic stipulations, semantic constraints, and target restrictions, and we call this character $\mathbb{C}$. When the technician makes another print, this act p_2 is sequenced with the previous acts e_1 and e_2 to make another representational act, $c_2 = e_1 \circ e_2 \circ p_2$, which is a reproduction of the act $c_1 = e_1 \circ e_2 \circ p_1$, and as such shares the character $\mathbb{C}$ with it. Another printing act then gives rise to another representational act $c_3 = e_1 \circ e_2 \circ p_3$, and so on until the final reproductive act, say, $c_{2875} = e_1 \circ e_2 \circ p_{2875}$. This way, multiple representational acts are reproduced, preserving the character $\mathbb{C}$. The class tok($\mathbf{C}$) of these acts is a reproductive class, and this makes up a mapmaking practice for downtown Chicago.

What happens if the map publisher decides to update the maps they publish, to reflect the recent change in the mapped region? Then a new sequence of an information-collecting act, a design act, and a map-printing act takes place, making up a representational act, say c_{2876}. If the reproductive process is of the kind that normally preserves $\mathbb{C}$, c_{2876} belongs to the same reproductive class as $c_1, c_2, \ldots, c_{2875}$ do. Then all subsequent acts reproducing c_{2876} are all members of this class tok($\mathbf{C}$), extending the map publisher's mapmaking practice.

Example 2: System of Scheduling Tables. Let us consider a case where representations are used in more private settings. If the manager of a shop regularly hears from the part-time workers to decide on who will work on what day of the next week, and she puts up a table of the next week's work schedule on the door of her office, that will construct a reproductive class of representational acts. The reproductively established character $\mathbb{C}$ is the satisfaction of syntactic stipulations (regulating, for example, what types of symbols can appear in a table cell), semantic constraints (regulating, for example, what types of symbols indicate "on" and what types "off"), and target restriction limiting a representational act only to the weekly work schedules of this particular shop. The manager's act of hearing and table-drawing conducted each week makes an individual representational act, where the represented object is the shop's work schedule during the week about which the manager hears from the workers, and the representing object is the particular table drawn by the manager. The act is reproduced every week, accumulating as the tokens in tok($\mathbf{C}$).

Unlike the case of *Downtown Chicago Map*, the arrangement of symbols in the produced table typically changes act by act, but these acts are still members of the reproductive class they are continuations of the same reproductive process that normally preserves $\mathbb{C}$. For the same reason, a representational act done by the sub-manager can be considered a member of the same reproductive class if he or she has copied the manager's table-drawing acts with respect to $\mathbb{C}$.

What if the sub-manager keeps using the same table format but makes a small change in syntactic and semantic constraints, such as using a check mark rather than a circle to indicate "on"? On the one hand, this act can be considered a member of a new reproductive class, which itself can grow if it is copied by further acts. The new class has a slightly different reproductively established character $\mathbb{C}'$ than the class tok($\mathbf{C}$) of the manager's original representational acts. On the other hand, this act is a partial reproduction of the manager's acts, which preserves most of the constraints on them. Thus, this act, as well as any further acts that copies it, can be considered members of a larger reproductive class that extends tok($\mathbf{C}$). This larger reproductive class can be modeled by a B&S model just as well, and since it still preserves a significant set of constraints, it probably merits a serious logical investigation. This reproductive class makes a higher *taxon* of representational acts, whose lower taxa tok($\mathbf{C}$) and the newly started reproductive class.

6 Conclusion

Thanks to all this philosophical discussion, we now have a better conceptual foundation on which we apply the B&S model. It was not clear from Barwise and Seligman's sketch what exactly their model is a model of—what exactly they conceive as a representation system. Our study reveals that it is, roughly, a reproductive class of individual representational acts that inherits a set of representational rules. Our initial test shows that this concept does a pretty good job in carving out a piece of reality that a B&S model is to capture.

Now, the B&S model features three local logics $\mathcal{L}_C$, $\mathcal{L}_S$, and $\mathcal{L}_T$ as its main ingredients. It lets us characterize various interesting properties of representation systems as conditions on $\mathcal{L}_C$, $\mathcal{L}_S$, and $\mathcal{L}_T$, and investigate the logical consequences of these properties. For example, Barwise and Seligman have shown that the properties such as free ride, over-specificity, and auto-consistency of diagrammatic representations can be captured and studied in this way. We believe that the triple of $\mathcal{L}_C$, $\mathcal{L}_S$, and $\mathcal{L}_T$ embeds many more interesting properties, but speculation aside, that work of defining such properties and investigating their consequences is mathematical in nature, since it is exclusively concerned with the model itself and the model in this case is a mathematical structure.

This entails that whatever result it obtains is fundamentally hypothetical—it states that a representation system has such and such properties assuming it has such and such $\mathcal{L}_C$, $\mathcal{L}_S$, and $\mathcal{L}_T$ as its core logic, source logic, and target logic. We do not have to be concerned with which system in the world has such a combination of logics, and for that matter, if there is such a system at all. In fact, it was not possible to be concerned with these matters, since we did not know exactly what a representation system is that a B&S model is a model of.

The philosophical explication in the present paper changes this situation. Now that we know what piece of reality a B&S model is to capture, the specifications of $\mathcal{L}_C$, $\mathcal{L}_S$, and $\mathcal{L}_T$ in a B&S model can be put in empirical test. We have an independent grasp of what the model is about, so we can put the model and the modeled object side by side and evaluate their fit. And when the fit is good, we can say, categorically, that *this* representation system has such and such properties, with a clear understanding of what "this system" refers to. As examples in Section 5 suggest, reproductive classes of representational acts inheriting a set of representational rules are abundant in both private and public media. The B&S model now seems ready for use in the exploration of their logical properties.

References

1. Barwise, J., Seligman, J.: Information Flow: the Logic of Distributed Systems. Cambridge University Press, Cambridge (1997)
2. Shimojima, A.: On the Efficacy of Representation. PhD thesis, Department of Philosophy, Indiana University (1996)
3. Shimojima, A.: Semantic properties of diagrams and their cognitive potentials (book in preparation)
4. Shin, S.J.: The logical status of diagrams. Cambridge U. P., Cambridge (1994)
5. Howse, J., Molina, F., Taylor, J., Kent, S., Gil, J.Y.: Spider diagrams: A diagrammatic reasoning system. Journal of Visual Languages and Computing 12, 299–324 (2001)
6. Miller, N.: Euclid and his twentieth century rivals: Diagrams in the logic of Euclidean geometry. CSLI Publications, Stanford (2007)
7. Millikan, R.G.: Language, Thought, and Other Biological Categories: New Foundation for Realism. The MIT Press, Cambridge (1984)
8. Millikan, R.G.: On clear and confused ideas: An essay about substance concepts. Cambridge U. P., Cambridge (2000)
9. Lewis, D.: Languages and language. In: Philosophical Papers, pp. 93–115. Oxford University Press, Oxford (1975/1985) (Originally published in 1975)
10. Barwise, J., Perry, J.: Situations and Attitudes. CSLI Publications, Stanford (1983/1999) (Originally published from MIT Press in 1983)
11. Barwise, J.: Logic and information. In: The Situation in Logic, pp. 37–57. CSLI Publications, Stanford (1989)

Logical and Geometrical Complementarities between Aristotelian Diagrams

Hans Smessaert[1] and Lorenz Demey[2]

[1] Department of Linguistics, KU Leuven
Hans.Smessaert@arts.kuleuven.be

[2] Center for Logic and Analytic Philosophy, KU Leuven
Lorenz.Demey@hiw.kuleuven.be

Abstract. This paper concerns the Aristotelian relations of contradiction, contrariety, subcontrariety and subalternation between 14 contingent formulae, which can get a 2D or 3D visual representation by means of Aristotelian diagrams. The overall 3D diagram representing these Aristotelian relations is the rhombic dodecahedron (RDH), a polyhedron consisting of 14 vertices and 12 rhombic faces (Section 2). The ultimate aim is to study the various complementarities between Aristotelian diagrams inside the RDH. The crucial notions are therefore those of subdiagram and of nesting or embedding smaller diagrams into bigger ones. Three types of Aristotelian squares are characterised in terms of which types of contradictory diagonals they contain (Section 3). Secondly, any Aristotelian hexagon contains 3 squares (Section 4), and any Aristotelian octagon contains 4 hexagons (Section 5), so that different types of bigger diagrams can be distinguished in terms of which types of subdiagrams they contain. In a final part, the logical complementarities between 6 and 8 formulae are related to the geometrical complementarities between the 3D embeddings of hexagons and octagons inside the RDH (Section 6).

Keywords: Aristotelian relations, square of oppositions, hexagon of oppositions, logical geometry, 3D visualisation, subdiagrams, complementarity, embedding.

1 Introduction

Aims of the Paper. In addition to using diagrams for the visual representation of individual formulae or propositions, logicians also use diagrams to visualize certain relations between formulae from some given logical system. For example, the relations of contradiction, contrariety, subcontrariety and subalternation which hold between a set of logical formulae, are standardly visualised by means of Aristotelian diagrams, such as the well-known square of oppositions. The latter has a rich tradition, originating in Aristotle's work on syllogistics, but it is also widely used by contemporary logicians to visualize interesting fragments of systems such as modal logic, (dynamic) epistemic logic and deontic logic. Furthermore, other Aristotelian diagrams beyond the traditional square have been studied in detail, the most widely known probably being the hexagon described

T. Dwyer et al. (Eds.): Diagrams 2014, LNAI 8578, pp. 246–260, 2014.

by Jacoby, Sesmat and Blanché [1,2,3]. In recent years, several three-dimensional Aristotelian diagrams have been proposed, such as the octahedron, cube or tetrahexahedron. One such 3D representation, namely the rhombic dodecahedron [4,5,6], henceforth referred to as RDH, visualises the Aristotelian relations between 14 contingent formulae and serves as the general frame of reference for the present paper. Our central aim is twofold, namely (i) to develop strategies for systematically charting the internal structure of the RDH and (ii) to study various complementarities between Aristotelian diagrams inside the RDH. In doing so, we provide a more unified account of a whole range of diagrams which have so far mostly been treated independently of one another in the literature.

The Embedding of Subdiagrams. The description of the internal structure of the RDH, and more in particular of the various types of complementarities, crucially relies on the idea that smaller diagrams occur inside bigger diagrams. These notions of subdiagram or diagram embedding/nesting have been studied for other types of diagrams as well, more in particular Euler diagrams [7], Venn diagrams [8], spider diagrams [9] or algebra diagrams [10]. The analysis proposed in the present paper is very much in line with the visual grammar or visual syntax approach developed by Engelhardt [11, p. 104] in that "various syntactic principles can be identified in graphics of different types, and the nature of visual representation allows for visual nesting and recursion [...] any object may contain a set of (sub-)objects within the space that it occupies. When this principle is repeated recursively, the spatial arrangement of (sub-)objects is, at each level, determined by the specific nature of the containing space at that level". Furthermore, the central role of the RDH as the overall Aristotelian diagram in this paper resembles that of the so-called 'top state' in work on the syntax and semantics of UML statecharts [12, pp. 327–328] which describes "the set of transitively nested substates of a composite state" and assumes that "in every statechart there is an inherent composite state called the top state which covers all the (pseudo) states and is the container of the states".

The Structure of the Paper. In Section 2 we first introduce the Aristotelian relations and the partitioning of logical formulae into 'pairs of contradictories' (PCDs). We then present the rhombic dodecahedron for the 3D visualisation of Aristotelian relations between 14 formulae. In the central part of the paper, i.e. Sections 3, 4 and 5, different types of (bigger) diagrams are distinguished on the basis of which types of subdiagrams they contain. In Section 3, three types of Aristotelian squares are characterised in terms of which types of PCDs they contain. In a next step, Section 4 defines different types of hexagons depending on which types of squares are embedded in them. Similarly, the distinction between the two sorts of octagons in Section 5 is based on differences in the nested hexagons. Section 6 then moves from the level of 2D visual representations to that of 3D visualisation: the different ways in which the 14 formulae can be partitioned into a hexagon (6 formulae) and an octagon (8 formulae) are related to the geometrical complementarities between the 3D embeddings of hexagons and octagons inside the RDH.

2 Aristotelian Relations in the Rhombic Dodecahedron

Aristotelian Relations. The traditional Aristotelian relations are defined as follows:

$$\begin{array}{llll} \varphi \text{ and } \psi \text{ are } \textit{contradictory} & \text{iff } \mathsf{S} \models \neg(\varphi \wedge \psi) & \text{and} & \mathsf{S} \models \neg(\neg\varphi \wedge \neg\psi), \\ \varphi \text{ and } \psi \text{ are } \textit{contrary} & \text{iff } \mathsf{S} \models \neg(\varphi \wedge \psi) & \text{and} & \mathsf{S} \not\models \neg(\neg\varphi \wedge \neg\psi), \\ \varphi \text{ and } \psi \text{ are } \textit{subcontrary} & \text{iff } \mathsf{S} \not\models \neg(\varphi \wedge \psi) & \text{and} & \mathsf{S} \models \neg(\neg\varphi \wedge \neg\psi), \\ \varphi \text{ and } \psi \text{ are in } \textit{subalternation} & \text{iff } \mathsf{S} \models \varphi \rightarrow \psi & \text{and} & \mathsf{S} \not\models \psi \rightarrow \varphi. \end{array}$$

Informally, two formulae are CONTRADICTORY when they *cannot be true together and cannot be false together*. They are CONTRARY when they *cannot be true together but may be false together* and SUBCONTRARY when they *cannot be false together but may be true together*. Finally, notice that SUBALTERNATION is not defined in terms of the formulae being *true together* or being *false together*, but in terms of *truth propagation*: there is a subalternation from φ to ψ when φ entails ψ but not vice versa.

Pairs of Contradictories. The 16 formulae (up to logical equivalence) from Classical Propositional Logic which can be built by means of unary or binary connectives and two propositional variables p and q, can be partitioned into the following 8 PAIRS OF CONTRADICTORIES (PCDs), namely 4 PCDs of type C and 4 PCDs of type O:[1]

$$\begin{array}{llllll} \text{PCDs of type C:} & \text{a. } (p \wedge q) & \text{b. } \neg(p \rightarrow q) & \text{c. } \neg(p \leftarrow q) & \text{d. } \neg(p \vee q) \\ & \text{a'. } \neg(p \wedge q) & \text{b'. } (p \rightarrow q) & \text{c'. } (p \leftarrow q) & \text{d'. } (p \vee q) \end{array}$$

$$\begin{array}{lllll} \text{PCDs of type O:} & \text{e. } p & \text{f. } q & \text{g. } (p \leftrightarrow q) & \text{h. } p \wedge \neg p \\ & \text{e'. } \neg p & \text{f'. } \neg q & \text{g'. } \neg(p \leftrightarrow q) & \text{h'. } p \vee \neg p \end{array}$$

Furthermore, any two formulae taken from the top row (a-d) in the PCDs of type C are contrary to one another, whereas any two formulae taken from the bottom row (a'-d') are subcontrary to one another. Notice that we will henceforth disregard the PCD containing the two non-contingent formulae in (h-h'), and focus on the 7 PCDs containing the 14 contingent formulae in (a-a') to (g-g').

The Rhombic Dodecahedron. Various isomorphic 3D visualisations have been proposed for the logical relations between the 14 contingent formulae above. Both the tetra-hexahedral representation of Sauriol [13] and the tetra-icosahedral representation of Moretti [14] take as their starting point the cube to which six pyramids are added, one on each face of the cube[2]. Sauriol makes use of 'obtuse' pyramids (whose angle between the base and each of the triangular faces is less than 45°), thus obtaining the convex polyhedron in Figure 1a. Moretti, on the other hand, makes use of 'acute' pyramids (whose angle between the base and each of the triangular faces is greater than 45°), thus obtaining the concave

[1] The rationale behind these abbreviations is explained in the next subsection.

[2] A radically different 3D representation, although fundamentally still isomorphic to the ones in Figure 1, is the 'double' tetrahedron of Dubois and Prade [15].

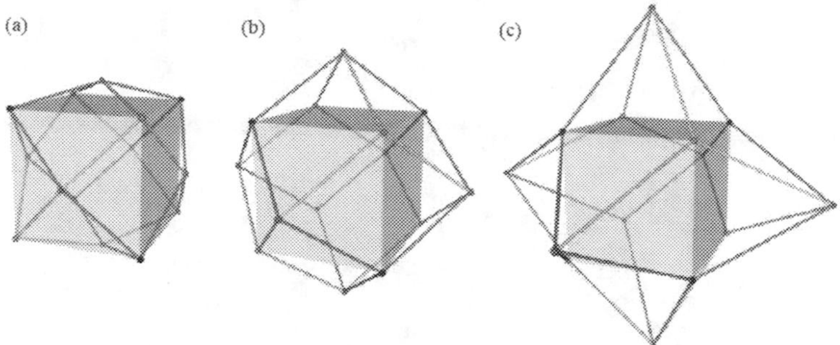

Fig. 1. (a) Sauriol's tetrahexahedron (b) the RDH (c) Moretti's tetraicosahedron

polyhedron in Figure 1c. The result in both cases is a polyhedron consisting of 24 triangular faces and 14 vertices (corresponding to the 14 formulae), namely the 8 vertices of the cube and the six pyramids' apices. The polyhedron proposed in Smessaert [5,6] and adopted in Demey [4], by contrast, is the RHOMBIC DODECAHEDRON (RDH) in Figure 1b. It can be considered as lying exactly in between the Sauriol structure in Figure 1a and the Moretti structure in Figure 1c, in the sense that the six pyramids added onto the faces of the cube are 'right' (i.e. having an angle between their base and each of their triangular faces of exactly 45°). As a consequence, each pair of triangular faces of adjacent pyramids falls in the same plane and constitutes a single rhombic face; an RDH thus consists of 14 vertices, but has 12 rhombic faces instead of 24 triangular faces.

Since the RDH is the polyhedral dual of the cuboctahedron (an Archimedean solid combining the properties of a cube and an octahedron [16]), it inherits this double connection, both with the cube and the octahedron. In particular the latter connection is absent from the Sauriol and Moretti structures, and can be seen as the major advantage of the RDH.

The crucial property of the visualisation in Figure 2 is that the Aristotelian relation of contradiction corresponds to central symmetry. In other words, each PCD corresponds to a diagonal through the center of the RDH. Two sets of diagonals can be distinguished: first of all, each formula from the contrariety set in (a-d), with a black label in Figure 2, constitutes a PCD diagonal with a formula from the subcontrary set in (a'-d'), with a grey label. In other words, the 8 vertices with the black and grey labels constitute the cube inside the RDH, and its diagonals represent the 4 PCDs of type C. Secondly, each of the three formulae in (e-g) constitutes a PCD diagonal with its negative counterpart in (e'-g'). Hence, the 6 vertices with a white label constitute the octahedron inside the RDH, the diagonals of which represent the 3 (contingent) PCDs of type O.

Notice that a similar rhombic dodecahedron is used in the visualisation of Zellweger [17]. In comparison to the Aristotelian RDH in Figure 2, however, Zellweger puts the two contingent formulae in (g-g') in the center of his RDH

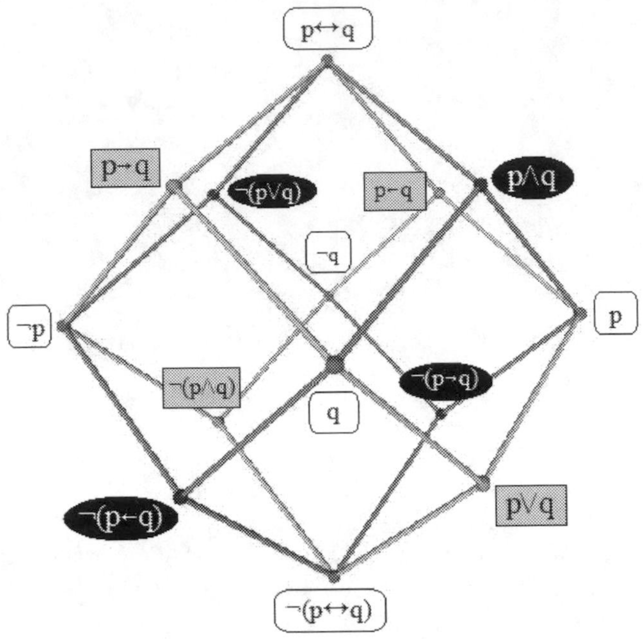

Fig. 2. The 3D visualisation of propositional connectives in the RDH

and the two non-contingent formulae in (h-h') at its top and bottom vertices[3]. Furthermore, in the Zellweger RDH, all 4 contrary formulae with the black label fall in one plane, all 4 subcontrary formulae with the grey label fall in one plane, and so do all 6 of the white label formulae. We argue elsewhere [18] that, although the Zellweger representation reflects the layered structure of the underlying Hasse diagram more directly, it is less suited for representing the Aristotelian relations.

3 Aristotelian Squares of Opposition

Three types of Aristotelian squares are distinguished in Figure 3. First of all, there are the *classical* squares in Figure 3a-b which have the diagonals for contradiction, the arrows going down for subalternation, the dashed line at the top connecting the contraries and the dotted line at the bottom connecting the subcontraries. There is a crucial difference between these squares in terms of which types of PCDs are involved. The *balanced classical* square in Figure 3a consists of two PCDs that are of the same type, i.e. of type C. The *unbalanced classical* square in Figure 3b, by contrast, consists of one PCD of type C (the diagonal from top left to bottom right) and one PCD of type O (the diagonal from bottom

[3] Although the non-contingent formulae in (h-h') are not explicitly represented in our RDH, they can be taken to coincide in its center.

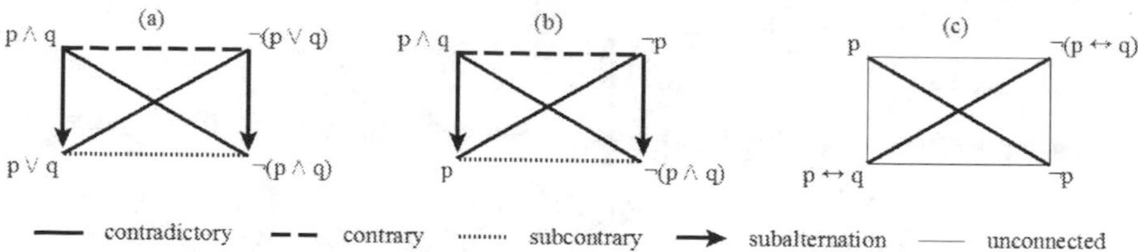

Fig. 3. Aristotelian Squares of Oppositions: (a) balanced classical (b) unbalanced classical (c) degenerate

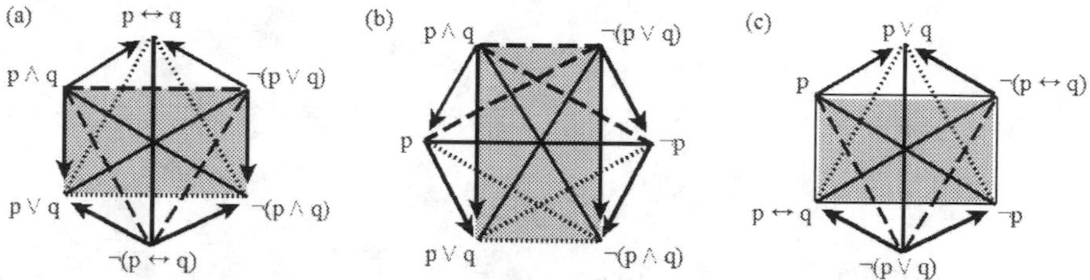

Fig. 4. Aristotelian Hexagons of Oppositions: (a) Jacoby-Sesmat-Blanché = JSB (b) Sherwood-Czezowski = SC (c) Unconnected-4 = U4

left to top right). The third square, in Figure 3c, is *degenerate* in terms of the Aristotelian relations holding between the 4 formulae. Only the diagonals for contradiction remain, whereas the 4 outer edges of the square do not represent any Aristotelian relation whatsoever: the 4 formulae are pairwise 'unconnected', i.e. logically independent [19]. Furthermore, the resulting configuration turns out to be balanced again: it consists of two PCDs of the O type[4].

4 Aristotelian Hexagons of Opposition

In this section, we first distinguish three types of Aristotelian hexagons in terms of which types of PCDs they consist of, namely the Jacoby-Sesmat-Blanché hexagon (JSB hexagon for short), the Sherwood-Czezowski hexagon (SC hexagon) and the Unconnected-4 hexagon (U4 hexagon). Secondly, each of these 3 types of hexagons is further characterized in terms of which types of squares it contains.

Three Types of Aristotelian Hexagons. The first two hexagons in Figure 4a-b illustrate the two by now standard ways in which a classical Aristotelian square can be extended or generalized to a hexagon [6]. The starting point in both cases is the balanced classical square of Figure 3a in the grey shaded area, consisting of two PCD diagonals of type C. Furthermore, in both cases the third PCD which is

[4] Another balanced degenerate square can be constructed using the four single-variable formulas p, $\neg p$, q and $\neg q$ (this square is embedded inside the Buridan octagon shown in Figure 8b). Next to these *balanced degenerate* squares, there also exist *unbalanced degenerate* squares. However, the latter play no role in the present paper.

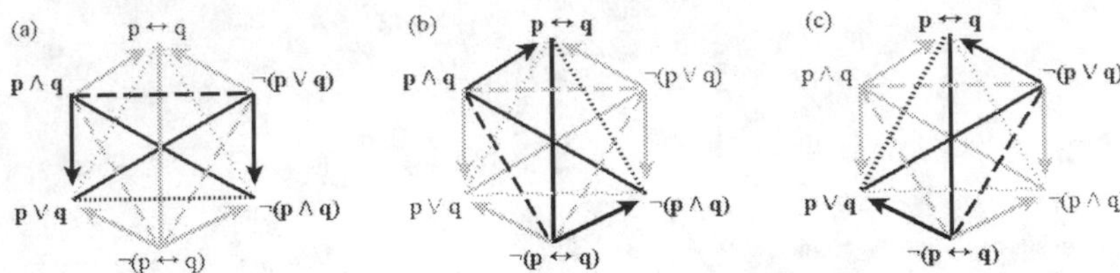

Fig. 5. The Aristotelian squares in the Jacoby-Sesmat-Blanché hexagon JSB

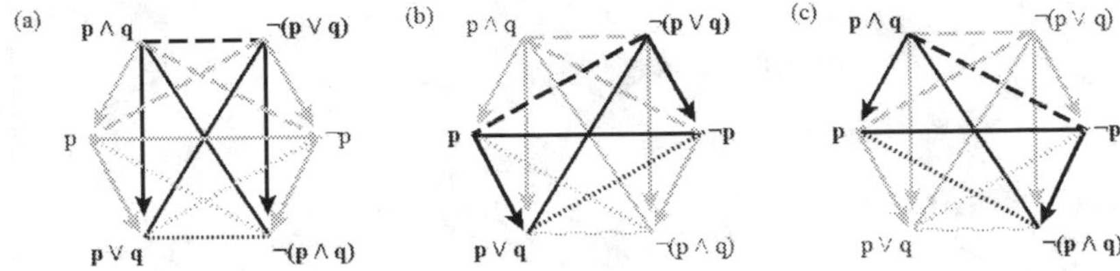

Fig. 6. The Aristotelian squares in the Sherwood-Czezowski hexagon SC

added is of type O. In the JSB hexagon of Jacoby [1], Sesmat [2] and Blanché [3] in Figure 4a the additional PCD consists of the disjunction of the square's upper two vertices and the conjunction of its lower two vertices, and is therefore added to the square 'vertically'. By contrast, in the SC hexagon of Sherwood [20,21] and Czezowski [22] in Figure 4b, the additional PCD consists of two formulae that are intermediate with respect to subalternation between the square's left two and right two vertices, and is therefore added to the square 'horizontally'. These different ways of inserting the third diagonal into the square result in two fundamentally distinct hexagonal constellations of Aristotelian relations. In the JSB hexagon in Figure 4a, the relations of contrariety and subcontrariety yield two triangles interlocking into a star-like shape inside the hexagon, whereas the arrows of subalternation constitute the outer edges of the hexagon and point from each vertex on the triangle of contraries to the two adjecent ones on the triangle of subcontraries. In the SC hexagon in Figure 4b, by contrast, the arrows of subalternation, which are all pointing downwards, constitute two triangles (of transitivity), whereas the contraries and subcontraries yield two ⋉ shapes instead of two triangles. With the third hexagon in Figure 4c, the starting point is also a balanced square in grey, but this time it is the degenerate square of Figure 3c, consisting of two PCD diagonals of type O. Adding a PCD of type C as the third diagonal in vertical position yields yet another Aristotelian configuration. Because of the presence of the 4 unconnectedness relations in the central square, this hexagon will be referred to as the Unconnected-4 or U4 hexagon[5]. In contrast

[5] At least two more types of hexagons can be defined: (i) the so-called 'weak' JSB hexagon in Moretti [14] and Pellissier [23] consists of 3 PCDs of type C but is isomorphic to the JSB hexagon in Figure 4a, (ii) the so-called Unconnected-12 or U12 hexagon consists of 3 PCDs of type O and, apart from the three diagonals of contradiction, exclusively contains 12 relations of unconnectedness.

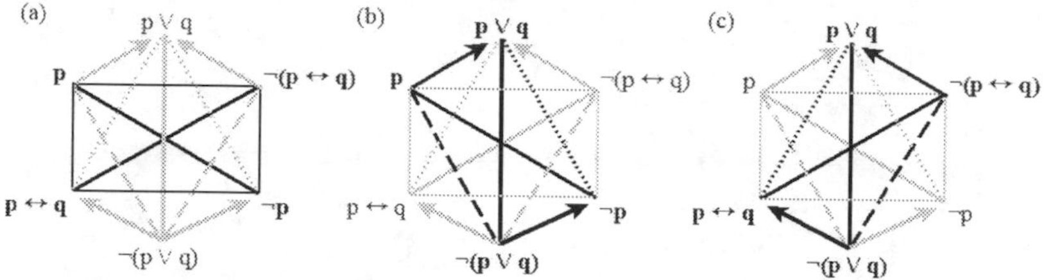

Fig. 7. The Aristotelian squares in the Unconnected-4 hexagon U4

to the JSB and SC hexagons in Figures 4a-b, U4 only contains 4 subalternation arrows instead of 6, and has V-shaped (instead of triangular) constellations for its contraries and subcontraries.

Squares Inside the Aristotelian Hexagons. We have just distinguished three types of Aristotelian hexagons in terms of which types of PCD diagonals they consist of. A second, closely related strategy for establishing a typology of hexagons is that of considering (i) which types of subdiagrams are embedded inside the bigger hexagonal diagram and (ii) in which way they are embedded. More in particular: any hexagon can be shown to contain 3 squares: since a hexagon consists of three diagonals, each of them can be left out in turn to yield a distinct square (consisting of 2 out of the 3 original diagonals). Looking at the overall constellations in Figures 5 to 7, we observe that all three types of hexagons contain one balanced square in the (a) diagram and two unbalanced classical squares in the (b-c) diagrams. A first difference, of course, is that with the JSB hexagon in Figure 5a and the SC-hexagon in Figure 6a the balanced square is the classical one, whereas with the U4 hexagon in Figure 7a it is the degenerate one. Secondly, although the JSB hexagon and the SC hexagon resemble one another as to which types of squares are embedded, they crucially differ as to the way in which the two unbalanced classical squares are embedded. In Figure 5b-c the embedding involves a rotation of 120° clockwise or counterclockwise (because of the triangular shape of the (sub)contraries), whereas in Figure 6b-c the embedding involves a rotation of only 30° (because of the ⋉ shape of the (sub)contraries). In Smessaert [6] this difference is argued to be due to the fact that the JSB hexagon is closed under the Boolean operations, whereas the SC hexagon is not. In the former case, the hexagon contains the meet and join of any of its pairs of formulae: each vertex on the triangle of contraries is the conjunction or meet of its two neighbours on the triangle of subcontraries, and vice versa, each vertex on the triangle of subcontraries is the disjunction or join of its two neighbours on the triangle of contraries. In the latter case, however, quite a number of pairs of vertices have a meet or join which does not belong to the hexagon (although their negation is always there due to the PCDs). Notice, by the way, that the U4 hexagon resembles the SC hexagon in not being closed under the Boolean operations either.

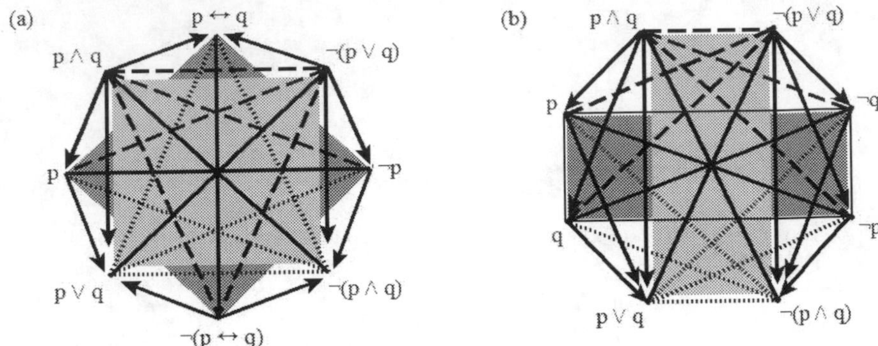

Fig. 8. Aristotelian octagons: (a) Béziau (b) Buridan

5 Aristotelian Octagons of Opposition

In this section two well-known types of Aristotelian octagons will be distinguished, namely the Béziau octagon [24] and the Buridan octagon [25,26]. They will be shown to differ from one another in terms of the types of hexagons that can be embedded into them as subdiagrams.

The Béziau Octagon versus the Buridan Octagon. If we adopt the original strategy for distinguishing diagrams, namely on the basis of the types of PCDs they consist of, the two octagons in Figure 8 turn out to be of the same type, since they both contain 2 PCDs of type C as well as 2 PCDs of type O. As a consequence, they can both be seen as combinations of one classical square (with the type C PCDs) in the lighter shade of grey and a degenerate square (with the type O PCDs) in the darker shade of grey. The crucial difference between the two octagons thus concerns the way in which these two squares are embedded into them. With the Béziau octagon in Figure 8a, the vertices of the two squares are strictly alternating on the outer edge, whereas with the Buridan octagon in Figure 8b, they are pairwise alternating. This results in two fundamentally distinct constellations of Aristotelian relations. If we focus on the 'triangular' components, the Béziau octagon on the left has two triangles of subalternation as well as an interlocking pair of triangles for contraries and subcontraries. Although the Buridan octagon on the right also contains 4 triangular shapes, they are all of the same type, namely two pairs of subalternation triangles.

Hexagons Inside the Aristotelian Octagons. So far, the two types of octagons were distinguished in terms of (i) different ways of interlocking the classical and the degenerate squares and (ii) different sets of triangular constellations of Aristotelian relations. The latter strategy for establishing a typology of octagons naturally leads to that of considering which types of hexagonal subdiagrams are embedded inside the bigger octagonal diagram. Any octagon can be shown to contain 4 hexagons: since an octagon consists of four diagonals, each of these can be left out in turn to yield a distinct hexagon (consisting of 3

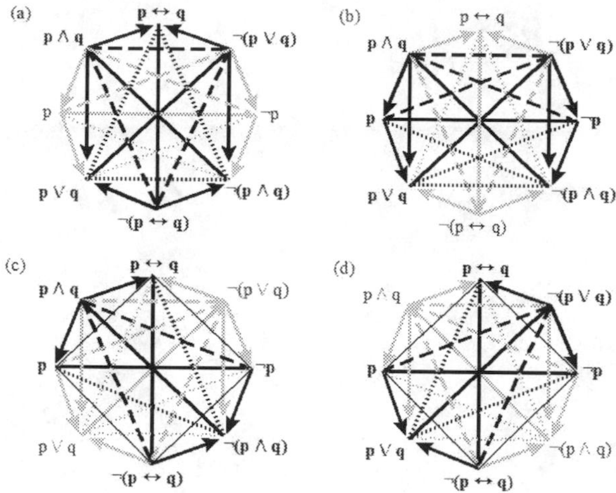

Fig. 9. The Aristotelian hexagons in the Béziau octagon: (a) JSB (b) SC (c-d) U4

out of the 4 original diagonals). Thus the Béziau octagon can first and foremost be considered as a combination of the JSB hexagon in Figure 9a and the SC hexagon in Figure 9b, with the outer edges completely defined by subalternation arrows. The Buridan octagon, by contrast, is fundamentally a combination of the two SC hexagons in Figure 10a-b, with all the subalternation arrows going downward and two pairs of interlacing ⋉ shapes for the contraries and the subcontraries. Notice that both with the Béziau octagon and the Buridan octagon the two remaining hexagons that can be embedded are of the U4 type. However, with the former in Figures 9c-d the U4 hexagons show up in their standard shape of Figure 4c, i.e. with the subalternation arrows along the edges and the (sub)contrary V-shapes on the inside, whereas with the latter in Figures 10c-d an alternative shape emerges for the U4 hexagons with more acute angles for the subalternation arrows and the (sub)contrary V-shapes.

6 Complementarities in the Rhombic Dodecahedron

Although in the previous three sections the Aristotelian diagrams have gradually become more complex, their visual representation remained two-dimensional, viz. from square to hexagon to octagon. In this section, however, we return to the complete set of 14 contingent formulae from Classical Propositional Logic (with the 7 PCDs introduced in Section 2) and their 3D visualisation inside the RDH. More in particular, we distinguish two different ways in which the 14 formulae can be partitioned into a hexagon (6 formulae) and an octagon (8 formulae) and relate them to the geometrical complementarities between the 3D embeddings of hexagons and octagons inside the RDH.

Complementarity between the JSB Hexagon and the Buridan Octagon. The two diagrams in Figure 11 reveal a first type of logical complementarity: if we

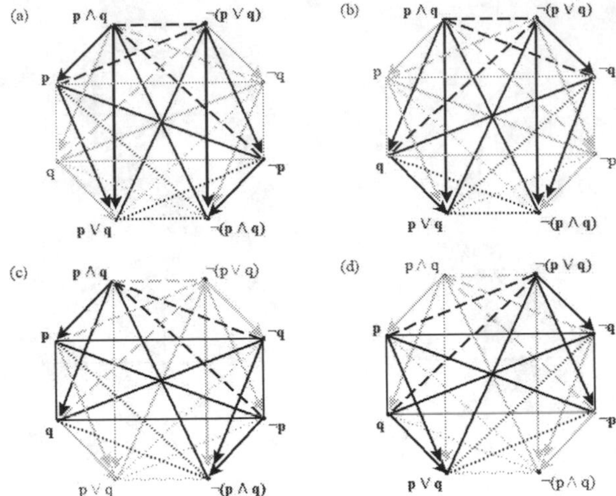

Fig. 10. The Aristotelian hexagons in the Buridan octagon: (a-b) SC (c-d) U4

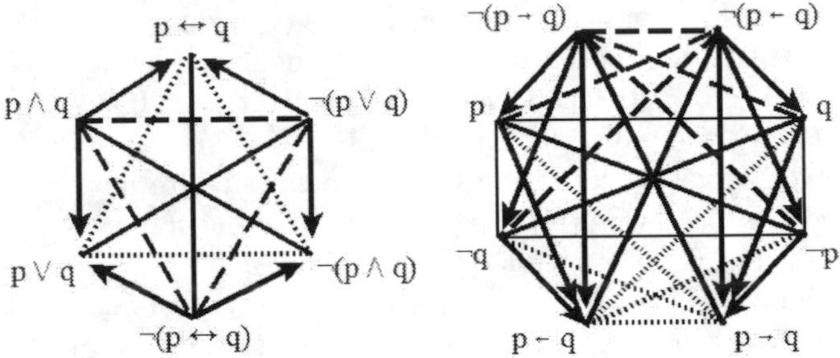

Fig. 11. Logical complementarity between a Jacoby-Sesmat-Blanché hexagon (left) and a Buridan octagon (right)

take 6 formulae whose Aristotelian relations constitute a JSB hexagon, the 8 remaining formulae yield an Aristotelian octagon of the Buridan type[6]. A number of authors [4,5,6,13,14] have demonstrated that there are exactly six different JSB hexagons embedded inside the RDH. On the left in Figure 12 we see that the embedding of a JSB hexagon constitutes a 2D plane which slices the 3D RDH solid in two equal parts. It can easily be shown that there are indeed exactly six planes that contain 6 out of the 14 vertices (i.e. 3 PCDs) of the RDH. One important result of the present paper is that, for each JSB hexagon in the RDH, the remaining 8 vertices yield a Buridan octagon, whose 3D embedding in the RDH is the solid visualised on the right in Figure 12. This object, which has no standard name in the literature on polyhedra, will be referred to as a RHOMBICUBE, the idea being

[6] Since a Buridan octagon is fundamentally a combination of two SC hexagons (see Figure 10a-b), this first logical complementarity can also be seen as holding between a JSB hexagon on the one hand and a pair of SC hexagons on the other.

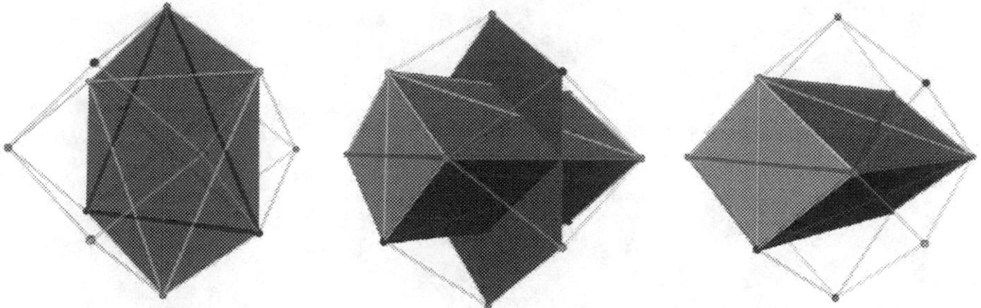

Fig. 12. 3D geometrical complementarity (middle) between a Jacoby-Sesmat-Blanché hexagon (left) and a rhombicube (right)

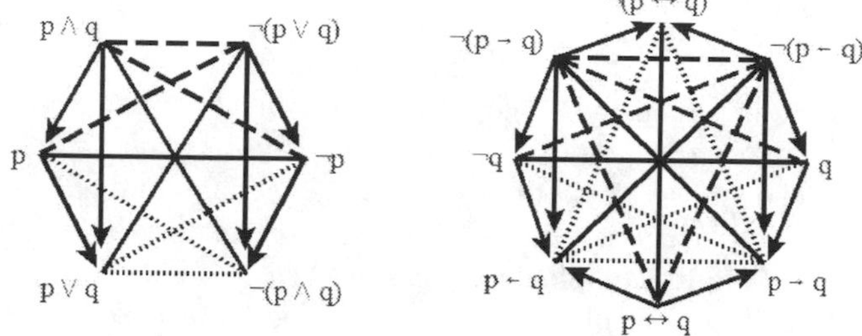

Fig. 13. Logical complementarity between a Sherwood-Czezowski hexagon (left) and a Béziau octagon (right)

that a cube is put on one of its edges and is squeezed at the top (edge) to the effect that its front and back faces turn from a square into a rhombic shape[7]. The logical complementarity between the JSB hexagon and the Buridan octagon in Figure 11 thus gets a very elegant counterpart in the 3D geometrical complementarity of the hexagon and the rhombicube in the middle of Figure 12.

Complementarity between the SC Hexagon and the Béziau Octagon. The operation of creating a partition of the 14 formulae can also be performed on the basis of the SC hexagon instead of the JSB hexagon. The two diagrams in Figure 13 thus reveal a second type of logical complementarity: if we take 6 formulae whose Aristotelian relations constitute an SC hexagon, the 8 remaining formulae yield an Aristotelian octagon of the Béziau type. As far as the embedding of an SC hexagon in a 3D polyhedron is concerned, the only proposal, to our knowledge, is that of Sauriol [13, p. 388], who embeds it into his tetrahexahedron. The left diagram in Figure 14 shows the 3D embedding of an SC hexagon in the RDH. This solid is a skew octahedron, i.e. it has 6 vertices and 8 triangular faces. Given that an SC hexagon can be seen as a

[7] The term does show up occasionally, either as an abbreviation for 'rhombicuboctahedron', which is a different, Archimedian solid, or else as a (less felicitous) alternative for the RDH itself.

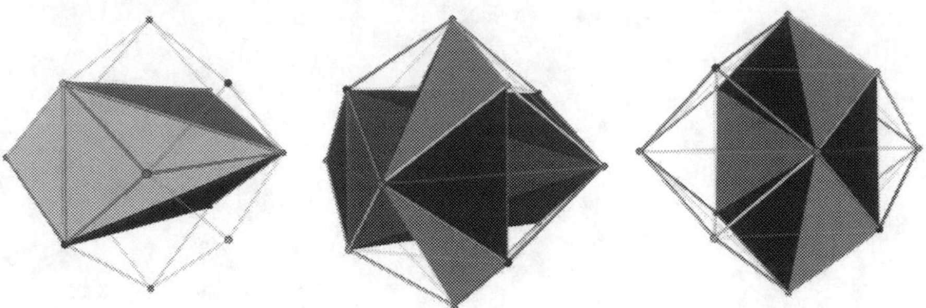

Fig. 14. 3D geometrical complementarity (middle) between a Sherwood-Czezowski hexagon (left) and a Béziau octagon (right)

Buridan octagon with one PCD diagonal left out (see Figure 10a-b), the skew octahedron in Figure 14 can be seen as a rhombicube with 2 vertices (or 1 PCD) 'sliced off'. Since there are exactly six rhombicubes embedded in the RDH (namely the complements of the six JSB hexagons), and each rhombicube contains two SC octahedra, it follows that there are twelve SC octahedrons inside the RDH. As for its complement, namely the 3D embedding of a Béziau octagon in the RDH, the result on the right in Figure 14 is a squeezed hexagonal bipyramid, which is a solid obtained by sticking together (base to base) two pyramids with a hexagonal base. As a consequence, the logical complementarity between the SC hexagon and the Béziau octagon in Figure 13 has as its counterpart the 3D geometrical complementarity of the octahedron and the hexagonal bipyramid in the middle of Figure 14. Although, from a strictly logical point of view, the complementarities in Figures 11 and 13 are on a par, there is a considerable difference, as far as the visual appeal is concerned, between the 3D geometrical complementarities in Figures 12 and 14. For example, the former has one central symmetry and three reflection symmetries, whereas the latter only has the central symmetry and one reflection symmetry. The main reason for this geometrical difference is that SC hexagons naturally come in pairs (as rhombicubes); the first complementarity respects this pairing, but the second cuts across it.

7 Conclusions and Prospects

The main aim of this paper has been to provide a more unified account of a range of Aristotelian diagrams, which are in general treated independently of one another in the literature. The central part was devoted to a general strategy for systematically charting the internal structure of the rhombic dodecahedron, which represents the Aristotelian relations in a Boolean closed set of 14 contingent formulae. Three families of squares are distinguished depending on the types of PCDs they consist of, 3 families of hexagons in terms of the types of embedded squares, and 2 families of octagons on the basis of the types of nested hexagons. In a final part two types of logical complementarities have been observed, namely (i) between a JSB hexagon and a Buridan octagon, and (ii) between an SC hexagon and a Béziau octagon. The difference in visual appeal of the corresponding 3D geometrical complementarities supports the claim that the former partition is more natural than the latter. The next step in this research project will be to provide an exhaustive typology (by means of combinatorial analysis) of the Aristotelian subdiagrams of the RDH. Central questions will be (i) how many families of octagons, decagons and dodecagons can be distinguished, and (ii) how many members does each family have (e.g. there are 6 strong JSB hexagons, but 12 SC hexagons inside the RDH).

Acknowledgements. The second author is financially supported by a PhD fellowship of the Research Foundation–Flanders (FWO).

References

1. Jacoby, P.: A Triangle of Opposites for Types of Propositions in Aristotelian Logic. The New Scholasticism 24(1), 32–56 (1950)
2. Sesmat, A.: Logique II. Les Raisonnements. Hermann, Paris (1951)
3. Blanché, R.: Structures Intellectuelles. Essai sur l'organisation systématique des concepts. Librairie Philosophique J. Vrin, Paris (1969)
4. Demey, L.: Structures of Oppositions in Public Announcement Logic. In: Béziau, J.Y., Jacquette, D. (eds.) Around and Beyond the Square of Opposition, pp. 313–339. Springer, Basel (2012)
5. Smessaert, H.: On the 3D visualisation of logical relations. Logica Universalis 3(2), 303–332 (2009)
6. Smessaert, H.: Boolean differences between two hexagonal extensions of the logical Square of Oppositions. In: Cox, P., Plimmer, B., Rodgers, P. (eds.) Diagrams 2012. LNCS (LNAI), vol. 7352, pp. 193–199. Springer, Heidelberg (2012)
7. Fish, A., Flower, J.: Euler Diagram Decomposition. In: Stapleton, G., Howse, J., Lee, J. (eds.) Diagrams 2008. LNCS (LNAI), vol. 5223, pp. 28–44. Springer, Heidelberg (2008)
8. Flower, J., Stapleton, G., Rodgers, P.: On the drawability of 3D Venn and Euler diagrams. Journal of Visual Languages and Computing (2013)
9. Urbas, M., Jamnik, M., Stapleton, G., Flower, J.: Speedith: A Diagrammatic Reasoner for Spider Diagrams. In: Cox, P., Plimmer, B., Rodgers, P. (eds.) Diagrams 2012. LNCS (LNAI), vol. 7352, pp. 163–177. Springer, Heidelberg (2012)
10. Cheng, P.C.-H.: Algebra Diagrams: A HANDi Introduction. In: Cox, P., Plimmer, B., Rodgers, P. (eds.) Diagrams 2012. LNCS (LNAI), vol. 7352, pp. 178–192. Springer, Heidelberg (2012)
11. Engelhardt, Y.: Objects and Spaces: The Visual Language of Graphics. In: Barker-Plummer, D., Cox, R., Swoboda, N. (eds.) Diagrams 2006. LNCS (LNAI), vol. 4045, pp. 104–108. Springer, Heidelberg (2006)
12. Jin, Y., Esser, R., Janneck, J.W.: Describing the Syntax and Semantics of UML Statecharts in a Heterogeneous Modelling Environment. In: Hegarty, M., Meyer, B., Hari Narayanan, N. (eds.) Diagrams 2002. LNCS (LNAI), vol. 2317, pp. 320–334. Springer, Heidelberg (2002)

13. Sauriol, P.: Remarques sur la théorie de l'hexagone logique de Blanché. Dialogue 7, 374–390 (1968)
14. Moretti, A.: The Geometry of Logical Opposition. Ph.D. thesis, University of Neuchâtel (2009)
15. Dubois, D., Prade, H.: From Blanché's Hexagonal Organization of Concepts to Formal Concept Analysis and Possibility Theory. Logica Universalis 6, 149–169 (2012)
16. Coxeter, H.S.M.: Regular Polytopes. Dover Publications (1973)
17. Zellweger, S.: Untapped potential in Peirce's iconic notation for the sixteen binary connectives. In: Hauser, N., Roberts, D.D., Evra, J.V. (eds.) Studies in the Logic of Charles Peirce, pp. 334–386. Indiana University Press (1997)
18. Demey, L., Smessaert, H.: The relationship between Aristotelian and Hasse diagrams. In: Dwyer, T., Purchase, H.C., Delaney, A. (eds.) Diagrams 2014. LNCS (LNAI), vol. 8578, pp. 215–229. Springer, Heidelberg (2014)
19. Smessaert, H., Demey, L.: Logical Geometries and Information in the Square of Oppositions. Submitted research paper (2014)
20. Khomskii, Y.: William of Sherwood, singular propositions and the hexagon of opposition. In: Béziau, J.Y., Payette, G. (eds.) New Perspectives on the Square of Opposition, pp. 43–60. Peter Lang, Bern (2011)
21. Kretzmann, N.: William of Sherwood's Introduction to Logic. Minnesota Archive Editions, Minneapolis (1966)
22. Czezowski, T.: On certain peculiarities of singular propositions. Mind 64(255), 392–395 (1955)
23. Pellissier, R.: Setting n-opposition. Logica Universalis 2(2), 235–263 (2008)
24. Béziau, J.Y.: New light on the square of oppositions and its nameless corner. Logical Investigations 10, 218–232 (2003)
25. Hughes, G.: The modal logic of John Buridan. In: Atti del Congresso Internazionale di Storia Della Logica: La Teorie Delle Modalitá, pp. 93–111. CLUEB, Bologna (1989)
26. Read, S.: John Buridan's Theory of Consequence and his Octagons of Opposition. In: Béziau, J.Y., Jacquette, D. (eds.) Around and Beyond the Square of Opposition, pp. 93–110. Springer, Basel (2012)

Logical Investigation of Reasoning with Tables

Ryo Takemura[1], Atsushi Shimojima[2], and Yasuhiro Katagiri[3]

[1] Nihon Univeristy, Japan
takemura.ryo@nihon-u.ac.jp
[2] Doshisha University, Japan
ashimoji@mail.doshisha.ac.jp
[3] Future University Hakodate, Japan
katagiri@fun.ac.jp

Abstract. In graphical or diagrammatic representations, not only the basic component of a diagram, but also a collection of multiple components can form a unit with semantic significance. We call such a collection a "global object", and we consider how this can assist in reasoning using diagrammatic representation. In this paper, we investigate reasoning with correspondence tables as a case study. Correspondence tables are a basic, yet widely applied graphical/diagrammatical representation system. Although there may be various types of global objects in a table, here we concentrate on global objects consisting of rows or columns taken as a whole. We investigate reasoning with tables by exploiting not only local conditions, specifying the values in individual table entries, but also global conditions, which specify constraints on rows and columns in the table. This type of reasoning with tables would typically be employed in a task solving simple scheduling problems, such as assigning workers to work on different days of the week, given global conditions such as the number of people to be assigned to each day, as well as local conditions such as the days of the week on which certain people cannot work. We investigate logical properties of reasoning with tables, and conclude, from the perspective of free ride, that the application of global objects makes such reasoning more efficient.

Keywords: mathematical logic, correspondence table, free ride.

1 Introduction

By "global objects," we mean those patterns or structures in diagrams that allow the extraction of higher-level information about the represented domain. Typical examples are a "cloud" consisting of multiple dots in a scatter plot that allows an estimation of the correlation strength of two variables [12], an "ascending staircase" made of multiple columns in a vertical bar graph that allows the observation of an increasing trend [13], and a group of adjacent contours lines in a topographical map that allows the identification of a characteristic landform of the terrain [11,9]. The extraction of such higher-level information has been variously called "macro reading" [20], "pattern perception" [7], "direct translation"

T. Dwyer et al. (Eds.): Diagrams 2014, LNAI 8578, pp. 261–276, 2014.

[13], and "cognitive integration" [14], and contrasted to the extraction of more concrete information from local objects, such as individual dots (scatter plot), bars (bar charts), and passing points of a contour line (geographical maps).

Given that the utilization of global objects significantly contribute to comprehension of graphical representations [5,6,20,13,21,10,7,14], it is natural to suspect its *inferential* advantage. That is, applying inferential operations on global components of a diagram may lead to a simplification or other positive change of inferential processes. Despite this prospect, few logical investigations have been conducted on what exact inferential advantages might be obtained by the utilization of global objects.

In our view, the paucity of logicians' interests in global objects is attributed to the difficulty of formally characterizing global objects—patterns and structures—that have been investigated in the graphics comprehension research cited above. The present paper tries to break the impasse and takes a few initial steps toward the logical explication of inferential advantages of utilizing global objects. For this purpose, we scale down the problem to the case of simple tabular representations. Simple as they may appear, rows and columns, as opposed to individual cells, can play the role of global objects and bring about a definite inferential advantage in certain natural inferential tasks. Also, rows and columns are simple enough to formally characterize, and we can define a heterogeneous logical system with tables, where the distinctive roles played by rows and columns are pinned down in the form of specific inference rules in the system. This allows us to compare logical systems with and without these inference rules, and to quantize the advantage of utilizing rows and columns on the basis of the complexity of proofs in the relevant logical system.

As it turns out, the inference rules in question involve "free rides" in the sense of Shimojima [15], and that is directly reflected in our results on computational complexity. Thus, the present work can be thought of as a quantitative analysis of the inferential advantage of free rides that nicely complements Shimojima's qualitative analysis.

In Section 2, we specify our reasoning with tables through an example. In Section 3, we define the syntax, semantics, and inference rules of our logic with tables (LT). We then investigate some logical properties, i.e., translation into a usual sentential system, soundness, and completeness of LT. In Section 4, we further discuss that our inference rules have multiple free rides, which make inference with tables more effective in a fragment characterized by our completeness theorem.

2 A Reasoning with Tables

Correspondence tables are one of the most basic graphical/diagrammatical representations, and have been applied in a variety of scenarios. Shimojima [17] studied the semantic mechanism of extracting information from a given table, and discussed the mechanism of derivative meaning. In addition to the extraction of information from given tables, we can use tables more dynamically to solve a given problem. This involves constructing a table and adding pieces of information, before manipulating, and finally reading the table as illustrated in the following example.

Example 1. Consider four people a, b, c, d who are scheduled to work separately on one of Monday, Tuesday, Wednesday, and Friday. The following constraints are known: (1) a works on Wednesday; (2) Neither b nor c can work on Monday; (3) On Friday, either c or d should work. Under these conditions, how we can arrange who works on which day?

Let us first consider this problem without using tables. Note that, in addition to conditions (1), (2), and (3), we know that:

(4) There is a one-to-one correspondence between the persons and the days.

First, condition (1) states that "a works on Wednesday." Thus, by (4), we find that "a does not work on Monday." Then, by combining this with (2) and (4), we find that "d works on Monday."

In the given situation, (3) is equivalent to the following (5) under (4):

(5) "a does not work on Friday, and b does not work on Friday."

Because we have already determined that "d works on Monday," (4) implies that "d does not work on Friday." As the above facts can be combined to give that "Neither a nor b nor d works on Friday," we find by (4) that "c works on Friday."

As for b, because we already know that "a works on Wednesday," "c works on Friday," and "d works on Monday," we have from (4) that "b works on Tuesday."

In this way, we are able to determine the working day of a, b, c, d.

Note that in the above reasoning, the condition (4) is necessary to derive any piece of information. Further note that there are various ways to solve the above problem. For example, in the above solution, we converted condition (3) with disjunction into (5) without disjunction. Alternatively, we could have divided (3) into two cases, and examined each case individually.

Next, let us solve the same problem using a correspondence table. We construct a table in which the rows are labeled according to the workers a, b, c, d, and the columns are labeled by days, M (for Monday), T (for Tuesday), W (for Wednesday), and F (for Friday). Based on the given conditions (1), (2), and (3), we insert $\bigcirc$ into each entry (x, Y) for which "x works on Y" holds, and insert $\times$ when "x does not work on Y" holds. Thus, we obtain the table T_0 in Fig. 1. Note that we applied condition (5) instead of the given condition (3).

In terms of tables, condition (4) is divided into the following two conditions:

(6) In each row, exactly one entry should be marked as $\bigcirc$, and the other entries should be $\times$;
(7) In each column, exactly one entry should be marked as $\bigcirc$, and the other entries should be $\times$.

Thus, from the fact that the (a, W)-entry is $\bigcirc$ and (6), we find that the (a, M), (a, T) entries are $\times$, as illustrated in T_1. Similarly, because the $(a, M), (b, M)$,

(c, M) entries are all ×, we find that (d, M) must be ○ by (7), as illustrated in T_2.

Hence, by successively applying (6) and (7), we finally get the determined table T_7. From this, we can read off any information about the working day of a, b, c, d.

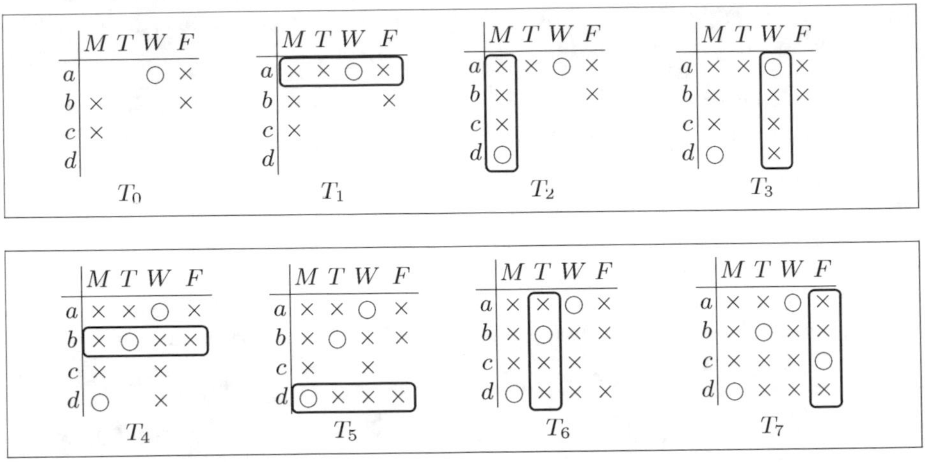

T_0

	M	T	W	F
a			○	×
b	×			×
c	×			
d				

T_1

	M	T	W	F
a	×	×	○	×
b	×			×
c	×			
d				

T_2

	M	T	W	F
a	×	×	○	×
b	×			×
c	×			
d	○			

T_3

	M	T	W	F
a	×	×	○	×
b	×		×	×
c	×		×	
d	○		×	

T_4

	M	T	W	F
a	×	×	○	×
b	×	○	×	×
c	×		×	
d	○		×	

T_5

	M	T	W	F
a	×	×	○	×
b	×	○	×	×
c	×		×	
d	○	×	×	×

T_6

	M	T	W	F
a	×	×	○	×
b	×	○	×	×
c	×	×	×	
d	○	×	×	×

T_7

	M	T	W	F
a	×	×	○	×
b	×	○	×	×
c	×	×	×	○
d	○	×	×	×

Fig. 1. Reasoning with a table in Example 1

Although all entries are either ○ or × in the above example, in general, some entries may not be determined. For example, if we remove condition (3), we obtain a partial table in which $(b, T), (b, F), (c, T), (c, F)$ remain indeterminate. Furthermore, although the number of ○ is fixed to be the same in every row and column, this is not necessarily the case. We do not assume such a restriction in our formalization of reasoning with tables. (See [19] for another example.)

By investigating the above example, we find that there are two types of condition in our problems. One is a "constraint over the framework of a given problem" (e.g., condition (4) above), and we call these *global conditions.* In view of tables, our global conditions are constraints over the number of ○ and × in every row and column. (Although there may be various types of global conditions, we here discuss a particular kind of them, which is formally defined in Definition 3 below.) The other type is a "specific condition for each object" (e.g., (1) above), and we call these *local conditions.* In view of tables, local conditions specify only particular entries. Our reasoning with tables is essentially conducted by combining global conditions and local conditions.

One of the remarkable facts is that, even if the given local and global conditions change, we are able to apply essentially the same strategy:

(I) We decompose, if necessary, the given conditions into local conditions (i.e., atomic sentences or their negation, such as "a does not work on Friday" and "b does not work on Friday") by applying logical laws (e.g., $(3) \wedge (4) \rightarrow (5)$).

(II) We construct a correspondence table using these local conditions (e.g., T_0).
(III) By applying global conditions, i.e., by exploiting constraints over the number of ○ or × in a row or column, we further insert ○ and × into the table.
(IV) Finally, we extract information from the table.

Although most of the given conditions in the above example are already local, more complex conditions may generally be given. In such cases, we frequently apply item (I) in the above procedure. Furthermore, a given condition, such as an implicational sentence, may be decomposed using basic information obtained after item (III). In these complex cases, we must repeat the whole procedure several times. Thus, a natural system of formalizing our reasoning with tables is a heterogeneous logical system combining tables and first-order formulas.

Reasoning tasks, such as that specified above, occur in simple scheduling problems, civil servant examinations in Japan, and in so-called logic puzzles, among others. Barker-Plummer and Swoboda [2] discussed similar problems. They formalized correspondence tables and their manipulations. Their system, consisting mainly of rules for case dividing and reduction to absurdity, is defined to be simple and have as few rules as possible. Conversely, we formalize our system based on each row and column as a global object so that they work effectively to solve a given problem.

3 Logic with Tables LT

In Section 3.1, we roughly review our HLT [19], while in Section 3.2, we define the syntax and semantics of LT, which is the table fragment of HLT. We introduce the inference rules of LT in Section 3.3 and then investigate logical properties of LT, that is, the translation of tables into formulas in Section 3.4, and the completeness theorem in Section 3.5.

3.1 A Heterogeneous Logic with Tables HLT

To formalize our heterogeneous reasoning with tables, we adopt many-sorted first-order logic, in which constants and variables of the usual first-order language are divided into two sorts: sorts of row and column. A **row-label** or **row-constant** (resp. variable) is denoted by a small letter a (resp. x), while a **col-label** (resp. variable) is denoted by a capital letter A (resp. X). (See, for example, [8] for many-sorted logic.) Then, by an **atomic formula** $B(a)$, or more formally $\bigcirc(a, B)$, we denote that "a and B are in a certain positive relation." Thus, sentences such as "a is B," "a matches B," and "a corresponds to B" are all expressed as $B(a)$. Atomic formulas and their negation without free variables are collectively called **(closed) literals**. Based on the atomic formulas, complex formulas are defined inductively as usual by using connectives $\wedge, \vee, \rightarrow, \neg, \forall, \exists$. Among such complex formulas, we distinguish "global formulas", as defined in Definition 3 below. Our correspondence tables are defined in Definition 1.

Semantics is defined as a particular case of the semantics of many-sorted logic, which is a natural generalization of the usual first-order set-theoretical semantics.

Since we are mainly concerned with the table fragment LT (Logic with Tables) of our Heterogeneous Logic with Tables HLT in this paper, we present definitions of the tables and global formulas without going into the detail for the heterogeneous system. See [19] for a formal description of the full HLT.

Although both HLT and LT are heterogeneous systems, on the one hand, in HLT, as well as manipulations of tables (modification of a table based on a given formula, see Section 3.3), the usual natural deduction rules for formulas and transfer rules between formulas and tables are defined. On the other hand, in the table fragment LT, only manipulations of tables are defined, and the other rules of HLT, in particular rules for formulas, are excluded. Thus, we here refer to only HLT as a heterogeneous system.

3.2 Syntax and Semantics of LT

First we define our tables.

Definition 1. A **table** T is an $m \times n$-matrix over symbols $\{\bigcirc, \times, b\}$; that is, a rectangular arrangement of the symbols, in which rows are labeled by distinct row-labels $a_1, \ldots, a_m$ and columns are labeled by distinct col-labels $A_1, \ldots, A_n$. In a specific representation of a table, we usually omit the symbol "b" and leave the entry blank. A table is said to be **determined** if there are no blank entries. Tables T_1 and T_2 are of the **same type** if their labels are the same.

According to the definition, no entry can be marked as $\bigcirc, \times, b$ at the same time.

Remark 1. A table T is abstractly defined as the function $T : \mathcal{R} \times \mathcal{C} \longrightarrow \{\bigcirc, \times, b\}$, where $\mathcal{R}$ (resp. $\mathcal{C}$) is some finite set of row-labels (resp. col-labels) of T.

As usual, any pair of tables, say T_1 and T_2, are identical if they have the same type, and if the $\bigcirc$, $\times$ marks of all entries in T_1 and T_2 are also identical. This is formally defined as follows.

Definition 2. Table T_1 is a subtable of T_2, written as $T_1 \subseteq T_2$, if:

- all row- and col-labels of T_1 are also those of T_2;
- for any (a_i, A_j)-entry in T_1: if it is $\bigcirc$ in T_1, it is also $\bigcirc$ in T_2, and if it is $\times$ in T_1, it is also $\times$ in T_2.

T_1 and T_2 are **(syntactically) equivalent** if $T_1 \subseteq T_2$ and $T_2 \subseteq T_1$ hold.

Note that, two specific tables that differ only in the order of their labels of rows and columns are equivalent.

Next, we define global formulas. To express sentences of the form "among n objects, there are exactly i objects that are A," we introduce a kind of counting quantifier, and write the sentence as $\exists^{i/n} x.A(x)$.

Definition 3. For fixed sets of row-labels $\mathcal{R} = \{a_1, \ldots, a_m\}$ and of col-labels $\mathcal{C} = \{A_1, \ldots, A_n\}$, the following forms of formulas are called **global formulas**: For any A and a,

$$\exists^{i/m} x \in \mathcal{R}.A(x), \quad \exists^{i/m} x \in \mathcal{R}.\neg A(x), \quad \exists^{i/n} X \in \mathcal{C}.X(a), \quad \exists^{i/n} X \in \mathcal{C}.\neg X(a).$$

If a set of labels is clear from the context, it is abbreviated as $\exists^{i/m} x.A(x)$.

Global formulas are simply abbreviations of the appropriate first-order formulas. For example, for some row-label a and col-labels $\mathcal{C} = \{A_1, A_2, A_3\}$, the global formula $\exists^{2/3} X \in \mathcal{C}.X(a)$ is an abbreviation of the following formula:

$$\Big(A_1(a) \wedge A_2(a) \wedge \neg A_3(a)\Big) \vee \Big(A_1(a) \wedge A_3(a) \wedge \neg A_2(a)\Big) \vee \Big(A_2(a) \wedge A_3(a) \wedge \neg A_1(a)\Big).$$

By $\mathcal{G}$, we denote a set of global formulas, and by **label sets of** $\mathcal{G}$, we mean the set consisting of label sets of every global formula of $\mathcal{G}$.

As for the semantics, informally speaking, $(a, B) = \bigcirc$ in a table T means that "a certain relation exists between a and B," or more specifically, $B(a)$ holds. In the same way, $(a, B) = \times$ in T means that $B(a)$ does not hold, i.e., the negation $\neg B(a)$ holds. $(a, B) = b$ in T means that it is not determined whether or not $B(a)$ hold. Although such an informal reading of a table can be formalized in an appropriate set-theoretical domain by applying the semantics of many-sorted logic, here we informally introduce the notion of models by avoiding technical details. See [19] for the details.

Let M be a set-theoretical domain. First-order formulas are interpreted as usual, and we write $M \models \varphi$ if formula φ has a **model** M, i.e., φ holds in M. Table T has a **model** M (written as $M \models T$) if the following holds:

- if $(a, B) = \bigcirc$ in T, then $B(a)$ holds in M;
- if $(a, B) = \times$ in T, then $B(a)$ does not hold, i.e., $\neg B(a)$ holds, in M.

Let $\mathcal{G}$ be a set of global formulas. We write $M \models \mathcal{G}, T$ if $M \models \mathcal{G}$ and $M \models T$, i.e., T has a model M in which $\mathcal{G}$ holds.

Since our tables and global formulas do not contain any variables, LT is essentially the propositional logic. Thus, our model M can be considered to be a set of literals, which hold in M.

Alternatively, a determined table T can also be regarded as a model in its own right, because we can define a model M from T as follows: $B(a)$ holds in M if $(a, B) = \bigcirc$ in T; and $B(a)$ does not hold in M if $(a, B) = \times$ in T (and either one is fine if (a, B) is blank or if there is no such entry in T). Conversely, if we have a model M of T, we are able to construct the determined instance T_M by applying the above definition in the opposite direction. Thus, the following are equivalent:

- T has a model;
- T can be extended to a determined table by consistently inserting $\bigcirc, \times$ marks into the blank entries of T.

The semantic consequence relation in our LT is defined as follows.

Definition 4. Let T_1 and T_2 be tables of the same type, and $\mathcal{G}$ be a set of global formulas whose label sets are those of T_1. T_2 is said to be a **semantic consequence** of $\mathcal{G}, T_1$, written as $\mathcal{G}, T_1 \models T_2$, if any model of $\mathcal{G}, T_1$ is also a model of T_2.

3.3 Inference Rules of LT

According to Plummer-Etchemendy [1], inference rules characteristic of heterogeneous systems are generally called *transfer rules*, and they allow the transfer of information from one form of representation to another. Typical rules in Hyperproof [3,4] are the Apply rule (from sentences to a diagram) and the Observe rule (from a diagram to sentences). Our inference rules, corresponding to the Apply rule, generally have the following form:

$$\frac{T \qquad \varphi}{T'}$$

where T and T' are our tables, and φ is a formula (more specifically, a literal or global formula). Following [3,4], we read this rule as: "we apply φ to amplify T to T'," "we extend T to T' by adding the new information of φ to T," or "φ justifies the specific modification of T to T'." More concretely, our heterogeneous HLT has the following rules, for example:

in **rule:** By applying $A(b)$, we extend table T in which the (b, A)-entry is blank, to table T' in which the (b, A)-entry is $\bigcirc$. Similarly for $\neg A(b)$.

row **rule:** By applying global formula $\exists^{i/n} X.X(a)$, we extend T in which exactly i entries of row a are $\bigcirc$, to T' in which the other entries of row a are $\times$. Similarly for a global formula of the form $\exists^{i/n} X.\neg X(a)$.

col **rule:** By applying global formula $\exists^{i/n} x.A(x)$, we extend T in which exactly i entries of column A are $\bigcirc$, to T' in which the other entries of A are $\times$. Similarly for a global formula of the form $\exists^{i/n} x.\neg A(x)$.

ext **rule:** This rule corresponds to the Observe rule in Hyperproof, and we extract information in a sentential form from a given table.

In addition to these transfer rules, our heterogeneous HLT has the usual natural deduction rules for first-order formulas. See [19]. Thus, the table fragment LT consists of the above *row* and *col* rules, which are formally defined below. Our rules are defined by specifying premises (a global formula and a table) and a conclusion (a table) for each rule.

Definition 5. Inference rules *row* and *col* of LT are defined as follows.

row$\times$ **rules** Premises: A global formula of the form $\exists^{i/n} X.X(a)$ (resp. $\exists^{i/n} X.\neg X(a)$), and a table T in which exactly i entries of row a are $\bigcirc$ (resp. b or $\times$) and the other entries of row a are b or $\times$ (resp. $\bigcirc$).

Conclusion: A table T' that is exactly the same as T except for the blank entries of row a, which are now $\times$.

row$\bigcirc$ **rules** Premises: A global formula of the form $\exists^{i/n} X.X(a)$ (resp. $\exists^{i/n} X.\neg X(a)$), and a table T in which exactly i entries of row a are b or $\bigcirc$ (resp. $\times$) and the other entries of row a are $\times$ (resp. b or $\bigcirc$).

Conclusion: A table T' that is exactly the same as T except for the blank entries of row a, which are now $\bigcirc$.

col× rules Premises: A global formula of the form $\exists^{j/m}x.A(x)$ (resp. $\exists^{j/m}x.\neg A(x)$), and a table T in which exactly j entries of column A are ○ (resp. b or ×) and the other entries of column A are b or × (resp. ○).
Conclusion: A table T' that is exactly the same as T except for the blank entries of column A, which are now ×.

col○ rules Premises: A global formula of the form $\exists^{j/m}x.A(x)$ (resp. $\exists^{j/m}X.\neg A(x)$), and a table T in which exactly j entries of column A are b or ○ (resp. ×) and the other entries of column A are × (resp. b or ○).
Conclusion: A table T' that is exactly the same as T except for the blank entries of column A, which are now ○.

The notion of proof in LT is defined inductively as usual in natural deduction.

Definition 6 (Provability in LT). Let T_1 and T_2 be tables, and $\mathcal{G}$ be a set of global formulas whose label sets are those of T_1. T_2 is **provable** from $\mathcal{G}, T_1$, written as $\mathcal{G}, T_1 \vdash T_2$, if there exists a proof of T_2 from the premises of $\mathcal{G}, T_1$.

Example 2. A proof in LT of Example 1 is given in Fig. 2.

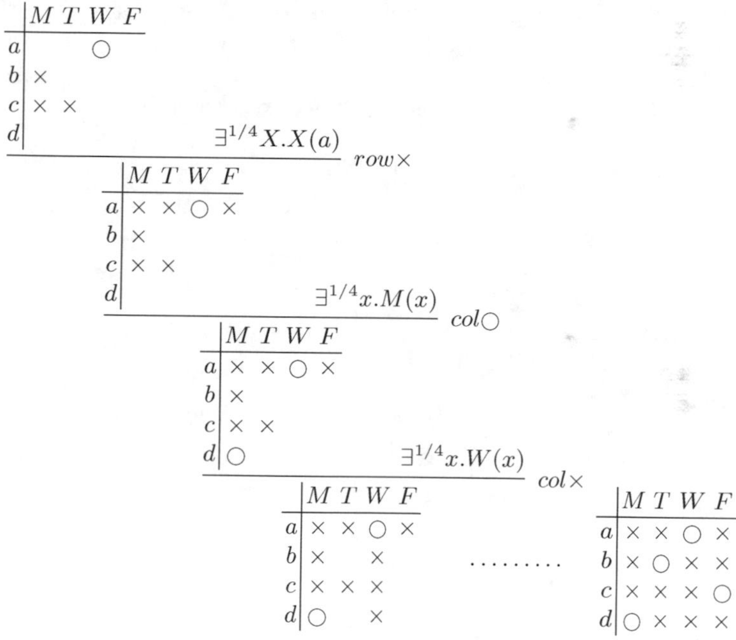

Fig. 2. A proof in LT of Example 1

3.4 Translation of LT

We investigate our tables in terms of the usual first-order language through logic translation of tables. Our tables are translated as follows.

Definition 7. A table T is **translated** into a conjunction of literals T° as follows:

$$T^\circ = \bigwedge\{A_i(a_j) \mid (a_j, A_i) = \bigcirc \text{ in } T\} \wedge \bigwedge\{\neg A_i(a_j) \mid (a_j, A_i) = \times \text{ in } T\}$$

Conversely, it is easily seen that, for any consistent conjunction of literals (without free variables), there exists a corresponding table. Thus, a table T and a consistent conjunction $\mathcal{L}$ of literals can be regarded as being interchangeable. Thus, by slightly abusing our notation, we sometimes write as $\mathcal{L} \subseteq T$.

Based on the translation of tables, *row* and *col* rules of LT are translated into combinations of natural deduction rules; see [19] for the details. The translation of an application of the *row*$\times$ rule is illustrated as follows.

Example 3. Let us consider the following application of the *row*$\times$-rule:

$$\dfrac{\begin{array}{c|ccc} & A_1 & A_2 & A_3 \\ \hline a & \bigcirc & \bigcirc & \end{array} \qquad \exists^{2/3} X.X(a)}{\begin{array}{c|ccc} & A_1 & A_2 & A_3 \\ \hline a & \bigcirc & \bigcirc & \times \end{array}}\ \mathit{row}\times$$

The premise table is translated into the formula $A_1 \wedge A_2$, while the conclusion table is translated into $A_1 \wedge A_2 \wedge \neg A_3$, where we abbreviate $A_i(a)$ as A_i. Note that $\exists^{2/3} X.X(a) := (A_1 \wedge A_2 \wedge \neg A_3) \vee (A_1 \wedge A_3 \wedge \neg A_2) \vee (A_2 \wedge A_3 \wedge \neg A_1)$. This application of the *row*$\times$-rule is translated into the following proof:

$$\dfrac{\dfrac{\exists^{2/3} X.X(a) \quad \dfrac{[A_1 \wedge A_2 \wedge \neg A_3]^1}{\neg A_3} \quad \dfrac{\dfrac{\dfrac{[A_1 \wedge A_3 \wedge \neg A_2]^1}{\neg A_2} \quad \dfrac{A_1 \wedge A_2}{A_2}}{\bot}}{\neg A_3} \quad \dfrac{\dfrac{\dfrac{[A_2 \wedge A_3 \wedge \neg A_1]^1}{\neg A_1} \quad \dfrac{A_1 \wedge A_2}{A_1}}{\bot}}{\neg A_3}}{\neg A_3}\,1 \qquad A_1 \wedge A_2}{A_1 \wedge A_2 \wedge \neg A_3}$$

Proposition 1 (Translation). *If $\mathcal{G}, T_1 \vdash T_2$ in* LT*, then $\mathcal{G}, T_1^\circ \vdash T_2^\circ$ in the natural deduction (without tables).*

By the above theorem of translation, soundness of LT is obtained through soundness of the usual natural deduction without tables.

Proposition 2 (Soundness of LT). *If $\mathcal{G}, T_1 \vdash T_2$ in* LT*, then $\mathcal{G}, T_1 \models T_2$.*

3.5 Completeness of LT

Let us now investigate the completeness theorem of LT. Unfortunately, LT is not complete with respect to our semantics; that is, for a given table T_1 and global formulas $\mathcal{G}$, there exists a table T_2 such that $\mathcal{G}, T_1 \models T_2$ but it cannot be obtained from T_1 with only *row* and *col* rules. Thus, by retaining our rules, we restrict our tables and global formulas as follows.

Informally speaking, a table T is **uniquely determinable** under $\mathcal{G}$, if every entry of T is uniquely determined semantically to either $\bigcirc$ or $\times$, without leaving any entry blank, regardless of any model M in which $\mathcal{G}$ holds.

Definition 8. Let T_M be the determined table defined by a table T and a model M. Let $\mathcal{G}$ be a set of global formulas whose label sets are those of T. T is said to be **uniquely determinable** under $\mathcal{G}$, if the following hold:

- $\mathcal{G}$ and T have a model M, i.e., $M \models \mathcal{G}, T$, and
- for any model N, if $N \models \mathcal{G}, T$ then $T_M = T_N$.

We further restrict our global formulas. We call a set $\mathcal{G}\mathbf{1}$ of global formulas **all-one-global formulas**, consisting of global formulas of the following forms: $\exists^{1/n} x.Y(x)$ for every Y and $\exists^{1/n} X.X(y)$ for every y, which implies, in terms of tables, that "for every row and column, exactly one entry should be ○."

The following is the main lemma used to prove our completeness.

Lemma 1 (Main lemma). *Let $\mathcal{G}\mathbf{1}$ be a set of all-one-global formulas. Let T be a table uniquely determinable under $\mathcal{G}\mathbf{1}$. Then, we have $\mathcal{G}\mathbf{1}, T \vdash T'$ in* LT *for some determined table T'.*

Proof (sketch). Let T' be a table obtained by applying as many *row* and *col* rules to $\mathcal{G}\mathbf{1}, T$ as possible. To show that this T' is the required determined table, assume to the contrary that T' contains some blank entries.

Let M be a model such that $M \models \mathcal{G}\mathbf{1}, T$. Then, by the soundness (Proposition 2) of LT, we have $M \models T'$. The determined table defined by T' and M is the same as that of T, which is denoted by T_M. Using this T_M, we construct another determined table T_N of T that is different from T_M. This contradicts the assumption that T is uniquely determinable under $\mathcal{G}\mathbf{1}$.

We illustrate our strategy for rewriting the given T_M in Fig. 3. We refer to each blank entry of T' using □ notation, and denote the entry in the determined T_M as $\boxed{\bigcirc}$ or $\boxed{\times}$. In each row and column of T_M, there are at least two boxed entries, because, if there were not, we would be able to apply a *row* or *col* rule to T'. In particular, there is another boxed entry other than $(a, A) = \boxed{\bigcirc}$ in column A of T_M, i.e., $(b, A) = \boxed{\times}$ in Fig. 3. Then, we replace the entire row a by row b. Note that, by the replacement, the positions of □ remains as before. After the replacement, if a ○-entry of row b, say (b, B), which appeared as the first row, becomes boxed, we terminate our rewriting process. Otherwise, we continue by replacing row b with another row, say c, which has not yet been replaced and $(c, B) = \boxed{\times}$. We continue the replacement until all ○-entries become boxed.

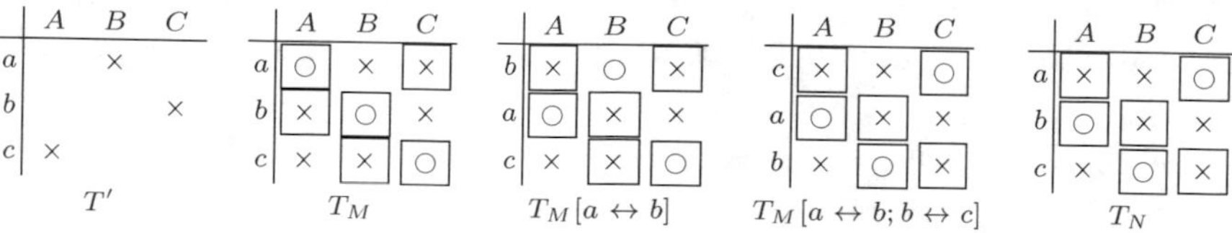

Fig. 3. Rewriting strategy

When the above rewriting process is complete, we obtain T_N by renaming labels of the resulting table so that the 1st, 2nd, 3rd, ... rows are labeled as $a, b, c, \ldots$, respectively. Then, it is clear that T_N and T_M are different only with respect to their boxed entries. ■

By the above main lemma, we obtain our completeness of LT.

Theorem 1 (Completeness of LT). *Let T_1, T_2 be tables of the same type, and $\mathcal{G}\mathbf{1}$ be a set of all-one-global formulas. Let T_1 be uniquely determinable under $\mathcal{G}\mathbf{1}$. If $\mathcal{G}\mathbf{1}, T_1 \models T_2$, then $\mathcal{G}\mathbf{1}, T_1 \vdash T_3$ in* LT *for some table T_3 such that $T_2 \subseteq T_3$.*

Proof. Let T_3 be a table obtained by applying as many *row* and *col* rules to $\mathcal{G}\mathbf{1}, T_1$ as possible. We show that $T_2 \subseteq T_3$. Assume to the contrary that $(a, B) = \bigcirc$ in T_2 and $(a, B) \neq \bigcirc$ in T_3 for some entry (a, B). (The same applies to the other case.) Since T_2 and T_3 are of the same type, and since T_3 is determined by Lemma 1, we have $(a, B) = \times$ in T_3. Then, for any model M such that $M \models \mathcal{G}\mathbf{1}, T_1$, we have $M \models B(a)$ and $M \models \neg B(a)$, which is a contradiction. ■

4 Effectiveness of Tables from a Free Ride Perspective

In Section 4.1, we introduce the notion of "free rides" of inference rules, and explain that our rules have multiple free rides. In Section 4.2, we further investigate the effectiveness of our rules in terms of the complexity of inference.

4.1 Free Rides of Inference Rules

The **free ride** property is one of the most basic properties of diagrammatic systems that provides an account of the inferential efficacy of diagrams. By adding a certain piece of information to a diagram, the resulting diagram somehow comes to present pieces of information not contained in the given premise diagrams. Shimojima [15] called this phenomenon *free ride*, and analyzed its semantic conditions within the framework of channel theory.

By slightly extending the notion of free ride, let us call diagrammatic objects, or translated formulas thereof, **free rides** if they do not appear in the given premise diagrams or sentences, but (automatically) appear in the conclusion after adding pieces of information to the given premise diagrams. The notion of free rides enables us to analyse the effectiveness of each inference rule. (Cf. [18].) Let us illustrate free rides of the *row*× rule by the following example.

Example 4. Let us consider the following application of the *row*× rule and its translation: By applying a global formula $\exists^{1/4}X.X(a)$, we extend table T_1 in which exactly one entry of row a is ○, to T_2 in which the other entries of row a are ×. In the translation, the double line represents the application of various rules.

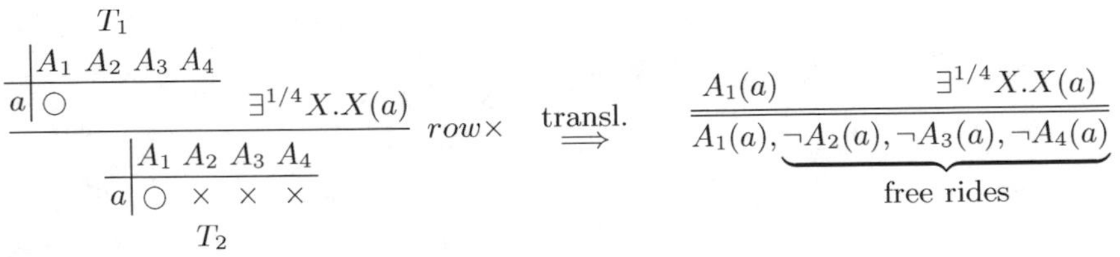

T_1

	A_1	A_2	A_3	A_4
a	○			

$\exists^{1/4}X.X(a)$ — *row*×

	A_1	A_2	A_3	A_4
a	○	×	×	×

T_2

transl. $\Longrightarrow$

$$\frac{A_1(a) \qquad \exists^{1/4}X.X(a)}{A_1(a), \underbrace{\neg A_2(a), \neg A_3(a), \neg A_4(a)}_{\text{free rides}}}$$

Note that pieces of information, $\neg A_2(a), \neg A_3(a), \neg A_4(a)$ in the conclusion do not appear in premise table T_1, which is translated into $A_1(a)$. Furthermore, they do not appear explicitly, or they are indeterminate, in the given global formula $\exists^{1/4}X.X(a) := (A_1 \wedge \neg A_2 \wedge \neg A_3 \wedge \neg A_4) \vee (\neg A_1 \wedge A_2 \wedge \neg A_3 \wedge \neg A_4) \vee (\neg A_1 \wedge \neg A_2 \wedge A_3 \wedge \neg A_4) \vee (\neg A_1 \wedge \neg A_2 \wedge \neg A_3 \wedge A_4)$. Note that the global formula does not imply any definite information about which entries are $\bigcirc$ and which are $\times$. If we extend T_1 to T_2 so that $\exists^{1/4}X.X(a)$ holds, all blank entries in row *a have to be* $\times$ in T_2 since there already exists one $\bigcirc$ in row a of T_1. Note that, in the extension of T_1 to T_2, we do not care about the location of $\bigcirc$ in T_1, because the same applies even if $(a, A_2) = \bigcirc$ and the other entries are blank in T_1. The only thing that matters is the number of $\bigcirc$ in row a in T_1, which corresponds to the number $1/4$ of the given global formula $\exists^{1/4}X.X(a)$. Then, we *freely* find that all blank entries of T_1 are $\times$ without checking each entry one by one. In other words, the pieces of information, $\neg A_2(a), \neg A_3(a), \neg A_4(a)$, are free rides of this *row*$\times$ as they do not appear explicitly in the premises.

As seen in Example 4, the *row*$\times$ rule has multiple free rides. By checking each rule of LT, we obtain the following proposition.

Proposition 3 (Free rides). *row and col rules of* LT *have multiple free rides.*

4.2 Complexity of Inference

We further investigate the multiple free rides of our inference rules in terms of complexity of inference. To this end, we consider the framework of the full heterogeneous HLT, which contains our LT as a subsystem. In the heterogeneous HLT, in addition to the rules for tables, we have usual natural deduction rules for first-order formulas. Furthermore, as shown by the translation (Proposition 1), everything provable by using tables in LT is also provable in HLT by using formulas instead of tables. Thus, we compare, in the framework of the heterogeneous HLT, complexities of inference both with and without the use of tables.

As usual, we formally define the complexity as the length, i.e., the number of formulas and tables, of a given heterogeneous proof in HLT. We consider the restricted logical consequence relation $\mathcal{G}\mathbf{1}, T_1 \vdash T_2$ (and its translation $\mathcal{G}\mathbf{1}, T_1^\circ \vdash T_2^\circ$), where T_1, T_2 are $n \times m$-tables, and T_1 is uniquely determinable under $\mathcal{G}\mathbf{1}$.

Let us first examine inference without using tables. In order to state each formula (literal) of T_2° corresponding to an entry in T_2, we need $n \times m$ formulas. Although we need zero steps to derive such a formula when it is given in the premise T_1°, it is difficult to estimate how many steps (definitely more than one) we need to derive the formula generally. (Cf. the natural deduction proof in Example 3.) Thus, we assume we need at least one step to derive all the formulas of T_2°. Then, we need $n \times m$ formulas to derive T_2°. Hence, we estimate the following number of steps, i.e., formulas, to prove $\mathcal{G}\mathbf{1}, T_1^\circ \vdash T_2^\circ$:

$$(n \times m) + (n \times m).$$

Next, we examine inference using tables. Our completeness (Theorem 1) implies the following theorem, which states that we are able to set an upper bound on the length of inference with tables.

Theorem 2 (Upper bound). *Let $\mathcal{G}1$ be a set of all-one-global formulas, and T be an $n \times m$-table that is uniquely determinable under $\mathcal{G}1$. Let $\mathcal{L}$ be a conjunction of literals whose labels (constants) are those of T. If $\mathcal{G}1, T \vdash \mathcal{L}$ in the heterogeneous* HLT*, then, with a proof of at most $n + m$ length, we have $\mathcal{G}1, T \vdash T'$ for some T' such that $\mathcal{L} \subseteq T'$.*

Proof. Let $\mathcal{G}1, T \vdash \mathcal{L}$ in HLT. Then, by the soundness of HLT, we have $\mathcal{G}1, T \models \mathcal{L}$. Also, by the completeness of LT (Theorem 1), we have, with only rules for tables, $\mathcal{G}1, T \vdash T'$ for some T' such that $\mathcal{L} \subseteq T'$. Note that the rules applicable to T are only *row* and *col* rules, and, by an application of one of these rules, one of the rows and columns is filled with $\bigcirc$ and $\times$ symbols. Since there are only $n + m$ rows and columns in T, we obtain T' within at most $n + m$ steps (tables). ■

In order to state each formula of the entries in T_2, i.e., to read table T_2, we need $n \times m$ formulas. Furthermore, by the above theorem, in order to derive such formulas using tables, we need at most $n + m$ tables. Hence, we estimate the following number of formulas and tables to prove $\mathcal{G}1, T_1 \vdash T_2$:

$$(n \times m) + (n + m).$$

Since the number of steps to state each formula is the same, the above comparison between inference with and without tables is summarized as the difference between $n + m$ and $n \times m$, or between $2n$ and n^2 more concisely.

The effectiveness of our system stems from the multiple free rides of our inference rules. In contrast to the rules for deriving each entry in a table one by one, our rules, having multiple free rides, derive multiple entries at once. The bigger a given table is, the more significant the difference between $2n$ and n^2 becomes. However, note that the above result is obtained in a restricted fragment, where global formulas are restricted to all-one-global formulas $\mathcal{G}1$, and where T_1 is uniquely determinable under $\mathcal{G}1$. In contrast, there is no such restriction on inference using natural deduction rules without tables, although we need n^2 steps to infer. In other words, the above theorem characterizes a fragment in which our tables work effectively. Thus, in practical applications, we can divide the solution to a given problem into two phases. The first phase consists of applying our *row* and *col* rules to a given table; while the other consists of applications of the usual natural deduction rules.

5 Conclusion and Future Work

We studied reasoning with tables in which local and global conditions are exploited. By regarding each row and column as a global object, we formalized our logic with tables LT, which is a subsystem of the heterogeneous logic with tables HLT. LT is shown to be complete with respect to the usual set-theoretical semantics (Theorem 1). Our inference rules, *row* and *col* rules, are designed to take full advantage of the global objects. Thus an inferential advantage of our tables is captured as multiple free rides of our rules (Proposition 3 and Theorem 2).

Our basic research can be extended in a variety of ways. Our correspondence tables and global objects (rows and columns) can be generalized, respectively. We here discuss the following among others.

- Our completeness of LT is restricted to the fragment in which only all-one-global formulas and uniquely determinable tables are considered. The theorem can be generalized by weakening this restriction or by introducing other inference rules. We intend investigating a more general completeness in the future.
- In addition to the free ride discussed in Section 4.1, there may be another kind of free ride in our system. From a local point of view, any addition of a symbol in an entry is just that—the addition of a symbol in that entry. From a global point of view, however, it means something additional, namely, the addition of a symbol in the *row* and *column* comprising the entry. As our inference rules act upon the number of symbols in a row or in a column, such an additional global effect can add up to trigger them and advance our reasoning. In fact, such effects can be seen abundantly in the example discussed in Section 2. We leave the formalization and analysis of this type of free ride to future work.

References

1. Barker-Plummer, D., Etchemendy, J.: A computational architecture for heterogeneous reasoning. Journal of Experimental & Theoretical Artificial Intelligence 19(3), 195–225 (2007)
2. Barker-Plummer, D., Swoboda, N.: Reasoning with coincidence grids–A sequent-based logic and an analysis of complexity. Journal of Visual Languages and Computing 22(1), 56–65 (2011)
3. Barwise, J., Etchemendy, J.: Heterogeneous Logic. In: Allwein, G., Barwise, J. (eds.) Logical Reasoning with Diagrams. Oxford Studies In Logic And Computation Series, pp. 179–200 (1996)
4. Barwise, J., Etchemendy, J.: Hyperproof: For Macintosh. The Center for the Study of Language and Information Publications (1995)
5. Bertin, J.: Semiology of Graphics: Diagrams, Networks, Maps. The University of Wisconsin Press, Madison (1973)
6. Bertin, J.: Graphics and Graphic Information. de Gruyter, W., Berlin (1981) (Originally published in France in 1977)
7. Cleveland, W.S.: The Elements of Graphing Data. Hobart Press, Summit (1994)
8. Enderton, H.B.: A Mathematical Introduction to Logic, 2nd edn. Academic Press (2000)
9. Gilhooly, K.J., Wood, M., Kinnear, P.R., Green, C.: Skill in map reading and memory for maps. Quarterly Journal of Experimental Psychology 40A, 87–107 (1988)
10. Guthrie, J.T., Weber, S., Kimmerly, N.: Searching Documents: Cognitive Processes and Deficits in Understanding Graphs, Tables, and Illustrations. Contemporary Educational Psychology 18, 186–221 (1993)
11. Kinnear, P.R., Wood, M.: Memory for topographic contour maps. British Journal of Psychology 78, 395–402 (1987)
12. Kosslyn, S.M.: Elements of Graph Design. W. H. Freeman and Company (1994)
13. Pinker, S.: A Theory of Graph Comprehension. In: Freedle, R. (ed.) Artificial Intelligence and the Future of Testing, pp. 73–126. L. Erlbaum Associates (1990)

14. Ratwani, R.M., Trafton, J.G., Boehm-Davis, D.A.: Thinking Graphically: Connecting Vision and Cognition During Graph Comprehension. Journal of Experimental Psychology: Applied 14(1), 36–49 (2008)
15. Shimojima, A.: On the Efficacy of Representation. Ph.D thesis, Indiana University (1996)
16. Shimojima, A.: Derivative Meaning in Graphical Representations. In: Proceedings of 1999 IEEE Symposium on Visual Languages, pp. 212–219 (1999)
17. Shimojima, A.: The Inferential-Expressive Trade-Off: A Case Study of Tabular Representations. In: Hegarty, M., Meyer, B., Narayanan, N.H. (eds.) Diagrams 2002. LNCS (LNAI), vol. 2317, pp. 116–130. Springer, Heidelberg (2002)
18. Takemura, R.: Proof theory for reasoning with Euler diagrams: a Logic Translation and Normalization. Studia Logica 101(1), 157–191 (2013)
19. Takemura, R.: A Heterogeneous Logic with Tables. In: Burton, J., Choudhury, L. (eds.) Proceedings of International Workshop on Diagrams, Logic and Cognition (DLAC 2013). CEUR Series, vol. 1132, pp. 9–16 (2014)
20. Tufte, E.R.: Envisioning Information. Graphics Press, Cheshire (1990)
21. Wainer, H.: Understanding Graphs and Tables. Educational Researcher 21, 14–23 (1992)

A Framework for Heterogeneous Reasoning in Formal and Informal Domains

Matej Urbas and Mateja Jamnik

University of Cambridge Computer Laboratory
{Matej.Urbas,Mateja.Jamnik}@cl.cam.ac.uk

Abstract. Heterogeneous reasoning refers to theorem proving with mixed diagrammatic and sentential languages and inference steps. We introduce a heterogeneous logic that enables a simple and flexible way to extend logics of existing general-purpose theorem provers with representations from entirely different and possibly not formalised domains. We use our heterogeneous logic in a framework that enables integrating different reasoning tools into new heterogeneous reasoning systems. Our implementation of this framework is MixR – we demonstrate its flexibility and extensibility with a few examples.

Keywords: interactive, heterogeneous, diagrammatic, theorem proving.

1 Introduction

Theorem provers generally use a sentential logical language with which they express formulae and construct proofs. However, human mathematicians typically use not only multiple but also *informal* representations such as diagrams or images within the same problem for different parts of the solution. Often, there may be parts of the problem that can be better (more intuitively or more efficiently) solved in one representation, language or theorem prover, and other parts in another. However, existing general-purpose theorem provers currently use sentential logics only and do not provide support for diagrammatic and heterogeneous reasoning. Moreover, tools for combining systems (e.g., OpenBox [1], Sledgehammer [2], Omega [6], HETS [9], Chiron [4]) do not allow augmentations to theorem provers that would enable flexible and heterogeneous mixing of formal as well as informal representations and reasoning steps within the same proof attempt. They also do not have the ability to integrate foreign or non formalised data into formulae of theorem provers, and reason with it natively. We analyse related work in more detail in Sec. 5.

Our goal is to enable *heterogeneous reasoning* (**HR**) [11], that is, reasoning with mixed representations and also with inference steps from different existing *sentential reasoners* (**SR**) as well as *diagrammatic reasoners* (**DR**). Furthermore, when logical formalisation of a particular representation (e.g., diagrams, images, or audio) is not tractable, we want to allow the embedding of such data in existing provers and still enable informal heterogeneous reasoning with these opaque objects within an otherwise formal proof. Our aim is to provide an HR

T. Dwyer et al. (Eds.): Diagrams 2014, LNAI 8578, pp. 277–292, 2014.

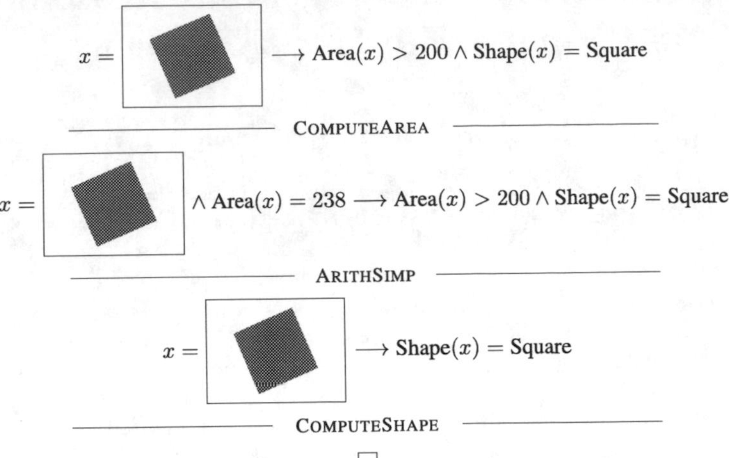

Fig. 1. A heterogeneous proof: it consists of three proof steps. The ComputeArea inference step is heterogeneous. It takes a bitmap image and extracts some information (the area of the square) which is expressed in the sentential language. The ArithSimp inference step is sentential. The ComputeShape is also a heterogeneous inference step. It extracts that the bitmap shape is a square and thus resolves the implication.

framework that enables the construction of heterogeneous proofs. Fig. 1 shows a simple example of a heterogeneous proof that we want to construct in an HR system. The idea is that the user can choose the most appropriate sentential and diagrammatic reasoners and representations, integrate them using our framework, and produce proofs where they can readily choose which parts of the proof they want to represent with which language, and which parts they want to prove with which reasoner.

Our work aims to model human flexible and informal reasoning with their plethora of representations and reasoning techniques. The applications of our work are targeted at tool developers whose sentential or diagrammatic theorem provers' power could be enhanced by bringing to them new and possibly informal representations and reasoning tools. Moreover, domain specific tools like, for example, those for image processing, circuit design, natural language processing, Venn and spider diagrams, which typically do not have access to reasoning engines, can utilize our framework to gain formal reasoning capabilities. Thus, the main contributions of our work are:

- A generic infrastructure for extending existing general-purpose theorem provers (**TP**) with heterogeneous reasoning – we call this *heterogeneous logic* and describe it in more detail in Sec. 2;
- A mechanism, called *placeholders*, for embedding foreign data into formulae of existing theorem provers: it is a crucial part of our heterogeneous logic;
- The MixR framework, which is an implementation of the heterogeneous logic and placeholders. It is a generic infrastructure for creating new HR systems.

MixR can integrate arbitrary existing TPs of any modality with each other into new HR systems – we describe the architecture of MixR in Sec. 3.

A tool developer can plug their chosen reasoners into MixR by writing MixR drivers for them. MixR, in turn, integrates them with each other into a new HR system. For example, we plugged Speedith [13] for spider diagrams and Isabelle [10] for sentential higher-order logic into MixR to create the Diabelli [12] HR system (see Sec. 4.1). MixR provides a user interface as well as an application programming interface (**API**) for drivers. Using the API, the drivers can share, translate and visualise formulae of various modalities. They may also apply foreign inference steps and query other drivers to invoke foreign reasoning tools.

MixR provides placeholders which store foreign data that is dealt with by external tools. This data is directly embedded into formulae of a prover which treats them as primitive objects that can be reasoned with its standard inference engine. When required, the reasoner can invoke external tools on this data to obtain new knowledge as the result. Our approach using placeholders removes the need for translations between representations which is particularly useful when no such translation is available or even possible (e.g., diagrammatic representations from CAD tools, images, and signal processing).

We demonstrate the generality and extensibility of MixR in Sec. 4 by presenting three examples: Diabelli for mixing spider diagrams and sentential higher-order logic, and two new prototype drivers: one for image processing, and another for natural languages. These show how to integrate tools and languages of different modalities with existing TPs. We also show that HR is achievable by merely extending general purpose TPs rather than creating entirely new ones.

2 Heterogeneous Logic

Our heterogeneous logic provides a generic infrastructure to formally or informally connect multiple logics as well as representations with each other. It serves as the foundation for building HR frameworks – MixR is its example implementation (see Sec. 3). We first define the basic concepts and then the logic of HR.

■ **Participating logics and reasoners** are integrated by our heterogeneous logic into a single system. Participating logics may either be diagrammatic, sentential, or of another modality. Participating reasoners provide languages, inference rules, theories, and proof structure. Goal-providing reasoners are *master* reasoners, and others are *slave* reasoners. Master reasoners must provide a language, proof obligations (goals) expressed in that language, inference steps, and a concept of a proof. Slave reasoners must only provide inference rules.

Reasoners may be formal and logical (e.g., a sentential general-purpose theorem prover, a formal diagrammatic prover), or informal (e.g., CAD software, image processing, or signal processing tools whose procedures for knowledge extraction have not been formally verified).

■ **Participating languages and inference rules** are provided by participating reasoners and can be diagrammatic, sentential, or of other modality. We denote the *set of all participating languages* with $\mathbb{L}$, and the *set of all participating* *inference rules* with $\mathbb{I}$. Our heterogeneous logic does not impose any restrictions on the syntax and semantic interpretations of these languages – they are left for the reasoner to provide. The languages can be formal or informal. For example, general-purpose TPs typically use formal logical languages and inference rules. Image processing software, however, uses informal images and informal image processing algorithms as its inference rules for extracting knowledge (e.g., a trend in a chart). Thus, the languages express and store data, while the inference rules (algorithms) extract knowledge.

■ **Heterogeneous formula** is a family of representations of the same original expression but possibly in different languages. A single representation is denoted with ϕ_a and is expressed in a particular language $\mathcal{L}_a \in \mathbb{L}$.

Definition 1. *A **heterogeneous formula** is a pair (ϕ_m, R), where ϕ_m is the main representation and R is a set of derived representations: $R = \{\phi_a, \phi_b, \ldots\}$.*

Every heterogeneous formula has a specific *main* representation ϕ_m from which others can be derived. These derived representations in the set R are in some semantic relation to the main representation ϕ_m which must be specified either by the translation procedure or by the inference rule that produced the representation. Our heterogeneous logic does not impose any restrictions on how these translations or inferences are defined (see below).

The languages of representations may be sentential, diagrammatic, or can contain multiple embedded foreign expressions. Representations must be assertions that can be evaluated to establish their truthfulness. The reasoner that provides the language defines and manages this notion of truth. Some languages (such as formal logics) have a clear definition of truth, while others (such as natural languages) lack formality. Nonetheless, many of these informal domains have notions of assertions that are compatible with the ones in formal TPs and can still be partly defined. We exploit this via placeholders, described below.

■ **Heterogeneous goal**, Γ, is a heterogeneous formula consisting of *premises* (heterogeneous formulae) and a *conclusion* (a single heterogeneous formula):

$$\Gamma = P_1, P_2, \ldots, P_n \Longrightarrow C$$

Each premise and the conclusion may be represented in a different language. Goals provide a standard way to exchange proof obligations and proofs between participating reasoners. Each reasoner can be invoked on parts of the goal which can be (re)presented in its language. Goals originate from a master reasoner. If they are expressed in one language, then they are homogeneous. If they are expressed in different languages or contain placeholders for non-formalisable or non-translatable languages, then they are heterogeneous.

■ **Placeholder** (or embedded foreign formula) is our novel concept that allows sentences of one language to be inserted directly (without translation) into a host formula of another language. Placeholders support embedding of representations that are foreign to a participating language (e.g., in Fig. 1 an image is embedded within a sentential logical formula). This is particularly useful when a translation of one, possibly informal, representation into another is not available or even possible. The novelty of placeholders is that they do not require translation or changing the syntax of the participating language. Instead, placeholders *encode* the foreign formula using the syntax of the host formula (for a concrete example of a placeholder, see Sec. 4.2). This differs from existing work on logical theories since they do not admit foreign data, they only use uninterpreted constants for building theories in their own, *single* language. We, in contrast, use uninterpreted constants to carry foreign data expressed in any language. This data becomes part of a chosen host language and theories can be built using it.

Definition 2. *A* ***placeholder*** $\pi[\mathcal{V}, \mathcal{L}_p, \phi_p]$ *is a formula in the language* $\mathcal{L}_c \in \mathbb{L}$*, where* ϕ_p *is a* payload formula*,* $\mathcal{V}$ *is a set of variables bound implicitly in* ϕ_p*, and* $\mathcal{L}_p \in \mathbb{L}$ *is the language in which* ϕ_p *is expressed.*

The role of the placeholder $\pi[\mathcal{V}, \mathcal{L}_p, \phi_p]$ *in a sentence of the language* $\mathcal{L}_c$ *is the same as that of a Boolean predicate* $P(v_1, \cdots, v_n)$ *where* $\{v_i \mid 1 \leq i \leq n\} = \mathcal{V}$*.*

Variables $\mathcal{V}$ provide the link between the host and foreign formulae. For example: using an external reasoner, we can deduce that for any x and any y the placeholder $\pi[\{x, y\}, \mathcal{L}_{nl}, \text{“x is smaller than y”}]$ implies that $x < y$ in the host theory. Note that universal quantification cannot be to the right of the implication, since x and y are bound in the placeholder. If no variables were listed in $\mathcal{V}$, for example, $\pi[\emptyset, \mathcal{L}_{nl}, \text{“x is smaller than y”}]$, then x and y are independent universally quantified free variables in the host theory, and now the placeholder implies that $x < y$ for any x and any y, a clear falsehood.

Placeholders have no defined interpretation in the host reasoner. They only carry data (i.e., the payload) throughout a proof while being embedded within a host formula. The data is given to external reasoners for knowledge extraction. The extracted knowledge is then embedded back into the proof hosted by the master theorem prover. The external reasoners can themselves be formal, but the entire step is still informal due to placeholder's informal embedding of data.

■ **Inference rules** $I \in \mathbb{I}$ are applied on goals to constitute inference steps. In general, an inference step takes as input a set of initial HR goals $\{\Gamma_1, \ldots, \Gamma_n\}$ and produces a new, transformed set of $\{\Gamma'_1, \ldots, \Gamma'_m\}$. The inference rule I must guarantee that the new goals *logically entail* the initial goals (see below).

The general form of an inference step in our heterogeneous logic is:

$$\frac{\Gamma'_1, \ldots, \Gamma'_m}{\Gamma_1, \ldots, \Gamma_n} \text{ rule: } I.$$

We use the backward reasoning notation, where transformed goals are on top and the original ones on the bottom. Inference rules may transform premises as well as the conclusion of a goal, thus both forward and backward (goal-directed) reasoning methods are supported in our heterogeneous logic.

Inference rules adhere to typical constraints of goal-directed reasoning: strengthening or information-preserving rules can be applied on the conclusion of a goal; weakening or information-preserving rules can be applied on the premises of a goal. Our heterogeneous logic defines weakening, strengthening, and information-preservation relative to other languages on the basis of translations.

Consistency: Our heterogeneous logic does not impose any restrictions on how the inference rules are constructed. Consequently, no guarantees can be given about the consistency and correctness of proofs in general (see *heterogeneous proof* below). Nonetheless, our logic can formally guarantee correctness of proofs under the following conditions: the entire proof must be hosted in a single master reasoner, the proof must start in the language of the master reasoner, all inferences (including external ones) must produce goals in the language of the master reasoner or the produced goals can be translated into that language. Under these conditions, the master reasoner can verify the correctness of each step (this process is called proof reconstruction). On the other hand, proofs that extract knowledge from placeholders are necessarily informal, as no translation of their content into the language of a master reasoner exists.

Heterogeneity: An inference rule is *heterogeneous* if it transforms goals containing formulae of different languages, otherwise it is *homogeneous*. Heterogeneous rules have to be written by the developer of a heterogeneous system (e.g., heterogeneous inference rules in Sec. 4.3 take an image and produce a sentential formula). Heterogeneity is also achieved by *translating* formulae of different languages to a target language before passing them to the inference rule.

■ **Translation procedure** takes an existing formula representation of a premise, conclusion, or entire goal, and produces another one in another language. These representations must be in a correct logical entailment relation with each other. Depending on the entailment of translation we distinguish between translations that are: *strengthening* (applicable on conclusions), *weakening* (applicable on premises) and *information-preserving* (applicable anywhere). If entire goals are translated, weakening applications must not be used.

Consistency and validity: Translations are formal when the logic of one language, say $\mathcal{L}_a$, has been formalised in the logic of another language, say $\mathcal{L}_b$. If these translations are used in a proof, then such a proof is formal if hosted and reconstructed by the master reasoner that uses the language $\mathcal{L}_b$. Our formalisation of spider diagrams in the sentential logic of Isabelle/HOL is one such formal translation: proofs are hosted and reconstructed in Isabelle/HOL [12]. Translations can be left informal when formalisation is not feasible. Clearly, in this case their soundness cannot be guaranteed.

■ **Logical entailment** between two formulae is *heterogeneous* if the two formulae are of different languages, and *homogeneous* otherwise. The heterogeneous logical entailment is defined for formulae resulting from applying inference steps or translations procedures as:

Definition 3. *Formula ϕ_a, expressed in language $\mathcal{L}_a \in \mathbb{L}$,* **entails** *formula ϕ_b of language $\mathcal{L}_b \in \mathbb{L}$ with respect to language $\mathcal{L}_c$ if there exists a direct translation t_1 from ϕ_a to ϕ_a' and another translation t_2 from ϕ_b to ϕ_b', where both ϕ_a' and ϕ_b' are expressed in language $\mathcal{L}_c$, and ϕ_b' can be deduced from ϕ_a' in a finite homogeneous proof, or if for any interpretation such that ϕ_a' is true, so is ϕ_b'.*

Heterogeneous entailment is formal with respect to language $\mathcal{L}_c$ if the translations t_1 and t_2 are formalised within the logic of $\mathcal{L}_c$ (e.g., our formalisation of spider diagrams in Isabelle/HOL in [12]). Homogeneous entailment is a special case of heterogeneous entailment, where the translation is an identity map.

■ **Heterogeneous proofs** are based on discharging proof obligations which take the form of heterogeneous goals. An HR proof thus consists of initial heterogeneous goals, followed by multiple applications of *heterogeneous inference rules* which produce new goals. The inference rules are applied until only tautological goals remain – they are thus discharged and the proof is completed.

Soundness: Proofs in our heterogeneous logic are hosted in a master reasoner. Foreign inference rules can be used on goals by extracting the goals from the master reasoner, passing them to the slave reasoner for inference, and then inserting the result back into the master reasoner's proof. Only a formal translation can provide soundness and completeness guarantees. So, if the new formula is *not* a placeholder (i.e., a translation into the host language exists), then the master reasoner can verify its soundness. Otherwise, the master reasoner can trust that the new goal is correct using the oracle principle, but clearly the proof cannot be guaranteed to be sound. Tools such as HETS [9] could be utilised to provide other existing translations, and thus formal guarantees for them.

3 Architecture and Design of MixR

MixR implements the heterogeneous logic introduced in Sec. 2: it manages formulae, their translations, inference rules and communicates these with participating reasoners. Tool developers, who are the users of the MixR framework, can integrate their own representations (formula formats), translation procedures, inference rules, visualisations, or entire participating reasoners using MixR by providing drivers which are pluggable components. This results in a new *HR system* for end users. We present three examples of drivers in Sec. 4. MixR was implemented in Java using the NetBeans platform.[1]

Fig. 2 shows the general architecture of MixR. A MixR-based system adds HR on top of participating reasoners. *Master reasoners* host the proof in their proof management infrastructure. MixR, however, takes proof obligations (goals) from master reasoners, translates them, visualises them, and lets users interact and explore them. MixR also facilitates communication of goals with other *slave reasoners*. Slave reasoners transform the goals with their inference rules. The result is communicated back to the originating master reasoner. In summary, MixR's role is that of a mediator between multiple master and slave reasoners.

MixR is a small core that implements a set of *heterogeneous logic components* (concepts from our heterogeneous logic), *driver contracts* (that facilitate integration of reasoners), and *general UI components* (which display pending proof obligations, enable user interaction and coordinate formulae visualisation).

[1] All MixR sources are available from `https://github.com/urbas/mixr`.

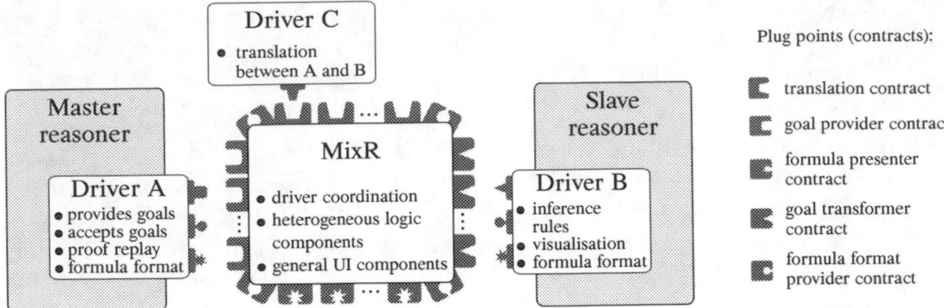

Fig. 2. The outline of MixR's architecture with hypothetical drivers. The central box represents MixR's core. It contains the implementation of heterogeneous logic components, general UI components, and driver plug-points. Drivers surround MixR's core and plug into it through the plug-points.

3.1 Heterogeneous Logic Concepts

Since heterogeneous proofs are hosted by master reasoners, it is the responsibility of master reasoners (and their drivers – see below) to implement the concept of heterogeneous proofs. MixR provides implementations for the other heterogeneous logic concepts: `FormulaFormat` (participating languages), `FormulaRepresentation`, heterogeneous `Formula`, heterogeneous `Goal`, `Placeholder`, heterogeneous `InferenceRule`, and `FormulaTranslator` (translation procedures). MixR also implements placeholders that provide the infrastructure for inserting foreign data into formulae of existing theorem provers (for examples, see Sec. 4). With these, specialised ad-hoc mixed systems do not have to repeat implementing or re-inventing placeholders. Instead, MixR allows the developer to focus on the rules and simply reuse the infrastructure.

3.2 Driver Contracts

MixR makes the functionality of every driver available to every other driver and also the user interface. To simplify and generalise all inter-driver communication, MixR defines a standard set of driver contracts: `GoalProvider`, `FormulaFormatProvider`, `FormulaPresenter`, `TranslationProvider`, and `GoalTransformer`.

The `GoalProvider` contract must be implemented by drivers that connect master reasoners to MixR. They communicate pending proof goals of the master reasoner to MixR. MixR also uses this contract to re-insert transformed goals back into the master reasoner.

Drivers providing new formula formats to MixR must implement the `FormulaFormatProvider` contract. They must provide a unique name of each format, a description and an API for manipulating the formulae of that format.

MixR gives users the choice of using sentential, diagrammatic as well as heterogeneous formulae. These are displayed to the user in a way most suitable for the given format. For this reason, drivers can implement the `FormulaPresenter` contract that facilitates the visualisation of formulae of particular formats.

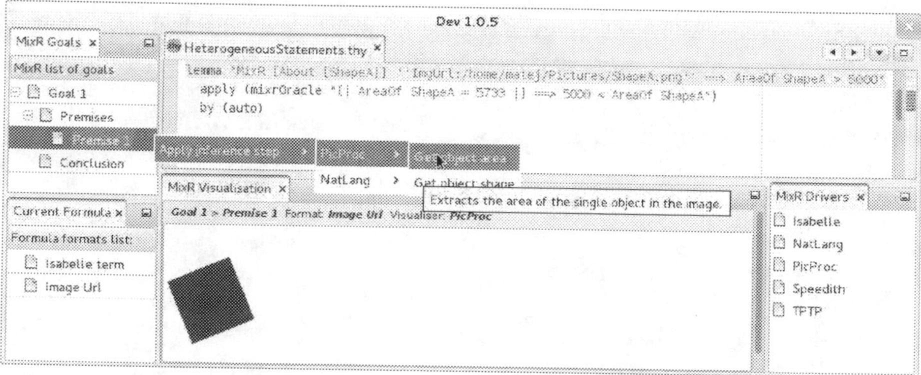

Fig. 3. The user interface of a MixR-based system that integrates PicProc with Isabelle

Translation procedures are integrated into MixR by drivers implementing the `TranslationProvider` contract. These drivers advertise all translations they contribute. MixR uses them automatically whenever a particular formula format is requested. The user may also invoke a translation on any formula interactively.

Drivers that transform goals with inference rules must implement the `Goal-Transformer` contract. They bring slave reasoner functionality into MixR.

3.3 General UI Components

MixR's general UI components present and control aspects that are common to the entire framework. Fig. 3 shows the user interface of a system created using MixR, which contains the following components (sub-windows):

- Top-left: The list of pending goals of the currently active master reasoner.
- Top-right: The proof script of the master reasoner.
- Pop-up menu under the mouse cursor: Shows how the user interacts with the list of pending goals (e.g., applying inference rules on the goals, translating them to a target representation, or selecting the part to be visualised).
- Bottom-centre: Visualises currently active or selected formulae.
- Bottom-left and bottom-right: Lists of all available reasoning languages, registered drivers, master reasoners, and slave reasoners.

This is how the interaction with inferences is carried out in MixR:

1. The user first selects the part of a goal on which to apply the inference rule. In Fig. 3 the user selected the first premise of the first goal (see top-left sub-window).
2. Next, the user chooses an inference rule by left-clicking the selected invocation target and choosing the desired inference rule (shown under the mouse cursor in Fig. 3).
3. Now, if the inference rule is an *interactive* one, the user may be asked for more input. In the example from Fig. 3 the selected rule *Get object area* requires no additional input.

4. Finally, the inference rule returns its result, which may be a formula foreign to the target master reasoner. If so, MixR tries to translate it. If no translation succeeds, the result is placed into a placeholder. MixR constructs a new goal with the result and passes it to the master reasoner's driver. The driver then inserts the goal into the master reasoner's proof script. In Fig. 3, the line starting with `apply` was inserted after the user invoked the inference.

4 Prototype Examples

We now demonstrate MixR's generality and its facilities to integrate diverse systems into heterogeneous reasoning tools. To integrate a new reasoner, we have to create a driver for it. The driver must implement a particular set of driver contracts to integrate the desired features into MixR. In the past, we used MixR to integrate the general purpose theorem prover Isabelle [10] and our diagrammatic prover Speedith [13] into an HR system Diabelli [12]. Here we present Diabelli, and two new drivers: *NatLang* and *PicProc*. *NatLang* integrates a mock sentential reasoner for informal natural languages. *PicProc* integrates an image processing visual reasoning tool. We use MixR to combine each with Isabelle that has the role of the master reasoner in this demonstration. These prototype drivers work on domains that are inherently hard to formalise. They showcase how such informal domains may still be integrated with a formal general purpose theorem prover. To give an idea of their complexity, the Speedith, Isabelle, NatLang and PicProc drivers each consist of between 400 and 600 lines of Java source code.

4.1 Diabelli

Diabelli is our HR system that allows users to construct heterogeneous proofs with spider diagrams mixed with the sentential higher-order logic language. It is a *system* resulting from plugging the Speedith [13] DR and the Isabelle [10] SR into the MixR *framework*. Diabelli does not utilise MixR's placeholder mechanism since Speedith's representation can be translated into Isabelle's, so Speedith's data is not foreign to Isabelle. Fig. 4 shows Diabelli's user interface with Isabelle/HOL sentential formulae mixed with Speedith's spider diagrams and spider-diagrammatic inference rules. The implementation details of the drivers for Speedith and Isabelle can be found in [12]. Diabelli provides a non-hypothetical, fully integrated, and non-trivial experimental evidence for how MixR contributes and aims to spur innovation within the community using and studying diagrams.

4.2 NatLang

The NatLang driver implements the contracts `FormulaFormatProvider` and `GoalTransformer`. It provides a single formula format, the `NatLangFormat`, which is a plain string with no restrictions or grammar. We use MixR's

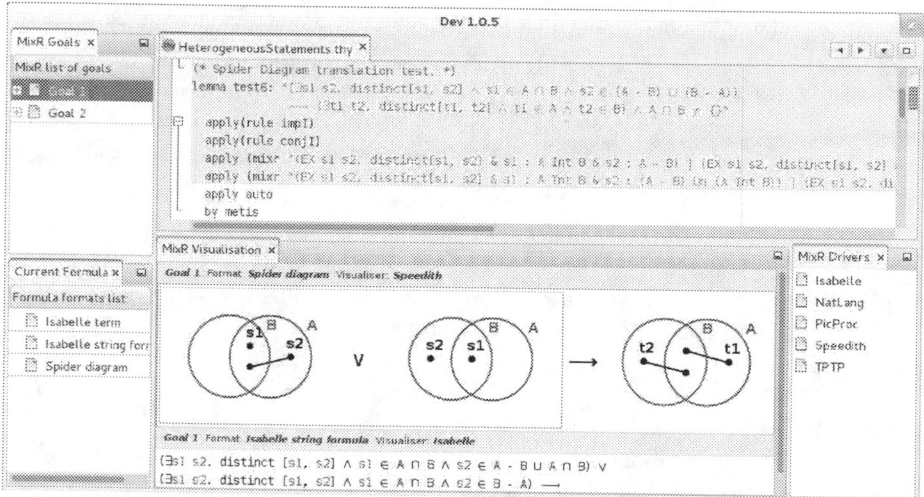

Fig. 4. Diabelli's main window. The top-right sub-window contains I3P's hosting proof script editor for Isabelle. Users may edit theories in the same way as in standalone I3P. Diabelli inserts heterogeneous proof steps into this proof script as the user applies them through Diabelli's GUI (by right-clicking on the goals in the top-left sub-window).

placeholders to insert NatLang expressions into Isabelle/HOL formulae. Here is an Isabelle/HOL formula containing one such placeholder:

```
(MixR [About[Teenager, Human]] "NatLang:Every teenager is a
    human.")
      ∧ (∀p. p ∈ Human ⟶ p ∈ Mortal)
⟶ Teenager ⊆ Mortal.
```

The predicate `MixR` is the placeholder. It contains the set of variables ($\mathcal{V} = \{$*Teenager, Human*$\}$) and a string (*payload*) which consists of the name of the foreign language ($\mathcal{L}_p =$ *NatLang*) and the actual foreign formula ($\phi_p =$ *Every teenager is a human*).

NatLang provides inference rules `HomogeneousInference` and `HeterogeneousInference`, which rely entirely on user-input. The homogeneous inference rule takes a NatLang expression and returns a NatLang formula which is re-embedded into the Isabelle/HOL formula as a placeholder. The heterogeneous rule takes a NatLang expression and returns an Isabelle/HOL formula, which it places into the Isabelle formula without a placeholder. Here is an example where *Teenager* $\subseteq$ *Human* has been inferred from the above NatLang sentence:

```
  (Teenager ⊆ Human ∧ (∀p. p ∈ Human ⟶ p ∈ Mortal)) ⟶
      Teenager ⊆ Mortal
```

The resulting goal may now be reasoned about with Isabelle's own theories and inferences. The homogeneous inference rule shows how a foreign inference step can be used within an Isabelle proof (i.e., both the input and the output formulae are foreign to the theorem prover). The heterogeneous inference rule, on the other hand, demonstrates how knowledge can be transferred between the natural language domain and the theorem prover's domain. Such reasoning is informal as a whole, but the informal parts are limited to small steps that link both domains. This mechanism provides more than the oracle principle: MixR extracts knowledge from the placeholder's data piecewise (as it's needed in the proof); the oracle principle is used to trust each such step of knowledge extraction, translation, and re-embedding.

4.3 PicProc

The PicProc example demonstrates the integration of images and image processing into sentential theorem provers. This example also utilises placeholders and the Isabelle master reasoner. Fig. 1 shows a heterogeneous proof that PicProc contributes to, and Fig. 3 shows the screenshot of the system resulting from plugging PicProc and Isabelle into MixR.

Similarly to the NatLang driver, PicProc provides a new formula format and two inference rules. The formula format is called `ImgUrl` and the provided inferences are `ComputeArea` and `ComputeShape`. The ImgUrl formulae are paths to bitmap images. The rule `ComputeArea` returns the area of the shape (the tilted square at the bottom of Fig. 3). The rule `ComputeShape` infers the type of the shape. This rule decides whether the shape is a circle, a triangle, or a square (based on the ratio between the area and the circumference of the shape).

Note that in Fig. 3 (top-right sub-window) the PicProc image is inserted into a `MixR` placeholder within the host Isabelle/HOL formula. The `About[ShapeA]` clause indicates that the placeholder talks about an object called `ShapeA`. Placeholder's variables $\mathcal{V}$ such as `ShapeA` must be extracted and listed by the driver (in this case the driver simply takes the name of the image file), otherwise Isabelle cannot know that the payload contains knowledge about them.

The PicProc driver also implements the `FormulaPresenter` contract which provides MixR with new visualisation capabilities. ImgUrl formulae found by MixR can now be passed to PicProc, which opens the image and displays it in MixR's visualisation panel (see the bottom-centre sub-window of Fig. 3).

5 Related Work and Evaluation

The topic of HR has been explored in a number of different contexts [11,1,12]. Here, we compare our work to existing logics and systems that provide some level of heterogeneity. We then evaluate its scalability and extensibility.

5.1 Logics

Chiron [4], context logic [3], and multilanguage hierarchical logics (MLHL) [5] are logics that provide heterogeneity on two levels: multiple reasoning paradigms with a single sentential language (Chiron and context logic), or single reasoning paradigm with multiple first-order sentential languages. In particular, these logics use purely sentential languages and establish entirely formal connections between all their components. For example, Chiron uses a *single* sentential language but allows switching between different reasoning paradigms (e.g., classical reasoning, set-theoretic reasoning, type-theoretic reasoning). Similarly, context logic uses a *single* sentential language, but its semantics are defined with contexts and multiple sets of statements. This enables context logic to encompass intuitionistic, classical, and predicate logic as its special cases. The goal of MLHL is similar – to enable reasoning in multiple sentential logics. Although MLHL allows multiple different languages, they are all confined to first-order logics.

Thus, Chiron, context logic, and MLHL have a different notion of heterogeneity to ours, and unlike MixR they also do not extend logics of existing theorem provers with HR, nor facilitate *formal and informal* HR.

5.2 Frameworks and Systems

■ **OpenBox** is the only other existing implementation of an HR framework [1]. It maintains its own proofs and uses external theorem provers to validate separate inference steps within its proofs. These provers may be diagrammatic or sentential, which makes OpenBox a heterogeneous framework. Unlike MixR, OpenBox does not extend existing reasoners with heterogeneous reasoning but rather utilises them to make its own proofs heterogeneous. This forces users of reasoners to abandon their existing work. In contrast, MixR reuses proofs and proof scripts of existing reasoners which means that users can retain their existing work and seamlessly extend it with HR. While separate formulae of OpenBox's proof can be of different modalities, a single OpenBox formula is of a single modality. This is in contrast to MixR where a single formula can contain embedded foreign formulae of different modalities. Furthermore, unlike MixR, OpenBox cannot embed foreign data into formulae of existing TPs.
■ **Sledgehammer** [2] is a component of the Isabelle [10] SR that passes translations of Isabelle statements to external reasoners for validation. It replaces external answers with corresponding Isabelle's inference steps. This purely sentential process requires formal translations and that obtained answers are fully expressible with Isabelle. Sledgehammer cannot process heterogeneous nor informal inference steps, and it cannot embedding foreign data into formulae.
■ **Omega** is a proof planning system [6] that uses different sentential reasoners, and is thus a homogeneous rather than a heterogeneous system. Unlike MixR, Omega imposes its own proof structure that is maintained by the blackboard mechanism. An external SR can only be invoked if the translation of the Omega formula into its representation exists. This is in contrast to MixR which can use placeholders to embed foreign data from external reasoners into formulae when no translation exists. Omega also does not extend existing reasoners in any way.
■ **HETS** (or *Heterogeneous Tool Set*) [9] differs from MixR in the use of the term "heterogeneous". In HETS it is used to refer to formal relations between multiple sentential logics. Unlike MixR, HETS is thus a purely sentential system, which produces formal translations between sentential logics. In contrast to MixR, which does not require translations, a logic cannot be used in HETS if it cannot be translated into other HETS logics.

5.3 Scalability and Extensibility

Current benchmark problem sets (such as TPTP) test the efficiency and scope of theorem provers in first-order or higher-order sentential logic. In contrast, our heterogeneous logic and framework aim to expand the vocabulary of existing TPs with foreign data, formulae, diagrams, informal reasoning, and foreign inference – none of these are included in benchmark sets. Thus, rather than quantitatively we qualitatively evaluate the extensibility of our heterogeneous logic and framework.

The *range of domains* that can be integrated using MixR was exemplified in Sec. 4 where we presented case-studies of Diabelli, NatLang and PicProc. These embed spider diagrams, natural language and image processing into an existing theorem prover. We now evaluate the *generality* of our framework by assessing if and which other reasoners can be plugged into it. We consider Cinderella [8], Hyperproof's Blocksworld [1] and Diamond [7] as representative DRs, and HOL4, HOL Light, Coq and Twelf as representative SRs.

Cinderella [8] (a reasoner for geometric constructions) and Diamond [7] (a reasoner for diagrammatic proofs on natural numbers) could both be master reasoners in MixR. Their drivers would have to implement UI editors for hosting diagrammatic proof scripts and visualisation procedures. Both, Cinderella and Diamond use automated sentential TPs to validate and verify their diagrammatic proof steps. These TPs could be integrated as slave reasoners in MixR. Cinderella could also be used to integrate the domain of geometry into other master reasoners such as Isabelle. This would require translation procedures between Cinderella's internal representation and Isabelle/HOL formulae.

Blocksworld expresses relations between 3D objects placed on a checkerboard. The reasoning about these relations is done within Hyperproof [1] using a first-order logic TP. Thus Blocksworld does not have a notion of a proof state, and can only be plugged into MixR as a slave reasoner. Its driver would have to integrate its visualisation, and also implement a translation procedure between Blocksworld's models and any existing master reasoner in MixR.

Interactive theorem provers HOL4, HOL Light, Coq, and Twelf all use textual proof scripts. Master reasoners need these to host proofs. All these provers enable reasoning with either oracles or axioms (required for informal reasoning). HOL4 and HOL Light provide the `mk_thm` command to introduce informal inferences; Coq provides the command `Axiom`; and Twelf supports axiomatic inference rules as part of its meta-theorem infrastructure. MixR's placeholders require uninterpreted constants and functions in the theorem prover's logic. HOL4 and HOL Light support uninterpreted functions through the `def_constant` command. Similarly, Coq provides the command `Variable` for this purpose. Finally, Twelf's fundamental approach to building theories is to start with uninterpreted symbols and provide inference rules for them later. Therefore HOL4, HOL Light, Coq, and Twelf could all be extended with our heterogeneous logic and thus plugged into MixR. This would require the developers to write additional translation procedures, communication channels between the theorem prover and MixR, and the integration of the theorem prover's text editor software with MixR. Theorem provers that are fully automated, such as Z3, ACL2, Vampire, and Spass lack proof scripts (or other theory- and proof-specification documents). Thus they can be integrated into MixR as slave reasoners but not as master reasoners.

6 Conclusion

In this paper, we presented a heterogeneous logic that marries formal logics with diverse and possibly informal representations. Our novel concept of placeholders enables existing logics to formally treat foreign representations within their own formulae. We implemented this logic in a heterogeneous framework MixR that facilitates a flexible integration of sentential or diagrammatic theorem provers and other formal and informal reasoners, representations and visualisations. These can be plugged into MixR via drivers to produce integrated HR systems. With MixR, we can explore in breadth and in depth the interaction between diagrammatic languages and formal sentential languages which invites multi-domain collaboration and exploration. Domains traditionally inaccessible to formal reasoners, can now be integrated and exploited. The developers of HR systems can be flexible about their system design: their choices may depend on issues surrounding efficiency, intuitiveness of proofs, or level of expertise of end users of resulting systems.

We presented three examples of integrated systems from diverse domains of spider diagrams, natural language and image processing. Plugging them into MixR alongside a general purpose theorem prover enabled for the first time to formally use the information from informal reasoners to construct proofs.

Many reasoning tools, representations and visualisation aids in AI exist mostly in isolation, specialised in their specific domains. Bringing them together in a simple, flexible and formal way allows them to contribute to the problem solving/theorem proving tasks. We believe this is desirable for several reasons: it better models what people do in problem solving, it allows developers to easily design systems that are flexible according to the needs of the end-users, and it enables us to take advantage of the existing powerful technology out there in a novel and sustainable way.

Acknowledgements. This work was supported by EPSRC Doctoral Training Grant and Computer Laboratory Premium Research Studentship (Urbas).

References

1. Barker-Plummer, D., Etchemendy, J., Liu, A., Murray, M., Swoboda, N.: Openproof - A flexible framework for heterogeneous reasoning. In: Stapleton, G., Howse, J., Lee, J. (eds.) Diagrams 2008. LNCS (LNAI), vol. 5223, pp. 347–349. Springer, Heidelberg (2008)
2. Böhme, S., Nipkow, T.: Sledgehammer: Judgement day. In: Giesl, J., Hähnle, R. (eds.) IJCAR 2010. LNCS (LNAI), vol. 6173, pp. 107–121. Springer, Heidelberg (2010)
3. Calcagno, C., Gardner, P., Zarfaty, U.: Context logic as modal logic: completeness and parametric inexpressivity. In: POPL, pp. 123–134. ACM (2007)

4. Farmer, W.M.: Biform Theories in Chiron. In: Kauers, M., Kerber, M., Miner, R., Windsteiger, W. (eds.) MKM/CALCULEMUS 2007. LNCS (LNAI), vol. 4573, pp. 66–79. Springer, Heidelberg (2007)
5. Giunchiglia, F., Serafini, L.: Multilanguage hierarchical logics, or: How we can do without modal logics. AI 65(1), 29–70 (1994)
6. Siekmann, J.H., et al.: Proof development with ΩMEGA. In: Voronkov, A. (ed.) CADE-18. LNCS (LNAI), vol. 2392, pp. 144–149. Springer, Heidelberg (2002)
7. Jamnik, M., Bundy, A., Green, I.: On Automating Diagrammatic Proofs of Arithmetic Arguments. JOLLI 8(3), 297–321 (1999)
8. Kortenkamp, U., Richter-Gebert, J.: Using automatic theorem proving to improve the usability of geometry software. In: Procedings of the Mathematical User-Interfaces Workshop, pp. 1–12 (2004)
9. Mossakowski, T., Maeder, C., Lüttich, K.: The Heterogeneous Tool Set, HETS. In: Grumberg, O., Huth, M. (eds.) TACAS 2007. LNCS, vol. 4424, pp. 519–522. Springer, Heidelberg (2007)
10. Paulson, L.C.: Isabelle - A Generic Theorem Prover. LNCS, vol. 828. Springer, Heidelberg (1994)
11. Shin, S.J.: Heterogeneous Reasoning and its Logic. BSL 10(1), 86–106 (2004)
12. Urbas, M., Jamnik, M.: Diabelli: A heterogeneous proof system. In: Gramlich, B., Miller, D., Sattler, U. (eds.) IJCAR 2012. LNCS (LNAI), vol. 7364, pp. 559–566. Springer, Heidelberg (2012)
13. Urbas, M., Jamnik, M., Stapleton, G., Flower, J.: Speedith: A diagrammatic reasoner for spider diagrams. In: Cox, P., Plimmer, B., Rodgers, P. (eds.) Diagrams 2012. LNCS (LNAI), vol. 7352, pp. 163–177. Springer, Heidelberg (2012)

The Second Venn Diagrammatic System

Renata de Freitas and Petrucio Viana

Institute of Mathematics and Statistics
UFF: Universidade Federal Fluminense, Niterói, Brazil
freitas@vm.uff.br, petrucio@cos.ufrj.br

Abstract. We present syntax and semantics of a diagrammatic language based on Venn diagrams in which a diagram is not read as a statement about sets, but as a set itself. We prove that our set of rules is sound and complete with respect to the intended semantics. Our system has two slight advantages in relation to the systems we usually encounter in the literature. First, the drawing of diagrams for terms is made inside the system, i.e., by a completely mechanical process based just on the rules of the system. Second, as a consequence, the validity of an inclusion is also verified inside the system and does not depend on any other means than those afforded by our set of rules. These characteristics are absent in the majority of the Venn diagrammatic systems.

Keywords: diagrammatic systems, Venn diagrams, soundness, completeness.

1 Two Approaches to Venn Diagrams

To start with, we consider a typical Venn Diagram:

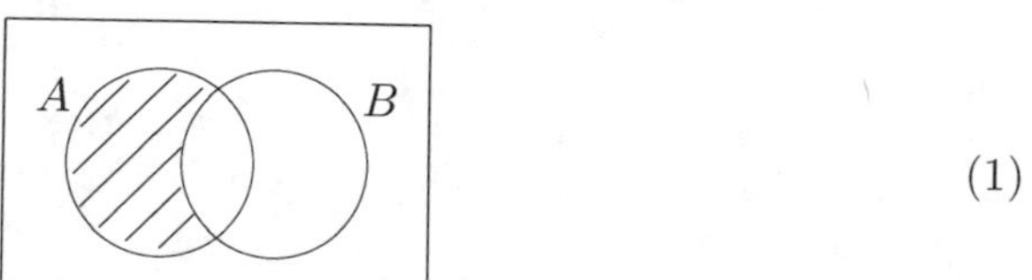

(1)

There are many ways to assign a meaning to this diagram. If we read it like Hammer [3] and Shin [4] do, then its meaning is *a proposition about sets*. More specifically, under this reading, the diagram means that the set A is a subset of the set B —in the usual symbolic language, $A \subseteq B$. References [3] and [4] formalize the syntax and semantics of Venn diagrams read as such and show how we can use them to prove *implications* as

$$A \subseteq B,\ \ B \subseteq C \implies A \subseteq C \tag{2}$$

By way of example, Figure 1 contains a proof, constructed in the Hammer and Shin framework, that statement (2) holds.

But some authors, particularly Stewart [5], have pointed out another *traditional* way of reading the diagram (1). In this second reading, its meaning is no

T. Dwyer et al. (Eds.): Diagrams 2014, LNAI 8578, pp. 293–307, 2014.

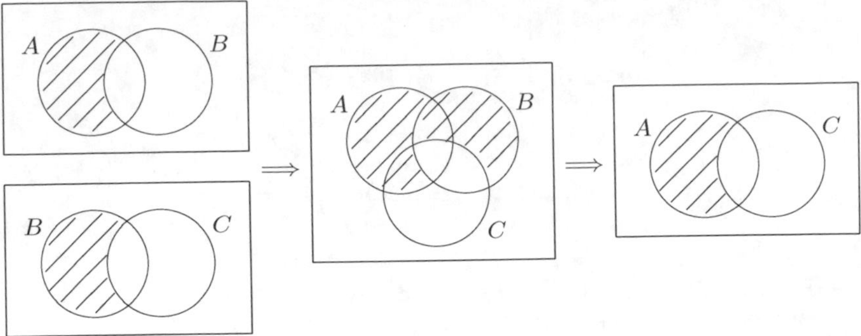

Fig. 1. A diagrammatic proof of $A \subseteq B,\ B \subseteq C \Longrightarrow A \subseteq C$

more a proposition on sets but instead a *set* itself. More specifically, according to this reading the diagram represents the set whose elements belong to A but not to B —in the usual symbolic language, $A \setminus B$. Now, instead of proving implications between Boolean inclusions, the natural statements we can use this diagram to prove are the *valid Boolean equalities* or *identities* —i.e., Boolean equalities that hold for any sets whatsoever— as, for example,

$$A \setminus B = (A \cup B) \setminus B \tag{3}$$

The *traditional* diagrammatic proof that (3) is an identity is just the diagram (1) itself, according to the remark that both the left hand side and the right hand side of the equality represent the same set as the diagram does. So, in this reading, the own diagram is a proof of the equality we are proving!

In the manner it is usually presented, the *traditional* use of Venn diagrams sketched above has a drawback: given two terms we want to compare through diagrams, we must know, in advance, how to draw the diagram representing the set each term represents. A similar observation also applies to the first approach: to prove an implication between inclusions, we must know how to draw a diagram having the same meaning of a given inclusion.

Our aim in this work is to present a diagrammatic formal system in which Venn diagrams are used to represent *sets*, instead of *propositions about sets*, and that may be applied to prove Boolean inclusions (and, consequently, equalities) from sets of Boolean inclusions (or even equalities) taken as hypotheses. Our system has two slight advantages in relation to the systems we usually encounter in the literature. First, the drawing of diagrams for terms is made inside the system, i.e., by a completely mechanic process based just in the rules of the system, without any appeal to the intuitive meaning of symbols. Second, as a consequence, the validity of an inclusion is also verified inside the system and does not depend of any other means than those afforded by our set of rules. These characteristics are absent in the majority of the Venn diagrammatic systems, where both the drawing of diagrams and the test for validities are heavily dependent on some acquaintance of the algebra of sets by the users part.

More specifically, we present a set of rules through which, for any given set of Boolean inclusions or equalities Σ and Boolean inclusion or equality involving Boolean algebraic terms X and Y, we are able to: (1) draw the diagrams of X and Y, (2) put these diagrams into a normal form, (3) compare the diagrams in normal form in order to decide whether the inclusion or equality is valid or not, (4) if not, we use the inclusions in Σ to transform the diagrams again and iterate this procedure to test if the inclusion or equality is a logical consequence of Σ or not.

In Section 2, we present an example illustrating some of the main characteristics of our system. In Section 3, we present syntax and semantics of the diagrammatic language. In Section 4, we present the rules of the system to prove the valid inclusions and show they are weakly sound and complete for the semantics presented in Section 3. In Section 5, we present an example illustrating the extension of the system to deal with reasoning from hypotheses. In Section 6, after introducing a new rule to the system, we show strong soundness and completeness. Finally, in Section 7, we present some conclusions and perspectives.

2 Proving an Identity

In this section, we illustrate the use of the system, by proving the identity (3).

In general terms, to test if $X = Y$ is a valid equality, we proceed as follows: (a) build the diagram of X, (b) put the diagram of X in normal form, (c) build the diagram of Y, (d) put the diagram of Y in normal form, (e) compare the normal forms.

We depart from two usual ways of drawing Venn diagrams, by adopting bullets instead of stripes to point to regions and by allowing non-atomic Boolean algebraic terms as labels of the curves. The way we use bullets to point out regions in a diagram resembles the way a "generic object" is used in an ordinary mathematical proof. Such an object x, which does not necessarily exist, usually appears in a proof of a given set-theoretic proposition, which starts from an assumption "Let x belongs to set A." In our system, the bullets in a diagram may be intuitively interpreted in the same way, that is, as pointing to some possible objects which do not necessarily exist, but whose belongness to some regions need to be considered to the overall reasoning being performed. In this sense, our use of bullets is completely different from the way shadows are used in the usual versions of Venn diagrams, i.e., shading usually expresses the emptiness of the corresponding region.

Figure 2 (a) shows how we put the diagram of $X = A \setminus B$ in normal form.

The sequence starts with the diagram of $A \setminus B$. The second diagram in the sequence is formed by two Venn diagrams, the upper one representing A, the lower one representing the complement of B, so that together they also represent $A \setminus B$. The third diagram in the sequence is also formed by two Venn diagrams,

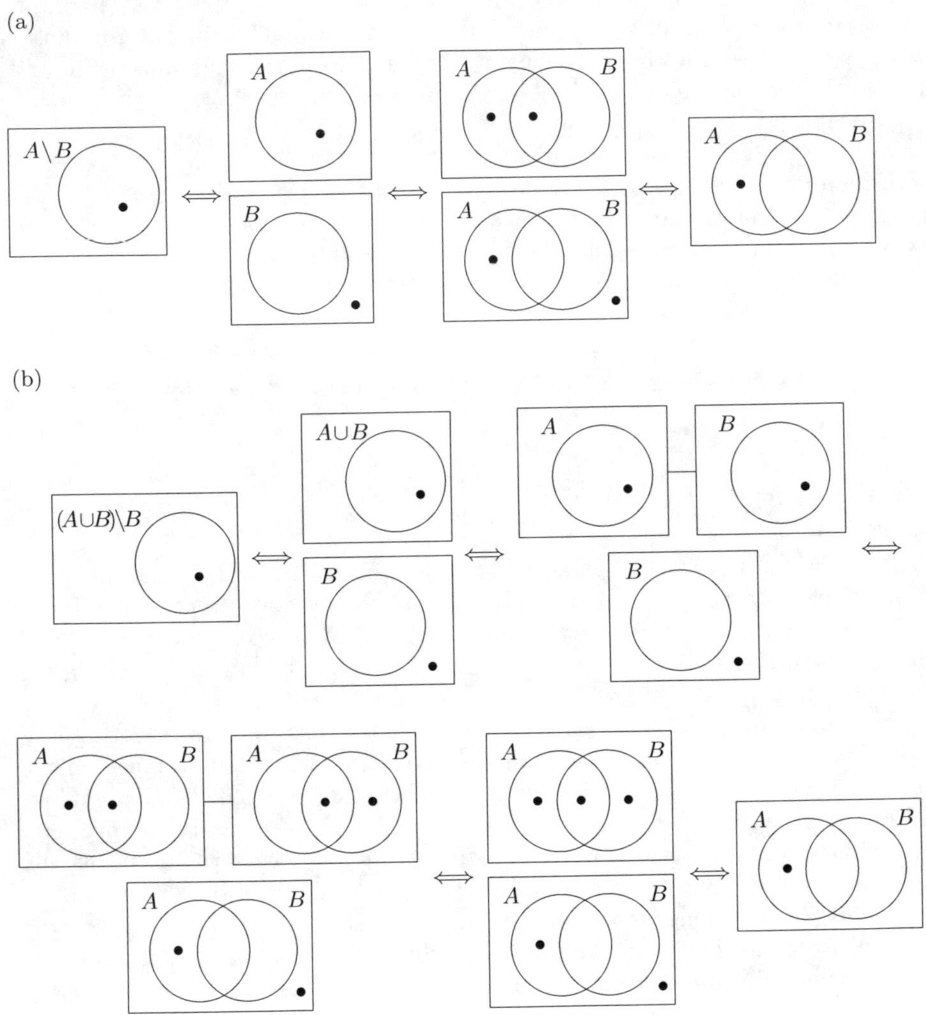

Fig. 2. (a) Putting the diagram of $A \setminus B$ in normal form. (b) Putting the diagram of $(A \cup B) \setminus B$ in normal form.

representing respectively A and the complement of B, now the difference is that both diagrams have the same set of labeling variables. As previously, the reading of any diagram in the sequence formed by the juxtaposition of more than one Venn diagram is as an intersection. So, the last diagram in the sequence is obtained from the third one by taking the common parts of the Venn diagrams that form it. This is the normal form of $A \setminus B$.

Figure 2 (b) shows how we put the diagram of $Y = (A \cup B) \setminus B$ in normal form, performing transformations similar to those displayed in Figure 2 (a). The main novelty is in the third and the fourth diagrams in the sequence. In both we have two Venn diagrams linked by a line. These stand for the unions of the sets represented by the linked Venn diagrams.

Figures 2 (a) and 2 (b) show that the normal form of the diagram of $A \setminus B$ and the normal form of the diagram of $(A \cup B) \setminus B$ are the same. So, based on them, we conclude that the equality (3) is, in fact, an identity.

3 Syntax and Semantics

In this section, we define the syntax and the semantics of the system. Since we are concerned with the basic logical properties of Venn diagrams, we adapt the approach taken in [1], where similar properties were investigated. We consider just Boolean inclusions, since equalities may be treated as double inclusions.

Let Var be a set of *variables for sets*, typically denoted by A, B, C, and Trm be the set of *Boolean terms* on Var, typically denoted by X, Y, Z. The elements of Var are atomic and the elements of Trm are molecular, given by

$$X ::= \mathsf{U} \mid \mathsf{O} \mid A \mid \overline{X} \mid X \cap X \mid X \cup X,$$

where U and O are constant symbols and A is an arbitrary variable. As usual, set difference $X \setminus Y$ is used as an abbreviation for $\overline{X} \cap Y$.

Given a finite set of terms $L = \{X_1, \ldots, X_n\}$, we denote by $\mathsf{MR}L$ the set of all 2^n sets $\{s_1, \ldots, s_n\}$, where $s_i = X_i$ or $s_i = \overline{X_i}$, for $1 \leq i \leq n$. In what follows, a set $\{s_1, \ldots, s_n\}$ will be denoted by the sequence $s_1 \cdots s_n$.

A *unitary diagram* is an ordered pair $d = (L, B)$, where $L \subseteq \mathsf{Trm}$ and $B \subseteq \mathsf{MR}L$ are finite sets. We call L the set of *labels* and B the set of *bullets* of d. A *compound diagram* is a set of unitary diagrams. A *multi-diagram* is a set of compound diagrams.

Given $X \in \mathsf{Trm}$, the *diagram of* X is $\Delta_X = \{\{(\{X\}, \{X\})\}\}$. Given a term X (set of terms L), we denote by $\mathsf{Var}X$ ($\mathsf{Var}L$) the set of variables occurring in X (in any term belonging to L). Given a multi-diagram Δ, we denote by $\mathsf{Var}\Delta$ is the set of variables occurring in the labels of the diagrams in Δ.

An *inclusion* is an expression of the form $\Delta_1 \subseteq \Delta_2$ where Δ_1 and Δ_2 are multi-diagrams. To simplify we refer to multi-diagrams simply as diagrams.

We interpret diagrams on models, according to the following ideas. Multi-diagrams are a formalization of the juxtaposed diagrams exemplified in Figures 2 (a) and 2 (b). So, the meaning of a multi-diagram is the intersection of the meanings of its components. Compound diagrams are a formalization of the linked diagrams exemplified in Figure 2 (b). So, the meaning of a compound diagram is the union of the meanings of its components. Unitary diagrams are a formalization of Venn diagrams having simple closed curves labeled by Boolean terms and some regions marked by dots. So, the meaning of a unitary diagram is the union of the meanings of its marked regions. If $M_1, \ldots, M_n$ are the meanings of the labels of the curves of the diagram, each region corresponds to an intersection $\pm M_1 \cap \cdots \cap \pm M_n$, where $\pm M_i$ is one of M_i or $\overline{M_i}$, for $1 \leq i \leq n$.

A *model* is a pair $\mathfrak{M} = (M, I)$, where M is a set and $I : \mathsf{Var} \to 2^M$. We ambiguously denote by I the natural extension of I to Trm, the set of all Boolean terms on Var, that agrees with I on the variables for sets and commutes with the Booleans. In particular, $I\mathsf{U} = M$ and $I\mathsf{O} = \emptyset$.

Let $d = (L, B)$ be a unitary diagram and $\mathfrak{M} = (M, I)$ be a model. The *meaning* of d in $\mathfrak{M}$ is

$$[\![d]\!]_{\mathfrak{M}} = \bigcup\{Ib : b \in B\},$$

where $Ib = \bigcap\{IX : X \in b\}$. Given a compound diagram D and a multi-diagram Δ, the meanings of D and Δ are defined as follows:

$$[\![D]\!]_{\mathfrak{M}} = \bigcup\{[\![d]\!]_{\mathfrak{M}} : d \in D\}$$

and

$$[\![\Delta]\!]_{\mathfrak{M}} = \bigcap\{[\![D]\!]_{\mathfrak{M}} : D \in \Delta\}.$$

An inclusion $\Delta_1 \subseteq \Delta_2$ is *true* in a model $\mathfrak{M}$ when $[\![\Delta_1]\!]_{\mathfrak{M}} \subseteq [\![\Delta_2]\!]_{\mathfrak{M}}$. When an inclusion is true in a model, we also say that the model *verifies* the inclusion. Validity, as usual, is defined as truth in all models. Two diagrams Δ_1 and Δ_2 are *equivalent* when $[\![\Delta_1]\!]_{\mathfrak{M}} = [\![\Delta_2]\!]_{\mathfrak{M}}$, for every model $\mathfrak{M}$. An inclusion $\Delta_1 \subseteq \Delta_2$ is a *consequence* of a set of inclusions Σ when it is true in every model that satisfies every inclusion in Σ.

4 Deductive Apparatus

In this section, we present a set of transformation rules to derive a diagram from another. Our set of rules is partitioned into rules to put diagrams in normal form and a rule to compare diagrams in normal form. We close the section proving weak soundness and completeness of our set of rules in relation to the semantics defined in Section 3.

Let Δ be a diagram and $L \subseteq \mathsf{Var}$. We say that Δ is in *normal form w.r.t.* L when $\Delta = \{\{(L, B)\}\}$, that is, the curves of the diagram are labeled exactly by the variables in L. If Δ_1 and Δ_2 are diagrams in normal form w.r.t. the same set of variables L, then Δ_1 is equivalent to Δ_2 iff $\Delta_1 = \Delta_2$. We call $\mathsf{NF}_L\Delta$ the diagram in normal form w.r.t. L that is equivalent to Δ.

The next set of rules is designed to put a diagram in normal form. Each one of these rules is presented as an equivalence between diagrams meaning that it can be applied to transform the diagram on the left hand side into the diagram on the right hand side, and vice-versa. As we see afterwards, given a diagram Δ

and a set L of variables containing the variables in Δ, by iterated applications of the rules, it is always possible to end up with $\mathsf{NF}_L\Delta$.

We present each rule followed by an example of its application. The examples illustrate the case when $\Delta = D = \emptyset$.

4.1 Rules to Put a Diagram in Normal Form

(U **in**) $\Delta \cup \{D \cup \{(\{\mathsf{U}\}, \{\mathsf{U}\})\}\} \Longleftrightarrow \Delta \cup \{D \cup \{(\emptyset, \{\emptyset\})\}\}$

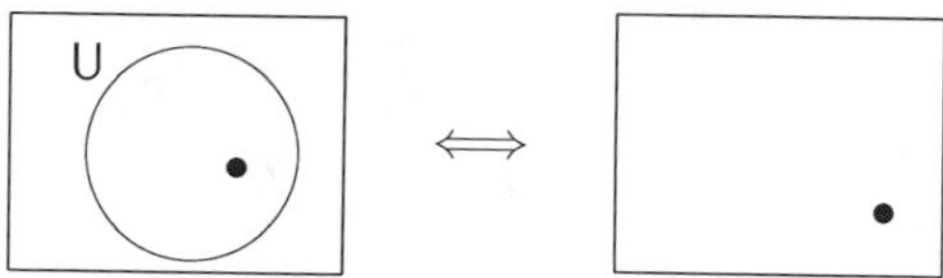

(U **out**) $\Delta \cup \{D \cup \{(\{\mathsf{U}\}, \{\overline{\mathsf{U}}\})\}\} \Longleftrightarrow \Delta \cup \{D \cup \{(\emptyset, \emptyset)\}\}$

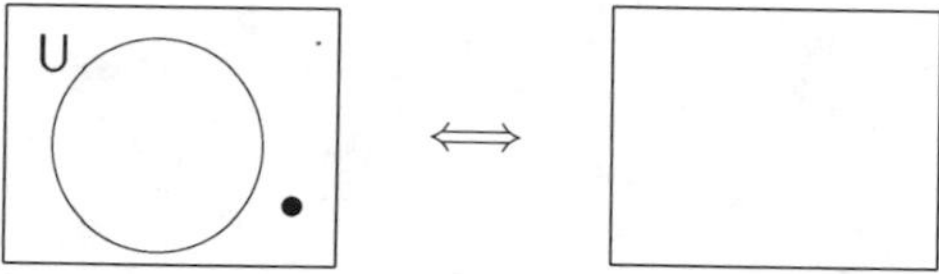

(O **in**) $\Delta \cup \{D \cup \{(\{\mathsf{O}\}, \{\mathsf{O}\})\}\} \Longleftrightarrow \Delta \cup \{D \cup \{(\emptyset, \emptyset)\}\}$

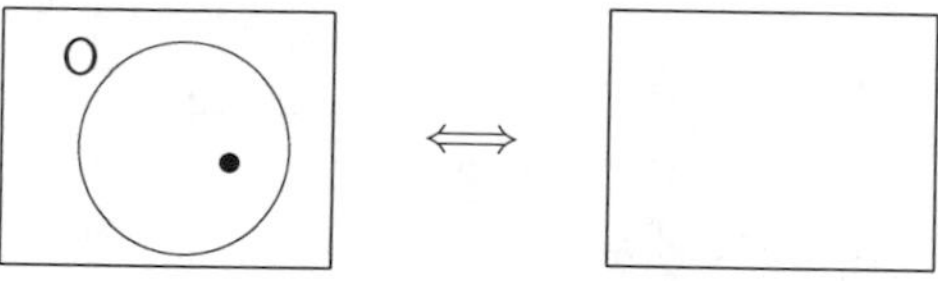

(O **out**) $\Delta \cup \{D \cup \{(\{\mathsf{O}\}, \{\overline{\mathsf{O}}\})\}\} \Longleftrightarrow \Delta \cup \{D \cup \{(\emptyset, \{\emptyset\})\}\}$

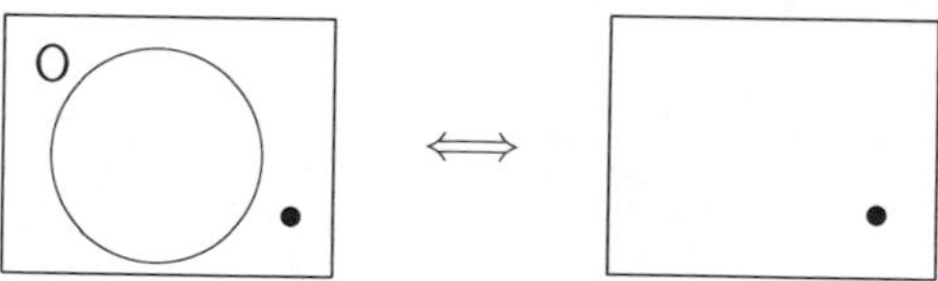

($\cap$ **in**) $\Delta \cup \{D \cup \{(\{X \cap Y\}, \{X \cap Y\})\}\} \Longleftrightarrow$
$\Delta \cup \{D \cup \{(\{X\}, \{X\})\}, D \cup \{(\{Y\}, \{Y\})\}\}$

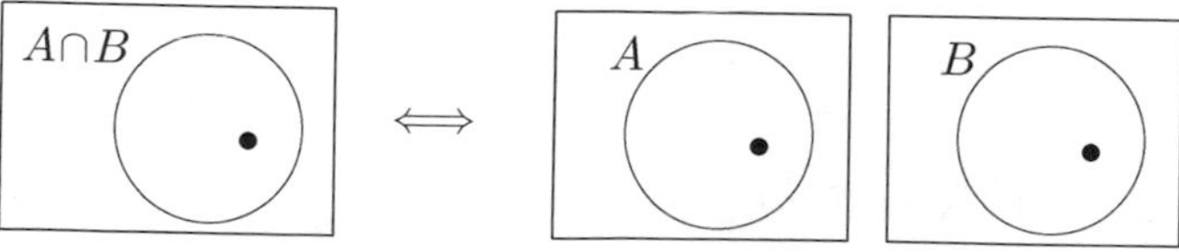

($\cap$ **out**) $\Delta\cup\{D\cup\{(\{X\cap Y\},\{\overline{X\cap Y}\})\}\}\Longleftrightarrow\Delta\cup\{D\cup\{(\{X\},\{\overline{X}\}),(\{Y\},\{\overline{Y}\})\}\}$

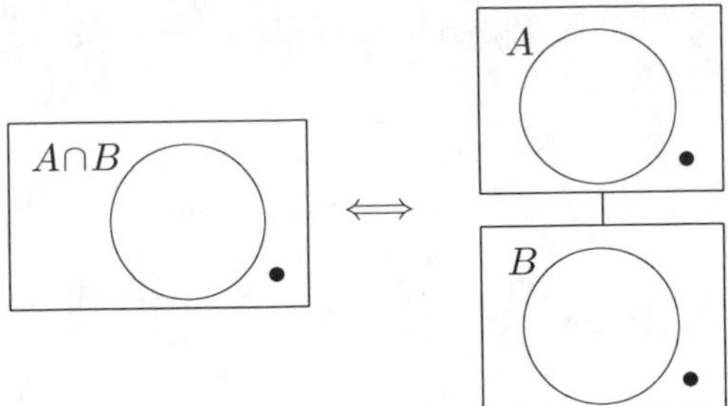

($\cup$ **in**) $\Delta\cup\{D\cup\{(\{X\cup Y\},\{X\cup Y\})\}\}\Longleftrightarrow\Delta\cup\{D\cup\{(\{X\},\{X\}),(\{Y\},\{Y\})\}\}$

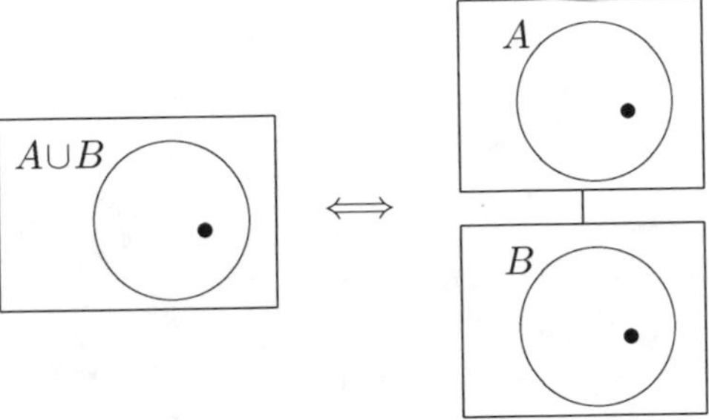

($\cup$ **out**) $\Delta\cup\{D\cup\{(\{X\cup Y\},\{\overline{X\cup Y}\})\}\}\Longleftrightarrow$
$\Delta\cup\{D\cup\{(\{X\},\{\overline{X}\})\},D\cup\{(\{Y\},\{\overline{Y}\})\}\}$

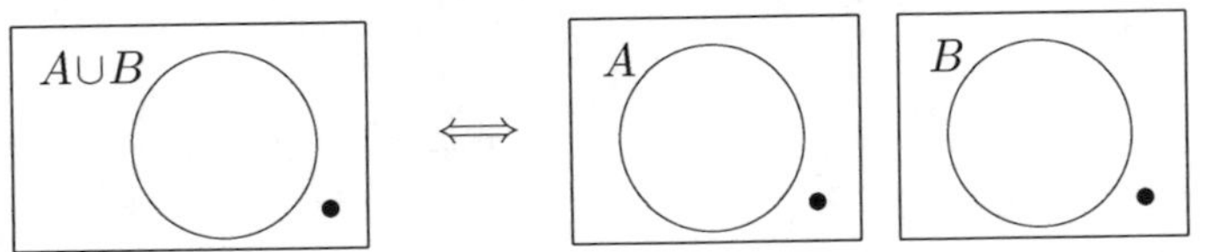

($-$ **in**) $\Delta\cup\{D\cup\{(\{\overline{X}\},\{\overline{X}\})\}\}\Longleftrightarrow\Delta\cup\{D\cup\{(\{X\},\{\overline{X}\})\}\}$

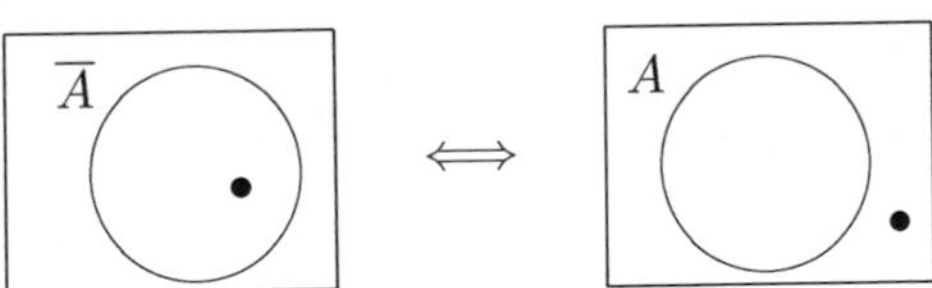

($-$ **out**) $\Delta\cup\{D\cup\{(\{\overline{X}\},\{\overline{\overline{X}}\})\}\}\Longleftrightarrow\Delta\cup\{D\cup\{(\{X\},\{X\})\}\}$

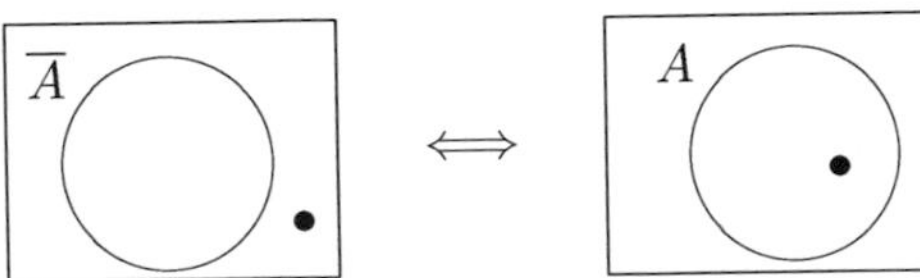

(Add Curve) $\Delta \cup \{D \cup \{(L, B)\}\} \Longleftrightarrow \Delta \cup \{D \cup \{(L \cup \{X\}, \{bX, b\overline{X} : b \in B\})\}\}$

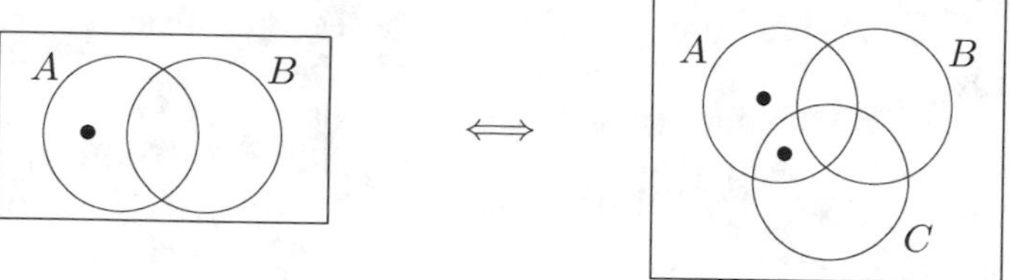

(Unify-c) $\Delta \cup \{D \cup \{(L, B_1), (L, B_2)\}\} \Longleftrightarrow \Delta \cup \{D \cup \{(L, B_1 \cup B_2)\}\}$

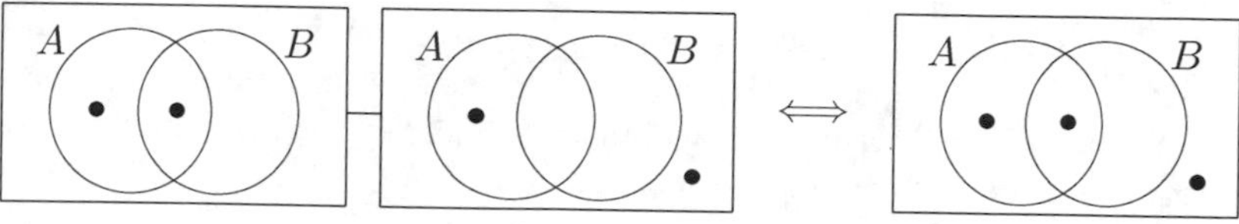

(Unify-m) $\Delta \cup \{\{(L, B_1)\}, \{(L, B_2)\}\} \Longleftrightarrow \Delta \cup \{\{(L, B_1 \cap B_2)\}\}$

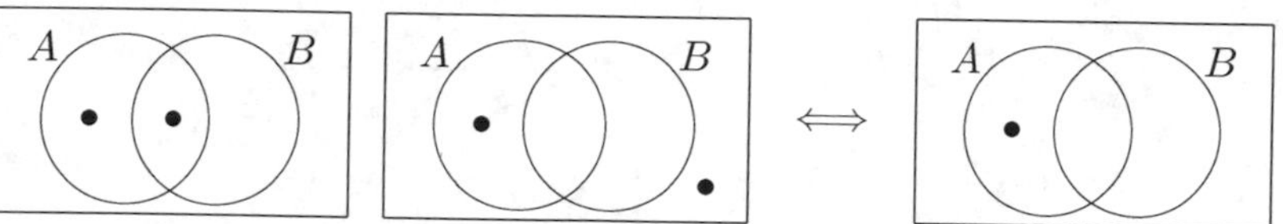

The following results justify the choice of the rules above.

Lemma 1. *If Δ' is a diagram obtained from a diagram Δ by some rule to put a diagram in normal form, then Δ and Δ' are equivalent.*

Theorem 1 (Normal form theorem). *For every diagram Δ and set of labels L' such that $\bigcup\{\mathsf{Var}L : d = (L, B), d \in D, D \in \Delta\} \subseteq L' \subseteq \mathsf{Var}$, there is a diagram Δ' such that Δ and Δ' are equivalent and Δ' is in normal form w.r.t. L'.*

Proof. Let Δ be a diagram and L' be a set of labels such that $\bigcup\{\mathsf{Var}L : d = (L, B), d \in D, D \in \Delta\} \subseteq L' \subseteq \mathsf{Var}$. By induction on the number of bullets in the unitary diagrams in Δ, applying rules **Unify-c** and **Add Curve**, we obtain a diagram Δ_0 such that each unitary diagram in Δ_0 has at most one curve, i.e.

$$d = (L, B) \in D \in \Delta_0 \implies L \text{ is a singleton}$$

By Lemma 1, diagrams Δ and Δ_0 are equivalent. By induction on the number of operators in the labels in Δ_0, applying rules **in/out**, we obtain a diagram Δ_1 such that

$$\bigcup\{L : d = (L, B), d \in D, D \in \Delta_1\} \subseteq L'.$$

By Lemma 1, diagrams Δ_0 and Δ_1 are equivalent. By applying Rule **Add curve**, we obtain Δ_2 and, for each $d = (L, B) \in D \in \Delta_2$, we have $L = L'$. By Lemma 1, diagrams Δ_1 and Δ_2 are equivalent. Since the set of labels of each unitary diagram in Δ_2 are the same, we can apply Rule **Unify-c** and obtain a diagram Δ_3 such that each compound diagram $D \in \Delta_3$ is a singleton. By Lemma 1, diagrams Δ_2 and Δ_3 are equivalent. Finally, by applying Rule **Unify-m**, we obtain a unitary diagram $\Delta' = \{\{(L', B')\}\}$ in normal form w.r.t. L'. By Lemma 1, diagrams Δ_3 and Δ' are equivalent. Hence, Δ and Δ' are equivalent.

4.2 Rule to Compare Diagrams in Normal Form

The next single rule is designed to the comparison of two given diagrams in normal form. It is presented as an implication between diagrams meaning that it can be applied to transform the diagram on the left hand side into the diagram on the right hand side, but not the other way around. As we see afterwards, if a diagram Δ' can be obtained from a given diagram Δ by an application of this rule, it follows that $\Delta \subseteq \Delta'$ is valid. We present the rule followed by an example of its application. As above, the example illustrates a case when $\Delta = D = \emptyset$.

(Add Bullets) $\Delta \cup \{D \cup \{(L, B)\}\} \Longrightarrow \Delta \cup \{D \cup \{(L, B \cup B')\}\}$

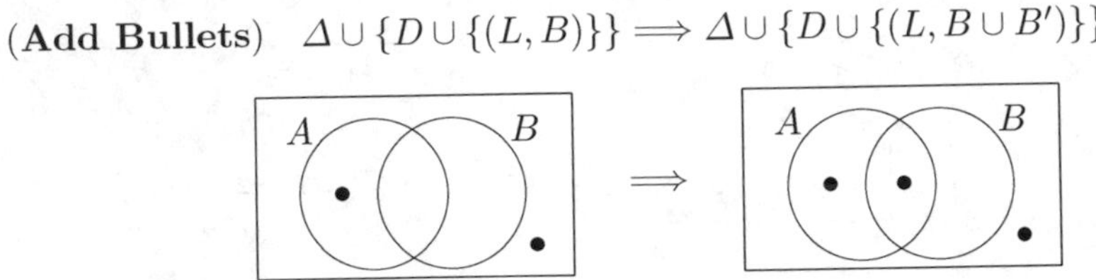

The following result justify the choice of the Rule **Add Bullets**.

Lemma 2. *If Δ' is a diagram obtained from a diagram Δ by application of the rule to compare diagrams in normal form, then $\Delta \subseteq \Delta'$ is valid.*

4.3 (Weak) Soundness and Completeness

We say that an inclusion $\Delta_1 \subseteq \Delta_2$ is a *theorem*, denoted by $\vdash \Delta_1 \subseteq \Delta_2$, when Δ_2 can be obtained from Δ_1 by a finite number (zero or more) of applications of the rules in Sections 4.1 and 4.2.

Theorem 2. *Let $\Delta_1 \subseteq \Delta_2$ be an inclusion. Then $\vdash \Delta_1 \subseteq \Delta_2$ iff $\Delta_1 \subseteq \Delta_2$ is valid.*

Proof. The proof of the sufficiency follows by Lemmas 1 and 2. To prove the necessity, suppose $\nvdash \Delta_1 \subseteq \Delta_2$. Let $\mathsf{NF}_{L'}\Delta_1 = \{\{(L', B_1')\}\}$ and $\mathsf{NF}_{L'}\Delta_2 = \{\{(L', B_2')\}\}$ where $L' = \bigcup\{\mathsf{Var}L : d = (L, B), d \in D, D \in \Delta_1 \cup \Delta_2\}$. By Theorem 1, we have $\vdash \Delta_1 \subseteq \mathsf{NF}_{L'}\Delta_1$ and $\vdash \mathsf{NF}_{L'}\Delta_2 \subseteq \Delta_2$. Now, since $\nvdash \Delta_1 \subseteq \Delta_2$, it is not possible to obtain $\mathsf{NF}_{L'}\Delta_2$ from $\mathsf{NF}_{L'}\Delta_1$ by a finite number (zero or more) of applications of **Add bullets**. So, in particular, there exists some $b \in L'$ such that $b \in B'_1$ but $b \notin B'_2$. Take the model $\mathfrak{M} = (M, I)$, such that $M = \{b\}$ and

$$IA = \begin{cases} \{b\}, \text{ if } A \in L' \text{ and } A \in b \\ \emptyset, \text{ otherwise} \end{cases}$$

It follows that $[\![\Delta_1]\!]_{\mathfrak{M}} = [\![\mathsf{NF}_{L'}\Delta_1]\!]_{\mathfrak{M}} = \{b\} \not\subseteq \emptyset = [\![\mathsf{NF}_{L'}\Delta_2]\!]_{\mathfrak{M}} = [\![\Delta_2]\!]_{\mathfrak{M}}$. That is, $\Delta_1 \subseteq \Delta_2$ is not valid.

As a corollary, we have that the problem of checking whether an inclusion between diagrams is valid or not is decidable.

Corollary 1. *Our set of rules provides an algorithm to decide whether a given inclusion between diagrams is valid or not.*

Proof. Given $\Delta_1 \subseteq \Delta_2$, apply the procedure outlined in the proof of Theorem 1 as follows. First, determine $\mathsf{NF}_{L'}\Delta_1 = \{\{(L', B'_1)\}\}$ and $\mathsf{NF}_{L'}\Delta_2 = \{\{(L', B'_2)\}\}$. Second, verify whether $B'_1 \subseteq B'_2$. If so, stop with yes; otherwise, stop with no.

5 An Extension to Deal with Hypotheses

In this section, we illustrate how the system can be extended to deal with reasoning from hypotheses, by solving half of the Exercise raised in [2], page 16, that is to prove the implication $(A \cap B) \cup C = A \cap (B \cup C) \implies C \subseteq A$.

In general terms, to test if an inclusion $X \subseteq Y$ follows or not from a set of inclusions Σ, we proceed as follows.

1. Construct $\mathsf{NF}_L\Delta_X$ and $\mathsf{NF}_L\Delta_Y$ where $L = \mathsf{Var}X \cup \mathsf{Var}Y$ and call these the *left diagram* and the *right diagram*, respectively.
2. If the right diagram can be obtained from the left one by applications of **Add bullets**, then stop with yes.
3. If not, take some hypothesis $X' \subseteq Y'$ in Σ and modify the left and the right diagrams as follows:
   (a) construct $\mathsf{NF}_{L'}\Delta_{X'}$ where $L' = \mathsf{Var}X \cup \mathsf{Var}Y \cup \mathsf{Var}X' \cup \mathsf{Var}Y'$ and call it $\Delta_1 = \{\{(L', B_1)\}\}$;
   (b) construct $\mathsf{NF}_{L'}\Delta_{Y'}$ and call it $\Delta_2 = \{\{(L', B_2)\}\}$;
   (c) calculate $B = B_1 \setminus B_2$;
   (d) construct the normal forms of the left and the right diagrams w.r.t. L' and call them $\Delta_l = \{\{(L', B_l)\}\}$ and $\Delta_r = \{\{(L', B_r)\}\}$, respectively;
   (e) construct the new (modified) *left diagram* as being $\Delta'_l = \{\{(L', B_l \setminus B)\}\}$ and the new *right diagram* as being Δ_r;

   and go back to step 2.

We start the application of the procedure by taking $\Sigma = \{(A \cap B) \cup C \subseteq A \cap (B \cup C), A \cap (B \cup C) \subseteq (A \cap B) \cup C\}$, $X = C$, and $Y = A$.

Steps 1 and 2 are executed by taking $L = \{A, C\}$.

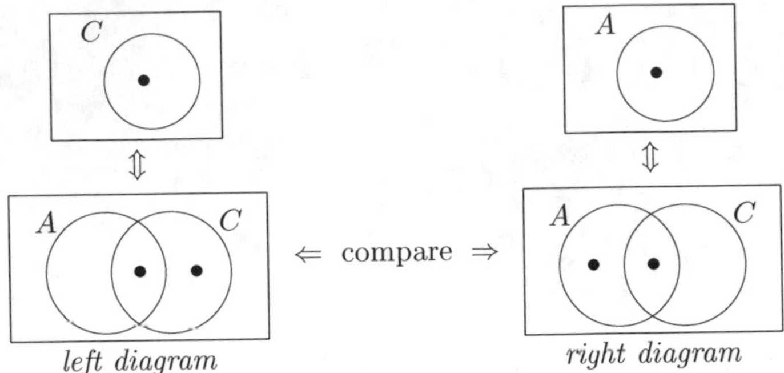

Step 3 is executed by taking $X' = (A \cap B) \cup C$ and $Y' = A \cap (B \cup C)$.

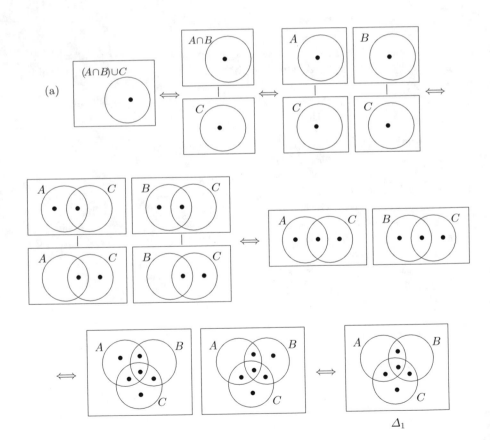

(a)

Δ_1

(b) The construction of Δ_2 is analogous.

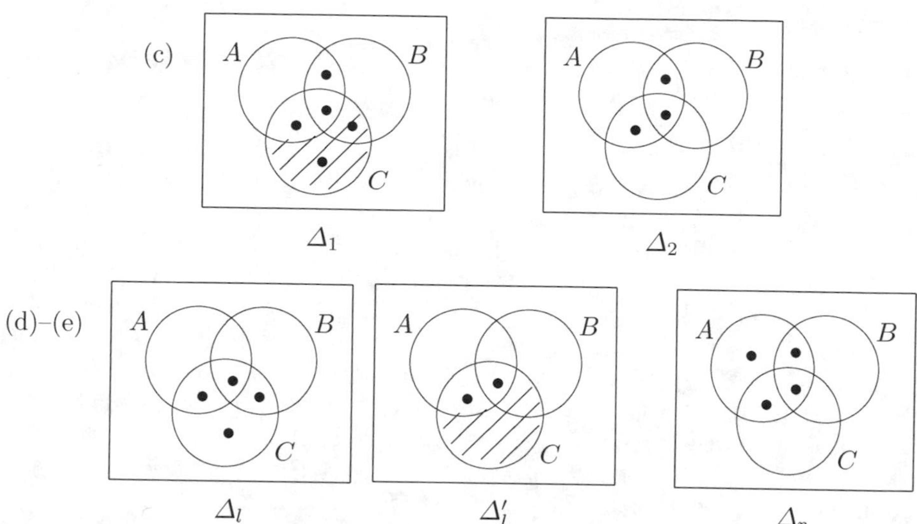

Go back to Step 2 and stop with yes.

6 Reasoning with Hypotheses

In this section, we complement the system, presenting a transformation rule to derive a diagram from another one, under a set of inclusions between diagrams taken as hypotheses. Together with the rules presented in Section 4, this new rule provides a strong sound and complete set of rules with respect to the semantics defined in Section 3.

The next single rule can be applied to transform a given diagram into another one, using one inclusion taken as hypothesis. It is presented as an implication between diagrams, meaning that it can be applied to transform the diagram on the left hand side into the diagram on the right hand side, but not the other way around. As we see afterwards, if a diagram Δ' can be obtained from a given diagram Δ by an application of this rule, given a hypothesis, it follows that $\Delta \subseteq \Delta'$ is a consequence of this hypothesis.

An example of application of this rule was already given in Section 5.

(Hyp) $\Delta \cup \{D \cup \{(L, B)\}\} \Longrightarrow \Delta \cup \{D \cup \{(L, B \setminus B')\}\}$,
if $B' = B_1 \setminus B_2$, where $\Delta_1 \subseteq \Delta_2$ is a hypothesis,
$\mathsf{NF}_L \Delta_1 = \{\{(L, B_1)\}\}$, and $\mathsf{NF}_L \Delta_2 = \{\{(L, B_2)\}\}$.

The following result justify the choice of the rule above.

Lemma 3. *If Δ' is a diagram obtained from a diagram Δ by application of the rule to derive from hypotheses, then $\Delta \subseteq \Delta'$ is a consequence of the set of hypotheses.*

We say that an inclusion between diagrams $\Delta \subseteq \Delta'$ is a *derivable from* a set of inclusions between diagrams Σ, denoted by $\Sigma \vdash \Delta \subseteq \Delta'$, when Δ' can be obtained from Δ by a finite number (zero or more) of applications of the rules in Sections 4.1, 4.2, and 6.

Theorem 3. *Let $\Delta \subseteq \Delta'$ be an inclusion and Σ be set of inclusions. Then $\Sigma \vdash \Delta \subseteq \Delta'$ iff $\Delta \subseteq \Delta'$ is a consequence of Σ.*

Proof. The proof of the sufficiency follows by Lemmas 1, 2 and 3. To prove the necessity, suppose $\Sigma \nvdash \Delta \subseteq \Delta'$. Consider an enumeration $\Delta'_1 \subseteq \Delta''_1, \Delta'_2 \subseteq \Delta''_2, \dots, \Delta'_n \subseteq \Delta''_n, \dots$ of the hypotheses in Σ. Let $\Delta_0, \Delta_1, \dots, \Delta_n, \dots$ be a sequence of diagrams defined as follows. For Δ_0, take $\mathsf{NF}_L\Delta$, where $L = \mathsf{Var}\Delta \cup \mathsf{Var}\Delta'_1 \cup \mathsf{Var}\Delta''_1$. For each $i > 0$, Δ_i is obtained from Δ_{i-1} by an application of rule **Hyp** using the hypothesis $\Delta'_i \subseteq \Delta''_i$. This can be done since Δ_{i-1} is in normal form w.r.t. $\mathsf{Var}\Delta_{i-1} \cup \mathsf{Var}\Delta'_i \cup \mathsf{Var}\Delta''_i$ and any diagram obtained by an application of **Hyp** to a diagram in normal form w.r.t. a set of variables is also in normal form w.r.t. this set. To continue the proof, we need the following constructions. Given the set of variables $\mathsf{Var} = \{A_1, A_2, \dots, A_i, \dots\}$, a unitary diagram $d = (L, B)$, and a bullet $b = s_{j_1} s_{j_2} \cdots s_{j_m} \in B$, we define $\mathcal{C}b$ as the set of infinite sequences $X_1 X_2 \cdots X_n \cdots$, where $X_i = s_{j_k}$, in case these is some j_k such that $s_{j_k} = A_i$ or $\overline{A_i}$, otherwise, $X_i = A_i$ or $\overline{A_i}$. Besides, we define $\mathcal{C}B$ as $\bigcup\{\mathcal{C}b : b \in B\}$. Now, taking B' as the set of bullets of Δ' and B_i as the set of bullets of the Δ_i, $i \in \omega$, we define the chain $\mathfrak{C}$ as

$$\mathcal{C}B_0 \setminus \mathcal{C}B' \supseteq \mathcal{C}B_1 \setminus \mathcal{C}B' \supseteq \cdots \supseteq \mathcal{C}B_n \setminus \mathcal{C}B' \supseteq \cdots$$

We have that $\bigcap \mathfrak{C} \neq \emptyset$. Otherwise, it would be possible to derive $\mathsf{NF}_L\Delta'$, where $L = \mathsf{Var}\Delta' \cup \mathsf{Var}\Delta_i$, from Δ_i by applying **Add bullets**. But this is not possible, since $\Sigma \nvdash \Delta \subseteq \Delta'$. Hence, take some $b \in \bigcap \mathfrak{C}$ and consider the model $\mathfrak{M} = (M, I)$ such that $M = \{b\}$ and $IA = \{b\}$ if $A \in b$, otherwise $IA = \emptyset$. By construction, we have that $\mathfrak{M}$ verifies each hypothesis in Σ, that $[\![\Delta]\!]_{\mathfrak{M}} = [\![\Delta_0]\!]_{\mathfrak{M}} = \{b\}$, and that $[\![\Delta']\!]_{\mathfrak{M}} = \emptyset$. That is, $\Delta \subseteq \Delta'$ is not a consequence of Σ.

As a corollary, we have a strong form of the Finite Model Property: every non-consequence of a set of hypotheses has a counter model whose domain is a singleton set.

Corollary 2. *Let $\Delta \subseteq \Delta'$ be an inclusion and Σ be set of inclusions such that $\Sigma \nvdash \Delta \subseteq \Delta'$. Then there is a model $\mathfrak{M} = (\{b\}, I)$ verifying every inclusion in Σ, whereas $[\![\Delta]\!]_{\mathfrak{M}} \nsubseteq [\![\Delta']\!]_{\mathfrak{M}}$.*

7 Conclusions and Perspectives

The work reported in this paper is clearly connected to two lines of development in the study of diagrammatic logics. The first one, taken by S.-J. Shin, E. Hammer, J. Howse, G. Stapleton — to mention just a few — explores the use of diagrams as propositions, and extends the Venn diagrams into a family of systems, presumably, up to expressive power of second order logic. In those papers which are relevant to our discussion, these authors investigated the expressive and proof powers of the systems considered, obtaining results similar to those here presented, but that go far beyond what we made, due to the scope of the systems to which their methods has already been applied. As a further research, we intend to investigate the extension of our approach to diagrams considered as terms to the other more expressive Venn diagrammatic logics, mainly spider diagrams, concept and generalized concept diagrams.

The second line that we want to mention, taken by the authors of this paper together with P.A.S. Veloso and S.R.M. Veloso, explores the use of diagrams as terms, and uses oriented graphs having two distinguished nodes and arcs labelled by variables to denote binary relations on a base set. In a serie of papers, we extend a system fitted just to express properties lacking negation up to the power of classical first order logic and, presumably, up to the power of intuitionistic logic. The ideas used to investigate the expressive and proof powers of these graphical systems are very similar to those we used to investigate Venn diagrams as terms. As a further research, we intend to investigate the close relationship it seems to exist between the Venn Diagrams presented here and those graphical systems investigated by us elsewhere.

The main idea applied to the completeness proof we presented here, that of putting the two diagrams in normal form and compare them, has a normal form to the proofs as a corollary. But as a referee to this paper pointed out, it is possible to establish some set theoretic identities without necessarily expressing each side in its normal form. As a further work, we plan to take this suggestion forward developing proof estrategies for set theoretic identities in our system, that do not take this normal form in consideration.

Acknowledgements. Research partially sponsored by Conselho Nacional de Desenvolvimento Científico e Tecnológico (CNPq) and Fundação de Amparo à Pesquisa do Estado do Rio de Janeiro (FAPERJ). The authors also thank the Programa de Engenharia de Sistemas e Computação da COPPE-UFRJ for material support during the production of this paper.

References

1. Burton, J., Stapleton, G., Howse, J.: Completeness proof strategies for Euler diagram logics. In: Chapman, P., Micallef, L. (eds.) Euler Diagrams 2012: Proceedings of the 3rd International Workshop on Euler Diagrams, Canterbury, UK, July 2 (2012)
2. Halmos, P.: Naive Set Theory. Springer, New York (1974)
3. Hammer, E., Danner, N.: Towards a model theory of Venn diagrams. J. Philos. Logic 25, 463–482 (1996)
4. Shin, S.-J.: The Logical Status of Diagrams. Cambridge University Press, Cambridge (1994)
5. Stewart, I.: The truth about Venn diagrams. Math. Gaz. 70, 47–54 (1976)

Diagrammatically Explaining Peircean Abduction

Flavio Zelazek

Department of Philosophy, Sapienza University of Rome, Italy
flavio.zelazek@gmail.com

Abstract. By using Euler diagrams, early Peircean abduction is explained as an inference based on the shrinkage of a class of properties; this renders it dual to inductive inference, which is based on the enlargement of a class of subjects. In fact, at a very general level these inferences can be interpreted as (category-theoretic) dual constructions, by representing them as commutative diagrams.

Keywords: abduction, induction, duality, Euler diagrams, commutative diagrams, category theory.

The question about how to characterize abduction, and its relation with induction, is a much debated one in philosophy and in Artificial Intelligence (cf. e.g. [1] for a survey).

An inference schema which is often used to represent abduction (for example in Abductive Logic Programming) is the following one, together with its propositional version:

$$\frac{\forall x\,(M(x) \to P(x)) \qquad P(s_1)}{M(s_1)}\;\textsc{Abd}_{\textsc{Q}} \qquad\qquad \frac{M \to P \qquad P}{M}\;\textsc{Abd}$$

On the other hand, the earliest definition of abduction given by Peirce [2] (2.424–425, 2.511) involves *multiple subjects and predicates*, and – as argued in [3] – characterizes abduction as the *dual* of induction; the resulting simplified inference schemata (forgetting terms and quantifiers) are the following:

$$\frac{S_1 \vee \cdots \vee S_j \to M \qquad S_1 \to P \qquad \vdots \qquad S_j \to P}{M \to P}\;\textsc{Ind}^* \qquad\qquad \frac{M \to P_1 \wedge \cdots \wedge P_k \qquad S \to P_1 \qquad \vdots \qquad S \to P_k}{S \to M}\;\textsc{Abd}^*$$

T. Dwyer et al. (Eds.): Diagrams 2014, LNAI 8578, pp. 308–310, 2014.

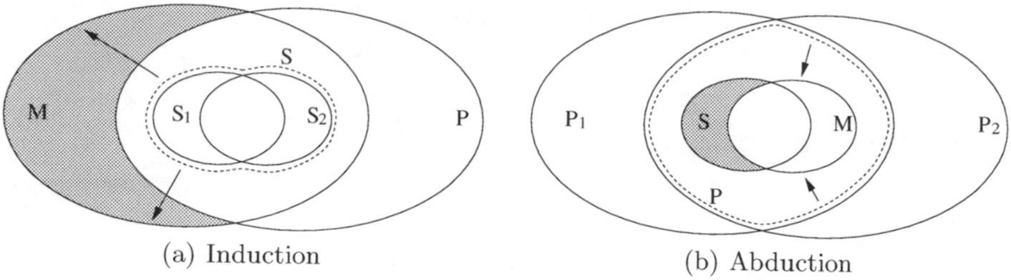

(a) Induction (b) Abduction

Fig. 1. Euler diagrams for induction and abduction

Now, the inferences (ABD$_Q$) and (ABD) are but limiting cases of (ABD*), just as *single-case induction* is a limiting case of (IND*).

Moreover, the schemata (IND*) and (ABD*) are the dual of each other inasmuch as the first involves an *increase in extension*, while the second involves an *increase in comprehension*;[1] which, if we consider just extensional (set-theoretical) notions, respectively correspond to the *enlargement of a (subject) class* and to the *shrinkage of a (predicate) class*, obtained by adding members to a union or, respectively, to an intersection of classes. So induction and abduction can be better represented as inferences between set-theoretical relations: simply replace $\rightarrow, \vee, \wedge$ with $\subseteq, \cup, \cap$.

The Euler diagrams depicted in Fig. 1(a) and in Fig. 1(b) (where the arrows indicate the enlargement or shrinkage of the classes S or P) help to understand how the increment of the considered subjects or properties (predicates) increases the degree of confirmation, or more generally the plausibility, of the two respective types of inferences, as explained in [3].

So, due to the multiplicity of predicate classes, (ABD*) conveys a a more general conception of abduction than the one represented by (ABD$_Q$) and (ABD). Moreover, due to the main property for which it is dual to induction, it better accounts for a kind of reasoning which is typical, e.g., of *clinical diagnosis*, where a single patient (subject) is checked for multiple medical signs (properties) in order to circumscribe, that is to shrink, the class of possible diseases.

Now, (IND*) and (ABD*) actually denote the same inference schema, if we take for granted the shifting to contrapositives.[2] This fact depends on the "extensional flattening" assumed here: subjects and properties cannot simply be exchanged by negation, for they have *different roles* in our inferential practices (indeed, Peirce maintained an *intensional* conception of predicates).

That said, in order to characterize abduction in a very general way, even a purely extensional one, another possibility is to consider the following commutative diagrams, in which objects are members of the power set $\mathcal{P}(U)$ for some

[1] Notice that Peirce uses the similar – but not equivalent – terms '*breadth*' and '*depth*': cf. [2] (2.407 ff.).

[2] In fact the contrapositive of (ABD*), with the $\neg P$'s and $\neg M$ renamed as S's and M, is classically equivalent just to (IND*); the same holds for the set-theoretical versions of the two schemata, considering the complements w.r.t the universe of discourse.

universe of discourse U, and arrows are inclusion functions (which map every element to itself):

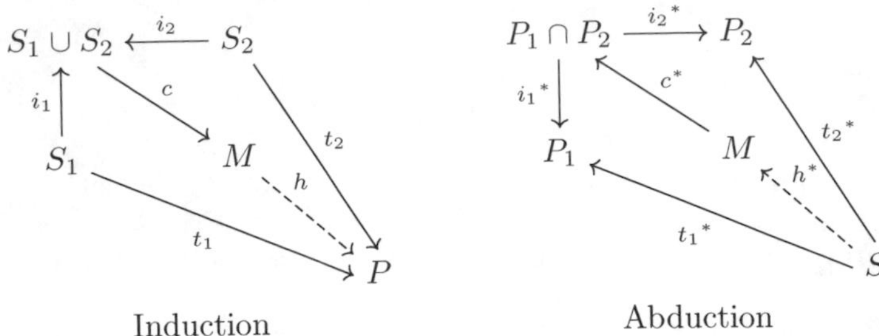

The starred functions in the rightmost diagram are again inclusions (as we have exchanged the S's and P's), so that we get two *dual constructions* in the same lattice $(\mathcal{P}(U), \subseteq)$, in which $P_1 \cap P_2$ and $S_1 \cup S_2$ are (in the category-theoretic sense) a *product* and a *coproduct*. But only some subset of *admissible tests*, or inclusions, is considered (e.g., one would test a raven for being black, but not a non-black thing for being a non-raven); then, once such tests are fixed, induction and abduction are identified each one by its peculiar (diagrammatic) "geometry" (i.e., configuration of the arrows, with the appropriate orientations). Conversely, once established the kind of inference under consideration, the classes involved are to be regarded as subjects or as predicates in accordance to the "place" they inhabit in either construction.

Finally, the various inclusion arrows have each a different epistemological status: c and c^* are *constraints* which have to hold empirically (the S's are tested in so far as they are M's, and the P's are verified in so far as they hold for M); all the t's are *tests*; h and h^* are *hypotheses* (typically, general the former, singular the latter); while the i's and i^*'s are just injections and projections.

References

1. Flach, P.A., Kakas, A.C.: Abduction and induction. Essays on their relation and integration. Applied logic series, vol. 18. Kluwer Academic Publishers, Dordrecht (2000)
2. Hartshorne, C., Weiss, P. (eds.): Collected papers of Charles Sanders Peirce. The Belknap Press of Harvard University Press, Cambridge (1960)
3. Zelazek, F.: Dual Aspects of Abduction and Induction. In: Lieto, A., Cruciani, M. (eds.) Proceedings of the First International Workshop on Artificial Intelligence and Cognition (AIC 2013). An Official Workshop of the 13th International Conference of the Italian Association for Artificial Intelligence (AI*IA 2013), Torino, Italy, December 3. CEUR Workshop Proceedings (2013)

Author Index